Gravitational Wave Data Analysis

NATO ASI Series

Advanced Science Institutes Series

A Series presenting the results of activities sponsored by the NATO Science Committee, which aims at the dissemination of advanced scientific and technological knowledge, with a view to strengthening links between scientific communities.

The Series is published by an international board of publishers in conjunction with the NATO Scientific Affairs Division

A	Life Sciences	Plenum Publishing Corporation
B	Physics	London and New York
C	Mathematical	Kluwer Academic Publishers
	and Physical Sciences	Dordrecht, Boston and London
D	Behavioural and Social Sciences	
E	Applied Sciences	
F	Computer and Systems Sciences	Springer-Verlag
G	Ecological Sciences	Berlin, Heidelberg, New York, London,
H	Cell Biology	Paris and Tokyo

Series C: Mathematical and Physical Sciences - Vol. 253

Gravitational Wave Data Analysis

edited by

B. F. Schutz

Department of Applied Mathematics and Astronomy,
University College, Cardiff, Wales, U.K.

Kluwer Academic Publishers

Dordrecht / Boston / London

Published in cooperation with NATO Scientific Affairs Division

Proceedings of the NATO Advanced Research Workshop on
Gravitational Wave Data Analysis
Dyffryn House, St. Nicholas, Cardiff, Wales, U.K.
6–9 July 1987

Library of Congress Cataloging in Publication Data

```
Gravitational wave data analysis : proceedings of the NATO advanced
  research workshop held at Dyffryn House, St. Nicholas, Cardiff,
  Wales, 6-9 July 1987 / edited by B.F. Schutz.
      p.    cm. -- (NATO ASI series. Series C, Mathematical and
  physical sciences ; vol. 253)
    Includes bibliographies and index.
    ISBN 978-94-010-7028-7
    1. Gravity waves--Congresses.   I. Schutz, Bernard F.  II. Series:
  NATO ASI series.  Series C, Mathematical and physical sciences ; no.
  253.
  QB341.G74 1988
  526'.7--dc19                                          88-13648
                                                            CIP
```

ISBN 978-94-010-7028-7

Published by Kluwer Academic Publishers,
P.O. Box 17, 3300 AA Dordrecht, The Netherlands.

Kluwer Academic Publishers incorporates the publishing programmes of
D. Reidel, Martinus Nijhoff, Dr W. Junk, and MTP Press.

Sold and distributed in the U.S.A. and Canada
by Kluwer Academic Publishers,
101 Philip Drive, Norwell, MA 02061, U.S.A.

In all other countries, sold and distributed
by Kluwer Academic Publishers Group,
P.O. Box 322, 3300 AH Dordrecht, The Netherlands.

Table of Contents

Preface ....   *vii*

Part 1: Sources of Gravitational Radiation

Sources of Gravitational Radiation
    by Bernard F. Schutz **3**

The Rate of Gravitational Collapse in the Milky Way
    by D. G. Blair   19

Gravitational Radiation from Rotating Stellar Core Collapse
    by L. S. Finn   33

Remarks on SN 1987a
    by Peter Kafka   55

Coalescing Binaries to Post-Newtonian Order
    by Andrzej Krolak   59

Part 2: Principles of Signal Processing

A Review of the Statistical Theory of Signal Detection
    by M. H. A. Davis   73

Radio Pulsar Search Techniques
    by A. G. Lyne   95

Sample Covariance Techniques in the Detection of Gravitational Waves
    by A. D. Irving   105

Part 3: Quantum Limits on Detectors

Parametric Transducers and Quantum Nondemolition in Bar Detectors
    by Mark F. Bocko *and* Warren W. Johnson   125

Squeezed States of Light
    by M. H. Muendel, G. Wagner, J. Gea-Banacloche *and* G. Leuchs   135

Part 4: Methods of Data Analysis in Gravitational Wave Detectors

Round Table Discussion — Gravitational Wave Detectors
    chaired by J. Hough   147

Spacecraft Gravitational Wave Experiments
by J. W. Armstrong   153

Gravitational Wave Experiments with Resonant Antennas
by G. Pizzella   173

Gravitational Antenna Bandwidths and Cross Sections
by J. Weber   195

Comparison of Bars and Interferometers: Detection of Transient Gravitational Radiation
by Daniel Dewey   201

Broadband Search Techniques for Periodic Sources of Gravitational Radiation
by Jeffrey Livas   217

Response of Michelson Interferometers to Linearly Polarized Gravitational Waves
of Arbitrary Direction of Propagation
by D. Fattaccioli   239

Data Analysis as a Noise Diagnostic: Looking for Transients in Interferometers
by Daniel Dewey   255

Data Acquisition and Analysis with the Glasgow Prototype Detector
by N. L. Mackenzie, H. Ward, J. Hough, G. A. Kerr, J. B. Mangan,
B. J. Meers, G. P. Newton, D. I. Robinson and N. A. Robertson   269

On the Analysis of Gravitational Wave Data
by S. P. Lawrence, E. Nahmad-Achar and B. F. Schutz   275

GRAVNET, Multiple Antenna Coincidences and Antenna Patterns for Resonant Bar
Antennas
by David Blair, Sergio Frasca and Guido Pizzella   285

Coincidence Probabilities for Networks of Laser Interferometric Detectors Observing
Coalescing Compact Binaries
by Massimo Tinto   299

Data Analysis Requirements of Networks of Detectors
by Bernard F. Schutz   315

Round-Table on Data Exchange
chaired by I. F. Corbett   327

Preface

The articles in this book represent the major contributions at the NATO Advanced Research Workshop that was held from 6 to 9 July 1987 in the magnificent setting of Dyffryn House and Gardens, in St. Nicholas, just outside Cardiff, Wales. The idea for such a meeting arose in discussions that I had in 1985 and 1986 with many of the principal members of the various groups building prototype laser-interferometric gravitational wave detectors. It became clear that the proposals that these groups were planning to submit for large-scale detectors would have to address questions like the following:

- What computing hardware might be required to sift through data coming in at rates of several gigabytes per day for gravitational wave events that might last only a second or less and occur as rarely as once a month?

- What software would be required for this task, and how much effort would be required to write it?

- Given that every group accepted that a worldwide network of detectors operating in co-incidence with one another was required in order to provide both convincing evidence of detections of gravitational waves and sufficient information to determine the amplitude and direction of the waves that had been detected, what sort of problems would the necessary data exchanges raise?

Yet most of the effort in these groups had, quite naturally, been concentrated on the detector systems. All the groups agreed that a chance to think primarily about the data analysis problem would be welcome. Moreover, no such meeting would be complete without the bar detector groups, who had already accumulated valuable experience in taking and analysing data (albeit at much lower data rates), and who would naturally also be involved in any observing network. The Workshop took shape from these requirements.

I undertook to organise the meeting, and in this I was very ably assisted by the other members of the Scientific Organising Committee: Peter Michelson, Guido Pizzella and Roland Schilling. In addition, Andrzej Krolak put in a great deal of work on the local organisation, without which the meeting would certainly not have succeeded. We are all very grateful to NATO for supporting the Workshop so generously.

The resulting volume of articles represents partly a resource for further investigations — such as the reviews by Davis and Lyne — and partly a snapshot of the state of our thinking and understanding in the summer of 1987. It is the first volume to be published on this subject, and I hope it will point the way toward future developments. None of us knows when the *first* gravitational wave will be observed in our detectors, but as the book shows, we are already looking beyond that momentous event to the establishment of *gravitational wave astronomy*, the regular detection and identification of gravitational waves from a great variety of different sources scattered throughout the universe.

Bernard Schutz

Cardiff, Wales

Part 1
SOURCES OF GRAVITATIONAL RADIATION

SOURCES OF GRAVITATIONAL RADIATION

Bernard F. Schutz
Department of Applied Mathematics and Astronomy
University College Cardiff
Cardiff, Wales, UK

ABSTRACT.   There is, not surprisingly, considerable uncertainty in the
theoretical   predictions   of   the   strength   and   rate   of   occurence   of
gravitational   wave   events   of   a   potentially   observable   strength   and
frequency.   Theory can at present only give an indication of what sources
might be seen, together with limits on their strength and event rate.   This
review   considers   both   the   kind   of   information   we   can   expect   from
observations   of   any   particular   class   of   sources,   and   the   degree   of
confidence we have in the observability of that class.   I begin with useful
formulae for estimating the detectability of different sources; in a given
detector, and provided optimal filtering can be used for sources with long
wave trains, then the relative detectability of two sources at the same
distance depends only on the total energy radiated and the dominant
frequency of the radiation.   I follow with a review of the most commonly
discussed   gravitational   wave   sources: gravitational   collapse,   coalescing
binaries,   pulsars,   rapidly   spinning   accreting   neutron   stars,   and   the
stochastic background.   I include a discussion in detail of two subjects
that have become interesting recently: the possibility that the rate of
gravitational collapse events in our own Galaxy is high, perhaps one every
one to three years; and estimates of bounds on the (highly uncertain) rate
at which binary coalescences occur at relatively large distances.   It is
very probable that a network of four detectors built to today's design
would detect at least a few events per year, and perhaps as many as
several hundred thousand.   The rate for systems containing at least one
black hole may be significant too.   I conclude with a discussion of the
kinds of astrophysical information that observations of coalescing binary
systems can provide: a value for Hubble's constant, information on the
neutron-star equation of state, tests of general relativity, surveys of the
mass distribution of the universe out to 0.5 Gpc or so, and (with further
technical development of the detectors) perhaps even values for the
deceleration parameter and the redshift at which star formation began.

1.   INTRODUCTION

The development of designs for ever more sensitive broad-band gravitational
wave detectors during the last five years has allowed theorists to re-

B. F. Schutz (ed.), Gravitational Wave Data Analysis, 3–17.
© *1989 by Kluwer Academic Publishers.*

examine the question of what sources future detectors would be likely to see; and the recent supernova in the Large Magellanic Cloud (SN1987a) has similarly reopened the question of how often relatively nearby events occur with a strength that would allow detection by the cryogenically cooled bar detectors that will soon be ready to do long-term observations. Because the subject of sources and detectors has been comprehensively reviewed very recently (Thorne 1987), I will give only a brief overview of the wide range of sources that theorists have studied, but I will include within it a more detailed discussion of the two points I have just referred to: nearby sources detectable by bars, and coalescing binary systems, which are theorists' current 'best bet' sources at large distances (greater than, say, 100 Mpc). But first I will address the 'detectability' of sources: what characteristics make one source easier to detect than another in a given detector?

2.   DETECTABILITY OF SIGNALS

Given the inevitable uncertainties associated with predicting what gravitational wave detectors will see, it is useful to have simple formulae that allow one to estimate the likely strength of the gravitational waves that would come from a hypothetical source, and the detectability of those waves. In this section I give some formulae and use them to make some interesting comparisons among possible sources.

2.1. Characteristics of the source.

The fundamental measure of the amplitude of the gravitational wave is h, the tidal strain produced by the wave in the distance between free particles. A useful back-of-an-envelope formula for h produced by nearly-Newtonian systems is (Schutz 1984)

$$h \lesssim \phi_{int}\phi_N/c^4, \tag{2.1}$$

where

$$\phi_N = GM/r$$

is the Newtonian potential of the source at the observer's distance (r), and ϕ_{int} is the typical value of the Newtonian potential *inside* the source. This expression provides an upper limit to the radiative field h, on the assumption that the motions giving rise to the radiation are driven by self-gravitational forces (ϕ_{int}). The upper limit is reached for systems with maximal asymmetry, but h will be well below the limit if the motions are nearly spherically symmetric.

The flux of energy in a gravitational wave is roughly

$$F \sim \frac{c^3}{16\pi G} |\dot{h}|^2 \sim \frac{\pi c^3}{4G} h^2 f^2, \tag{2.2}$$

where f is the wave's frequency. For a simple wave of dominant frequency f and duration τ (and number of cycles n = fτ), we find that the total radiance (energy per unit area) is

$$R \sim F\tau \sim \frac{\pi c^3}{4G} h^2 fn \sim \frac{\Delta E}{4\pi r^2} \, . \qquad (2.3)$$

2.2. Filtering wide-band data

If the effective bandwidth of the signal is smaller than that of the detector, then the signal to noise ratio (S/N) will be improved by optimal filtering (Davis 1988). The best filter is one which is as long as the signal. A useful estimate of the *best* amplitude signal-to-noise ratio is provided by the formula

$$\frac{S}{N} \sim \frac{hn^{\frac{1}{2}}}{\sigma} \sim (\frac{4R}{\pi f})^{\frac{1}{2}} \frac{1}{\sigma} \, ,$$

where σ is the noise in a bandwidth equal to f, and n and R are as above. Using our expression for the radiance, we get

$$\frac{S}{N} \sim \frac{1}{\sigma \pi r} (\frac{\Delta E}{f})^{\frac{1}{2}}. \qquad (2.4)$$

This signal to noise ratio is independent of n because we have assumed that we can integrate over all n cycles. This may, of course, be impractical for some sources, but our estimate still gives the best we can expect.

We can now compare the detectability of two different sources at the same distance, observed by different detectors:

$$\frac{(S/N)_1}{(S/N)_2} \sim \frac{\sigma_2}{\sigma_1} \lceil \frac{\Delta E_1 f_2}{\Delta E_2 f_1} \rceil^{\frac{1}{2}}. \qquad (2.5)$$

The frequency dependence of this formula may be strongly affected by the way that σ depends on f. For example, a simple interferometric detector (no reclycling) that is optimized to detect waves of frequency f has σ ∝ $f^{3/2}$, while a recycling interferometer has σ ∝ f (Thorne 1987).

2.3. Examples

Consider first the *supernova* SN1987a. Suppose that a neutron star was formed with a large enough rotation rate to render it unstable to the gravitational wave instability (see Schutz 1987 and references therein). This would mean a spin period of the order of 1 ms. Then the star would be a source of 'spindown radiation' until it loses enough angular momentum to stabilize itself. Which would be easier to detect: the initial burst or the subsequent spindown radiation? Suppose, relatively conservatively, that the burst produced an energy $\Delta E_1 \sim 0.01\ M_\odot$ at $f \sim 1$ kHz. The spindown for the m=2 and m=3 instabilities takes place very rapidly, so observations begun a week or so after the event might see the m=4 instability. Supposing that the star spins down from a period of 1.3 ms to 1.4 ms in 3 months, it would radiate 0.001 $M_\odot$ at a frequency of, say, 1 kHz (very uncertain). If the two kinds of radiation are looked for with the same detector, then Eq.(2.5) tells us that

$$\frac{(S/N)_{\text{moderate supernova}}}{(S/N)_{\text{spindown}}} \sim 3. \tag{2.6}$$

Given the uncertainties about these estimates, it is clear that the spindown radiation may in fact be as detectable as the initial burst, especially if its frequency is lower than our estimate of 1 kHz. And if detectors had been looking for the spindown from the m=2 mode to m=3 in the first second or so after the burst, they might have seen it more easily than the burst itself. This is a possibility that should be borne in mind in future observations.

A second example is the *comparison of a supernova explosion and a coalescing binary at the same distance*, when observed by a recycling interferometer, in which the noise σ scales as f (Thorne 1987). If we take a supernova of $\Delta E \sim 0.01\ M_\odot$ at a frequency of 1 kHz, and for the coalescing binary system we use $f \sim 100$ Hz and $\Delta E \sim 0.01\ M_\odot$ (a typical value for the orbital energy radiated as the system is tracked from 100 Hz to 200 Hz), then we find that

$$\frac{(S/N)_{\text{supernova}}}{(S/N)_{\text{coalescing binary}}} \sim \frac{1}{30}. \tag{2.7}$$

This equation explains why coalescing binaries have replaced supernova explosions as the theorists' choice for the most likely source to be detected. Even if in a given volume of space the rate at which coalescences occur is as small as 10^{-3} of the rate of such supernovae, we may see far more coalescing binary systems because we can detect events in a volume of space 27,000 times larger.

3.   A CATALOG OF SOURCES

In this section I will list and briefly describe the principal sources of gravitational radiation that are currently favored by theorists. Although considerable effort has been devoted to studying possible sources, it should be borne in mind that these lists are essentialy predictions, and the quality of our predictions is limited both by our imaginations and by the availablility of information on potential sources from astronomical observations in the *electromagnetic* spectrum. It would therefore be very surprising if gravitational wave observations did not turn up some completely unanticipated sources. The reader will find a more detailed discussion of many of these sources in Thorne (1987).

3.1. Gravitational collapse

Since gravitational collapse is the trigger for supernova explosions, the supernova rate gives at least an assurance that if we can 'see' as far as the Virgo cluster then there will be a few *potentially* observable events per year. There are two major uncertainties: (i) the degree of nonsphericity in a typical gravitational collapse, and (ii) the possiblility that the collapse rate is much higher than the supernova rate, with the majority of collapse events being electromagnetically 'quiet'. I will return to (ii) below. The problem of nonsphericity hinges in large part on rotation: if the collapsing core rotates moderately rapidly, the collapse will 'pancake' in an axisymmetric manner; if there is even more rotation, rapidly growing nonaxisymmetric instabilities may be excited, leading to a core that looks like a tumbling cigar or that even fissions into two or more pieces.

Numerical calculations of *rotating axisymmetric* gravitational collapse can now be done with reasonable accuracy, but the last word has by no means yet been heard. Müller (1982) finds that if the collapse forms a 1 $M_\odot$ neutron star, then typically only 10^{-7} $M_\odot$ are released in radiation, with a frequency of about 1 kHz. Piran & Stark (1986) show that if a 10 $M_\odot$ black hole forms, then one can expect perhaps 10^{-3} $M_\odot$ in gravitational radiation, at 1.5 kHz. Taking the collapse to be 15 Mpc away (the Virgo cluster) gives on these estimates

$$h \sim 5 \times 10^{-24} \left(\frac{15 \text{ Mpc}}{r} \right) \quad \text{for neutron stars, and}$$

$$h \sim 4 \times 10^{-22} \left(\frac{15 \text{ Mpc}}{r} \right) \quad \text{for black holes.}$$

On the other hand, if a collapse occurs in our galaxy, at a distance of 15 kpc, then the amplitudes are 10^3 times larger. This gets close to the sensitivity of present bar detectors. I will return below to the question of how likely a nearby gravitational collapse event might be. The waveform may be quite variable if a neutron star is formed, but Piran & Stark (1986) find that the waveform when a black hole forms is remarkably insensitive to the initial conditions, essentially because is is a superposition of the lowest few normal modes of the hole. This means that filtering might improve the detectablility of black-hole events by perhaps a factor of 2.

More promising for detection is a collapse which has enough rotation to go *nonaxisymmetric*. Here we do not yet have accurate numerical calculations to guide our estimates, although several groups are preparing attacks on this problem. We can get an optimistic estimate by using Eq.

(2.1) and taking $\phi_{int} \sim 0.1$ and $M \sim 1 \, M_\odot$ for a neutron star, and $\phi_{int} \sim 0.3$ and $M \sim 10 \, M_\odot$ for a black hole. These lead to

$$h \sim 5 \times 10^{-22} \left(\frac{15 \text{ Mpc}}{r} \right) \qquad \text{for neutron stars (optimistic),}$$

$$h \sim 1 \times 10^{-20} \left(\frac{15 \text{ Mpc}}{r} \right) \qquad \text{for black holes (optimistic).}$$

It is hard to predict the waveforms from present calculations, but it may turn out here again that black-hole formation could produce a relatively standard waveform. If the collapse leads to the formation of a strongly nonaxisymmetric, tumbling core, then Ipser & Monagan (1984) suggest that the waveform could last as long as 30 cycles at about 1 kHz; however, their estimate of the amplitude of the waves is rather smaller than the optimistic one above, so that their predicted signal is just about as detectable as a burst at 1 kHz with the optimistic amplitude.

What are the chances of a collapse in our own Galaxy? If a collapse occurs at one fifth the distance to the recent supernova SN1987a, and if the present generation of bar detectors are on the air at the time, it will stand a good chance of being detected, particularly as one would probably also have a neutrino signal in coincidence. Until recently the accepted view has been that the collapse rate in our Galaxy should be similar to the pulsar birthrate or the supernova rate in nearby galaxies, *i.e.* about once per 30 years. But Blair (1988) argues, on the basis of a model for pulsar beaming that seems increasingly to be supported by observations, that the pulsar birthrate may be much higher, perhaps up to 1 every one to three years.

How can this be reconciled with the supernova rate? One possiblility is that not all collapses to neutron stars produce supernovae. (Because of the softness of the equation of state in the neutronisation regime, it is not possible to form a neutron star by quasistatic contraction of a white dwarf: neutron stars will always be formed by collapse.) Indeed, one might argue that there could be an *anti-correlation* between the emission of gravitational radiation and the production of a visible expanding cloud of gas. The maximum emission of gravitational radiation will come from rapidly rotating collapse, in which the collapsed star deforms into a tumbling ellipsoid or fissions into two or more smaller pieces temporarily. In such a situation, it may be harder to produce the rebound shock that drives off the envelope: with a larger ratio of surface area to volume such configurations may not trap neutrinos as effectively, and the more turbulent hydrodynamic processes may convert more of the released binding energy into heat rather than kinetic energy. Moreover, it may even be that the supernova rate *is* more like one every few years than one every thirty, but that most supernovae are not seen, either because the supernova luminosity function allows for a large number of less-luminous explosions, or because most supernovae occur in obscuring molecular clouds, or because supernovae are simply missed because they are transient events.

If collapses are really more common than has been assumed until recently, then -- as W. Fairbank has stressed -- the present generation of bar detectors has a real chance of making the first observations of gravitational waves: they will be the only detectors capable of sustained operation for the next five to ten years. It must therefore be of high priority to establish a network of bar detectors committed to observations rather than to further technical development, detectors that will be on the air all the time. The establishment of GRAVNET (Blair, Frasca, & Pizzella 1988) is a most welcome development.

3.2. Coalescing binary systems

As should be clear from our discussion of detectability in §2, coalescing binary systems are very promising, essentially because they emit large amounts of energy from the decay of their orbits at a relatively low frequency over a long time. For example, two 1.4 $M_\odot$ neutron stars in circular orbits around each other will have an orbital frequency of 50 Hz -- and therefore a gravitational wave frequency of 100 Hz -- when their orbital radius is 150 km; the system still has a couple of seconds (!) to live before the radiation decay of the orbit causes the stars to collide or to tidally disrupt one another. The first discussion of these systems as sources of gravitational radiation was by Clark & Eardley (1977), but they only estimated the burst of radiation from the coalescence event. Our appreciation ofthe importance of the radiation from the orbit itself is due to Kip Thorne (see Thorne 1987). The signal we can expect (Thorne 1987; Krolak & Schutz 1988; Krolak 1988a,b) is conveniently expressed in terms of an r.m.s. average <> over the orientations of the detector and the source (but see Thorne 1987 for a better definition of a 'typical' signal strength). Consider a system consisting of two objects with total mass M × $M_\odot$ and reduced mass μ × $M_\odot$, at a distance r × 100 Mpc, whose Newtonian radiation comes off at the frequency f × 100 Hz. One finds that

$$\langle h \rangle = 1.02 \times 10^{-23} \mu \, M^{2/3} f^{2/3} r^{-1} . \qquad (3.1)$$

Corrections due to possible magnetic interactions or tidal effects are negligible; for the post-Newtonian corrections, see Krolak (1988a,b). The rough estimate given by Eq. (2.1) is in fact very good here, since all the mass is moving nonspherically.

The timescale for decay of the orbit is given by

$$\tau = f/\dot{f} = 7.97 \, \mu^{-1} M^{-2/3} f^{-8/3} \text{ sec.} \qquad (3.2)$$

Because the orbital decay accelerates so quickly, the actual lifetime is 3/8 of this timescale. Notice that, since both h and τ are measureable, one can take their product to find r: the masses M and μ drop out. This ability to measure r is almost unique in astronomy, and it forms the basis of a new method of determining Hubble's constant (Schutz 1986a, 1986b). I will explore other consequences below. The long timescale means that the the number n of cycles in §2.2 is large: 0.6 fτ, to be exact. So if a system consisting of two 1.4 $M_\odot$ stars is picked up when f = 100 Hz, its S/N can be improved by a factor of $n^{1/2} \sim 20$ over the S/N for a burst at 100 Hz. Compounding this with the fact that a recycling detector will be 10 times more sensitive at 100 Hz than at 1 kHz means that this signal will be 200 times more detectable than a 1 kHz burst with the same h; put another way, if the binary were at 15 Mpc, it would be as detectable as a gravitational collapse that produced h ~ 1.8×10^{-20}. As we concluded earlier, this means that we can see coalescing binary systems much further away.

The key question, then, is how often do coalescences occur relative to the supernova rate? Observations of pulsars in our Galaxy give us a clue, but one whose interpretation is still very uncertain. We know of one precursor of a coalescing binary system, namely the famous 'binary pulsar' PSR 1913+16. It consists of two neutron stars that will coalesce in less than 10^8 years. Although its eccentricity is large now, by the time it reaches an observable state it will have circularized. Given one such system out of about 400 known pulsars, the simplest estimate is that the coalescence rate is 1/400th of the supernova rate in any volume. This leads to a figure of about 3 per year out to a distance of 100 Mpc (Clark, et al, 1979). This is the rate that Tinto (1988) adopts in concluding that a network of four interferometers could observe 2000 coalescences per year. But this rate assumes that the binary pulsar is a typical pulsar, and that there is a steady state, with coalescences balancing births of coalescence precursors. I shall now discuss a number of uncertainties in this rate.

(i) *The binary pulsar is not typical.* There are now 7 known pulsars in binary systems, only one of which is a precursor to a coalescing binary system. But three of the five fastest pulsars are in binaries. Current thinking is that these are old neutron stars that have been spun up by accretion from the companion, and there are indications that they may have lifetimes up to 1000 times longer than most pulsars. This would depress the inferred coalescence rate by a factor of 100 to 1000.

(ii) *Small number statistics.* Since we see only one precursor system, there is some chance that we are lucky somehow, and that the real fraction of pulsars in precursor systems is 10 or even 100 times lower. With equal probability (on Poisson statistics) it could be 3 or 10 times higher. It is somewhat reassuring to note that there are 7 pulsars in binaries. Of these, PSR 1913+16 has the shortest orbital decay time, but two other systems (0655+64 and 1831–00) have likely gravitational-wave decay times that are of the order of a Hubble time. Our one precursor, therefore, may not be particularly unusual. I shall allow a (1 σ) factor of 10 less or 3 more for small-number statistics. So far, therefore, we have a coalescence rate of between 3×10^{-4} and 0.1 per year out to 100 Mpc.

(iii) *There are selection effects in pulsar statistics.* These effects can go both ways. Since binary pulsars are preferentially short-period pulsars, and since people are looking hard for short-period pulsars (Lyne 1988), coalescence precursors may in fact be over-represented among all pulsars. However, there are severe observational selection effects against seeing short-period pulsars in binaries, because the Doppler shifting of the pulsar period smears out the sharp peak in the power spectrum that pulsar searchers look for (Lyne 1988). It may be fair to expect that binary coalescence precursors are actually more common than current observations would suggest. Quantifying this is, however, very difficult. I shall allow a factor of 1 to 10 for this, raising our rate to between 3×10^{-4} and 1 per year out to 100 Mpc.

(iv) *There may be no steady state, no equality between the coalescence rate and the precursor birth rate.* This is because we are dealing with systems that have lifetimes of the order of a Hubble time, so that the coalescence rate may depend as much on what systems were formed in the initial phase of star formation as it does on the present rates. Population II stars in spirals and ellipticals may well contribute to the coalescence rate, while they do not contribute to the (Type II) supernova rate. Since the average rate of star formation in our galaxy since its formation is probably some five times its present rate, and since including elliptical galaxies doubles the number of stars we have been considering, this effect would increase the coalescence rate by a factor of 10, provided that the distribution of coalescence times τ (number of systems in a given coalescence time interval) is flat, independent of τ. If it rises with τ, e.g. as τ^α for $\alpha>0$, then the correction factor is larger. Only if large numbers of systems with short coalescence times are formed would the present rate be dominated by recently-formed stars. The statistics for τ for the binaries in Table I are not very good, but they show no evidence for a preference for small τ. Accordingly, I will allow a factor of 10 to 50 for this correction, leading to a rate between 3×10^{-3} and 50 per year to 100 Mpc.

(v) *The gravitational collapse rate may far exceed the supernova rate.* Recent theories of the evolution of pulsar beaming have suggested that the rate of formation of pulsars has been underestimated, and it may be as high as 1 per three years in our Galaxy, compared to a supernova rate of 1 per 30 to 60 years (Blair 1988). This would raise the rate of coalescence by a factor of 10. Allowing for the possibility that this is not correct, we get a rate between 3×10^{-3} and 500 per year to 100 Mpc.

(vi) *There may be a significant number of black-hole coalescing binary systems.* We do not have enough information from pulsar statistics to estimate how many binaries contain a neutron star and a black hole of, say, 10-15 $M_\odot$ with a sufficiently short coalescence time. However, we can use the statistics of X-ray sources to give us a clue to how many there may be, because precursor systems probably pass through a stage in which they look like binary X-ray sources. Of the binary X-ray sources we can see, a few percent contain black holes of substantial mass (Hayakawa 1986). It may therefore be that in any volume of space, the rate of coalescence involving black holes may be a few percent of the rate for 'ordinary' two-neutron-star coalescence. From Eqs.(3.1) and (3.2) one can deduce that a system containing a 10 $M_\odot$ black hole will have a S/N some about twice as large as one with two neutron stars (Schutz 1986b), and so will be visible in a volume of space some 10 times larger. This may partly compensate the smaller event rate, so that these systems might increase the observed coalescence rate. Similarly, if two-black-hole systems coalesce at a few percent of the black-hole-neutron-star rate, they will be equally important: with an improvement of S/N of about 3 over the one-black-hole system, they will be visible in a volume of space 25 times as large. The observed event rate for black holes might be a substantial fraction of the two-neutron-star event rate. Of course, we may see none of these systems: it may be that companions of black holes never themselves evolve to black holes or even neutron stars. The effect of this on our estimated event rate is to raise the upper limit by a factor of 2, giving between 3×10^{-3} and 1000 per year for a system that can see two-neutron-star coalescences out to 100 Mpc.

Given all these uncertainties, Tinto's figure of 2000 events per year becomes something between 2 and 6 × 10⁵ per year for a network of four detectors. It may be that only gravitational wave observations will show us where in this range the true rate lies. Nevertheless, the fact that by adopting the most pessimistic view on each point we still obtain a lower bound of a few per year is encouraging: there are very strong grounds for expecting networks of interferometers to detect coalescing binaries with great regularity.

3.3. Pulsars

Since pulsars have non-axisymmetric magnetic fields, they will inevitably be sources of gravitational radiation. But the amplitude of the gravitational waves produced by this mechanism is too small to be observable in the near future. However, we know little about the details of the structure of neutron stars, and in particular about the topography of their crusts. It is possible that some pulsars have lumps (slight ellipticity) of mass frozen in as the star cools and contracts. If the ellipticity is δ (the fractional distortion of the radius of the star) then h will be estimated by a variant on Eq.(2.1):

$$h \sim v_{rot}^2 \delta \phi_{int} \phi_N / c^6, \tag{3.3}$$

where v_{rot} is the equatorial rotational velocity of the star. For a star of mass 1.4 $M_\odot$ and radius 10 km, emitting gravitational radiation at frequency f (twice the rotational frequency of the star), we have

$$h \sim 6 \times 10^{-22} \delta \left(\frac{f}{100 \text{ Hz}}\right)^2 \frac{10 \text{ kpc}}{r}. \tag{3.4}$$

An upper limit on δ is set by the spindown rate of the pulsar: gravitational radiation cannot be carrying away more energy than the loss of rotational kinetic energy. For the *Crab pulsar*, this limit is $\delta < 10^{-3}$. Current observational limits by the Tokyo group (Owa *et al* 1986) are 3 × 10⁻²² on h and about 0.3 on δ. It is expected that current designs for LIGOs will (when resonant recylcing is implemented) reach perhaps $\delta \sim 10^{-6}$.

3.4 Wagoner radiation

The gravitational radiation instability mentioned in §2.3 can have a more spectacular effect than simply spinning a neutron star down rapidly. If the star is accreting from a binary companion, then it may acquire enough angular momentum to reach an instability point for a mode with a reasonably short growth time, say the m=4 mode. Further accretion will cause this mode to grow and to radiate angular momentum away, until a balance is reached between accreted and radiated angular momentum: the star then will remain a fixed-frequency 'beacon' on the accretion timescale. The intensity of this radiation will be proportional to the accretion rate, but so will be the intensity of X-rays from the source, so that the gravitational wave luminosity and the X-ray luminosity will be proportional. Wagoner (1984) showed that a source whose X-ray flux is F_x and whose gravitational radiation comes off with frequency f will have amplitude

$$h \sim 2 \times 10^{-27} \left(\frac{300 \text{ Hz}}{f}\right)\left(\frac{F_x}{10^{-8}\text{erg cm}^{-2} \text{ s}^{-1}}\right)^{\frac{1}{2}}. \qquad (3.5)$$

Notice that this estimate does not contain the distance to the object: this is taken care of automatically in the decrease of F_x with distance. Since there are many sources of X-rays with fluxes above 10^{-10} erg/cm²s, and since LIGOs may well be capable of reaching h $\sim 10^{-28}$ with resonant recycling and long observing runs, it should be possible to search for these sources. However, the signal-to-noise ratio will be degraded unless the frequency f is known in advance. One way this might be observable is as a low-amplitude modulation of the X-ray flux. Present data do not show any modulations that would be candidates for this effect, but there is a proposal to fly an X-ray observatory that would have considerably more sensitivity for such effects than any previous satellites have had (Wood, *et al* 1986). If this satellite is built and discovers modulation in any source, that source will be a good candidate for the Wagoner radiation.

3.5 Stochastic background

The stochastic background of gravitational waves is any 'confusion-limited' wave field, where the time between receiving different and unrelated wave trains is less than the length of each wave train. Such a background must be sought by looking for correlations between independent detectors: it will appear as an irreducible, correlated 'noise' in the antenna. The detectors must be nearer than about half a wavelength of the radiation if they are to have optimum sensitivity. Michelson (1987) has studied this problem in detail. For bar detectors, correlations between detectors on the same site give the best sensitivity, provided seismic disturbances (also correlated) can be eliminated. For LIGOs, observing above 100 Hz, baselines between proposed European detectors would be ideal, but the proposed American detectors are rather far apart.

Possible sources of a stochastic background in the frequency range 10–1000 Hz include cosmic strings (Vachaspati & Vilenkin 1985; Brandenburger, *et al* 1986), primordial waves (reviewed by Thorne 1987), and binary coalescence and black hole formation in an early generation of stars ('Population III': see Bond & Carr 1984). At lower frequencies, close binary systems in our Galaxy and various phase transitions in the early universe may contribute.

It is usual to characterise the strength of the background at a frequency f by giving Ω_g, the energy density in the waves in a bandwidth Δf = f as a fraction of the cosmological closure density. One of the firmest predictions of Ω_g is given by cosmic strings. Strings provide an attractive model for seeds of galaxy formation; with only one free parameter -- the mass per unit length μ of the strings -- they can explain a range of features of the observed masses and clustering properties of galaxies. Such seeds give off gravitational radiation, and for the best choice of μ one finds that $\Omega_g \sim 10^{-7}$ at observable frequencies. According to Michelson (1987), cryogenic bar-type detectors can reach this limit below 200 Hz, and

LIGOs may be able to get down as far as 10^{-12} at 50 Hz. It is clear that the testing of various theories for the stochastic background is one of the most interesting jobs for detectors being designed now.

4.   ASTROPHYSICAL INFORMATION FROM COALESCING BINARIES

If coalescing binaries can be observed by LIGOs, we will learn more than just that gravitational waves exist.   This is because, if the waves are observed with enough detectors, the waveform can tell us the absolute *distance* to the source.   Since direct distance measurements are rare in astronomy, I will devote this section to mentioning some of the possibilities that this opens up.

4.1. The distance to a coalescing binary

The key to measuring the distance is in Eqs.(3.1) and (3.2), which give the mean $\langle h \rangle$ of the waves and the timescale τ for the change in their frequency.   The masses of the stars enter these expressions in the same way, so that their product is

$$\langle h \rangle \tau = 8.13 \times 10^{-23} f^{-2} r^{-1} \text{ sec.} \qquad (4.1)$$

This is independent of the masses.   Since the frequency is known, a measurement of $\langle h \rangle$ and τ determines the distance r to the source (given in this equation in units of 100 Mpc).   Before we go on to see what we can learn from r, we need to address two points: First, can one measure $\langle h \rangle$, the r.m.s. orientation average of h, for any individual system?   Second, how secure and reliable is the model of the binary system that leads to Eqs.(3.1) and (3.2)?
    On the measurement of $\langle h \rangle$, it is clear that this is not directly measured.   But if it is possible to measure the actual orientation of the source and its direction on the sky, then one can reconstruct $\langle h \rangle$ from this information.   Now, a network of four detectors can determine this information (Schutz 1986a).   It is not hard to see how the direction to the source can be inferred from time-delay information; the orientation of the source on the sky and the inclination of its orbital plane can be determined by measuring the polarisation of the wave.
    On the reliability of the model, Krolak (1988a,b) has considered several possible complications to the basic model of a point-mass Newtonian binary in a circular orbit with quadrupole gravitational radiation reaction. By the time an orbit decays to the point where its period is a few milliseconds, it will be essentially perfectly circular: radiation reaction eliminates any initial eccentricity.   Tidal effects between neutron stars have a negligible effect on the orbit until they begin actually to exchange mass.   Magnetic effects are similarly negligible.   Post-Newtonian corrections to the orbit can be significant, at around the 5-10% level, but these are not really a complication.   In fact if they are large enough to influence the observations, then they can be solved for from the data, and this will give independent information that determines the individual masses of the stars.   It seems, therefore, that the model is exceedingly robust and

that the distances inferred from it are likely to be essentially free from systematic error.

4.2 What we may be able to infer from observations

As Tinto (1988) shows, a network of detectors may be able to see a sizeable fraction of all coalescing binary events out to 1.5 Gpc or even further. The following discussion of the implications of this is based upon Krolak & Schutz (1988).

(i) *Hubble's constant*. If the event rate is high enough to give several events per year from within 100 Mpc, then Schutz (1986a) has shown that an accurate determination of Hubble's constant is possible. This is because such events have a signal-to-noise ratio sufficiently large that a determination of the distance to the system accurate to a few percent is possible, and of the direction to ±3°. If a search of this window reveals an optically visible supernova-like event from the actual coalescence of the two stars, then by measuring the redshift of the 'host' galaxy one can determine H_0 to a few percent, much better than we know it today. In the absence of any optical counterparts, a statistical method will suffice in the long run (after 10-20 events are recorded).

(ii) *Standard mileposts*. If an optical identification can be made as described above, the determination of the distance to the host galaxy provides a milepost for the calibration of a variety of distance measures and intrinsic luminosity measurements.

(iii) *Masses of neutron stars and black holes*. It is clear that, unless we detect the post-Newtonian corrections to the radiation, we can measure only the combination $\mu\, M^{2/3}$, which I shall call the mass parameter of the system. Provided that the event rate is high enough to generate good statistics on the mass parameter (a few hundred events per year would probably suffice), one may be able to infer the maximum mass of neutron stars. The existence of massive black holes in binaries would be clear if we observe large values of the mass parameter. If the post-Newtonian terms can be detected, even more direct information will emerge.

(iv) *Mass distribution in "local" region of the universe*. Events out to, say, 500 Mpc (z about 0.1 - 0.2) can be used to test for isotropy, homogeneity, superclustering, and the existence of voids. Binary coalescences ought to be randomly distributed among stars (whose distribution may or may not coincide with that of bright galaxies), so they should be ideal tracers of the stellar population.

(v) *Tests of general relativity*. The information from an observation by four detectors will test Einstein's predictions regarding gravitational wave polarization: four detectors overdetermine the solution for the wave, so any inconsistency among them would be evidence for polarization states not allowed by general relativity (Schutz 1986a). If an event is accompanied by the kind of optical counterpart discussed above, then a measurement of the speed of propagation of gravitational waves is possible: since the optical emission should brighten up within a day after the gravitational waves leave, the arrival at Earth of the two emissions with a day or so separating them would be evidence that the gravitational waves travel at the speed of light to an accuracy of better than one part in 10^9. Coalescence events provide other tests as well: the post-Newtonian terms predicted in Eqs. (1) and (2) might be independently detectable in radiation at *four* different frequencies, and must be consistent with one another. A two-black-hole coalescence event would be the strongest evidence one could imagine for the existence of black holes: not only would it identify the holes, but it would also test the predictions of general relativity regarding their radiation in the strong-field limit. (By the time observations are possible, numerical calculations should have determined these predictions to a high accuracy: Miyama *et al* 1986.)

(vi) *Neutron-star equation of state.* By following a coalescing binary system until the S/N drops to about 1, one may be able to get information about the dynamics of the actual coalescence of the two objects: when mass transfer begins, whether the coalescence is accompanied by collapse to a black hole, etc. Coupling this with numerical simulations should constrain the neutron-star equation of state. Bursts of gamma rays may also be detectable in coincidence with coalescence events (Blinnikov, *et al,* 1984). At the great distances involved, this could improve the above test of the speed of the waves by a factor of 10 or more in accuracy.

If LIGOs can improve their sensitivity by using technology that develops over the next ten years, then observations can be pushed to larger distances. Let us suppose (rather arbitrarily) that an improvement in sensitivity of a factor of 10 is possible, perhaps by a combination of higher laser power, better mirrors, longer baselines, the use of squeezed light (Leuchs 1988), and active seismic isolation. Neutron-star binaries can then be seen to a luminosity distance of 8 Gpc, which corresponds to a redshift larger than 1. The event rate will not be a problem with such a volume of galaxies: even on the most pessimistic assumptions, there would be thousands of events per year. Black-hole binaries would be seen at essentially any redshift, and those at redshifts of order unity would have S/N > 20. Such observations then open up a new set of possibilities: the measurement of the deceleration parameter and the mass density of the universe, the detection of differences in the formation rates of neutron stars due to chemical evolution, gravitational lensing of coalescing binary events, and the observation of the redshift at which star formation begins. See Krolak & Schutz (1987) for details.

These kinds of observations make a strong case for continuing the effort to develop LIGOs, even if that effort requires ten or more years of work on large systems to get them to their ultimate potential. If bar detectors are the observing systems for the near future, LIGOs seem to be the gravitational wave observatories of the future. At present, general relativity is important to astrophysics as an interpreter of observations made primarily in the electromagnetic spectrum. When observations like these begin to be made, general relativity will enter a new epoch as a provider of new and qualitatively different observational data to astronomy.

REFERENCES

Blair, D. 1988, this volume.
Blair, D., Frasca, S., & Pizzella, G. 1988, this volume.
Blinnikov, S. I., Novikov, I. D., Perevodchikova, T. V., & Polnarev, A. G., *Sov. Astron. Lett.,* **10**, 177 (1984).

Bond, J. R., & Carr, B. J. 1984, *Mon. Not. R. astr. Soc.*, **207**, 585.

Brandenburger, R.H., Albrecht, A., & Turok, N. 1986, *Nucl. Phys. B*, **277**, 605.

Clark, J. P. A., & Eardley, D. M. 1977 *Astrophys. J.*, **215**, 315.

Clark, J. P. A., van den Heuvel, E. P. J., & Sutantyo, W. 1979, *Astron. Astrophys.*, **72**, 120.

Davis, M. 1988, this volume

Hayakawa, S. 1986 in Gravitational Collapse and Relativity, eds. H. Sato & T. Nakamura (World Scientific, Singapore).

Ipser, J. R., & Monagan, R. A. 1984, *Astrophys. J.*, **282**, 287.

Krolak, A. 1988a, this volume.

Krolak, A. 1988b, in preparation.

Krolak, A., & Schutz, B. F. 1987, *Gen. Rel. Gravit.*, **19**, 1163.

Leuchs, G. 1988, this volume.

Lyne, A. 1988, this volume.

Michelson, P. F. 1987, *Mon. Not. R. astr. Soc.*, **227**, 933.

Miyama, S., Nagasawa, M., & Nakamura, T. 1986, in Gravitational Collapse and Relativity, eds. H. Sato & T. Nakamura (World Scientific, Singapore).

Müller, E. 1982, *Astron. Astr.*, **114**, 53.

Piran, T., & Stark, R. F. 1986, in *Dynamical Spacetimes and Numerical Relativity*, ed. J. Centrella (Cambridge University Press), pp 40-73.

Schutz, B. F. 1984, *Am. J. Phys.*, **52**, 412.

Schutz, B. F. 1986a, *Nature*, **323**, 310.

Schutz, B. F. 1986b, in Gravitational Collapse and Relativity, eds. H. Sato & T. Nakamura (World Scientific, Singapore).

Schutz, B. F. 1987, in *Gravitation in Astrophysics*, eds. B. Carter & J. B. Hartle (Plenum, New York).

Thorne, K.S. 1987, in *300 Years of Gravitation*, eds. S.W. Hawking and W. Israel (Cambridge University Press).

Vachaspati, T., & Vilenkin, A. 1985, *Phys. Rev. D*, **31**, 3052.

Wagoner, R. V. 1984, *Astrophys. J.*, **278**, 345.

Wood, K. S., Michelson, P. F., Boynton, P., Yearian, M. R., Gursky, H., Friedman, N., and Dieter, J. 1986, *A Proposal to NASA fo an X-Ray Large Array (XLA) for the NASA Space Station* (Stanford University, Palo Alto).

THE RATE OF GRAVITATIONAL COLLAPSE IN THE MILKY WAY

D.G. Blair
Physics Department
University of Western Australia
Australia   6009

ABSTRACT.   Data for historical supernovae, external galaxy supernovae,
and radio pulsars are considered in order to estimate the rate of
gravitational collapse events of interest to gravitational radiation
astronomers.   It is shown that the rate of historical supernovae has been
under estimated, and is at least 1 per 10 years in our galaxy.   The rate
of supernovae in external galaxies is consistent with the historical
rate, but the radio pulsar rate, of 1 per 4 years obtained from
geometrical evolution, is an order of magnitude larger than expected from
the supernova rate.

1.INTRODUCTION

The present generation of gravitational radiation antennas have a
strain sensitivity $h \sim 10^{-18}$ at frequencies $\sim$ 1 kHz.[1,2].   Over the next
few years improvements should bring this sensitivity down to $h \sim 10^{-19}$.
This range of sensitivities is sufficient to see relatively low
efficiency transient gravitational radiation events within our galaxy
($\varepsilon \sim 10^{-2}$-10^{-4}) but is insufficient to detect even high efficiency events
at the Virgo cluster.[3]
Traditional estimates of the rate of gravitational collapse based on
supernova statistics give an event rate of 1 per 30-50 years.   Here we
look again at the available data, which includes not just supernovae, but
also pulsars, in order to obtain more reliable limits.   Two major factors
are identified as leading to errors in obtaining event rates.   The first
is the failure to allow for the true surface density distribution of
sources in the galaxy.   Strong evidence supports an exponential disc,
whereas previous studies assume a uniform disc.   The second applies to
radio pulsars: it is impossible to estimate their birthrate without
proper understanding of their evolution.   The recent confirmation of
pulsar geometrical evolution[4,5,6] allows a new and more rigorous
estimate of the pulsar birthrate through knowledge of the time evolution
of the beaming factor.

B. F. Schutz (ed.), Gravitational Wave Data Analysis, 19–32.
© 1989 by Kluwer Academic Publishers.

None of the new estimates for gravitational collapse rates are
independent. Exponential disc data is obtained from studies of the
pulsar population, and from supernova studies. Historical supernovae
must be analysed in the context of the exponential disc. The population
of pulsars is estimated from integration over the exponential disc galaxy
based on local observations in our own.

In section 2 we review data for the surface density of supernovae in
external galaxies, and for the surface density of pulsars and massive
stars in the Milky Way. The data is shown to be basically consistent,
supporting an exponential surface density distribution with a scale
length of 7.5 kpc in our galaxy. Based on this surface density
distribution we derive a curve for the effective area sampled as a
function of horizon from the Solar system. In section 3 we examine the
historical supernova rate in the light of the exponential disc model.

In section 4 we review Tammann's analysis of the rate of supernovae
in external galaxies while applying corrections for selection and
possible errors in galactic classification. [7]

In section 5 we review radio pulsar evolutionary models, and derive
the radio pulsar birthrate.

In section 6 we conclude by emphasising that the rate of
gravitational collapse in our galaxy is probably much higher than
previously estimated, but that a significant discrepancy exists between
the supernova rate and the pulsar birthrate, which may only be resolved
by hypothesising that pulsars form from lower mass systems by silent
collapse. [8]

2. THE SURFACE DENSITY

The surface density of radio pulsars has been estimated by Lyne,
Manchester and Taylor [9]. It shows a clear exponential form, with a
scale length of about 7.5 kpc. There is an apparent "hole" towards the
galactic centre, however, where errors also become large. Recent data
shows however [10] that the exponential form continues to within 2 kpc of
the galactic centre. Figure 1 shows the old pulsar data, along with data
for the surface density of massive stars, as well as a 7.5 kpc scale
length exponential for comparison. [11]

Surface density curves can be derived from data given on supernovae
in external galaxies. Figure 2 shows the surface density for supernovae
plotted from data given by McCarthy [12] (figures 2(a) and 2(b) and
derived from data given by Tammann [7] (Figure 2(c)). The galactic radius
estimate is based on Hubbles constant H=100 km/sec/Mpc, for Figure 2(a)
and 2(b) [12], whereas in figure 2(c) the radius scale assumes an optical
radius of 14 kpc, and is based on data for supernovae in 33 face-on Sc
galaxies. [7]

All the data shows clear evidence for exponential distributions.
The supernova data indicates that they are centrally concentrated, much
as Lyne's new pulsar data indicates.

The discrepancy between the scale lengths in Figure 1 and 2 are
likely to be due to a combination of errors in the distance scale for our
galaxy, errors in the Hubble constant and errors in estimating the
optical radius of Sc galaxies. It is worth noting that the supernova

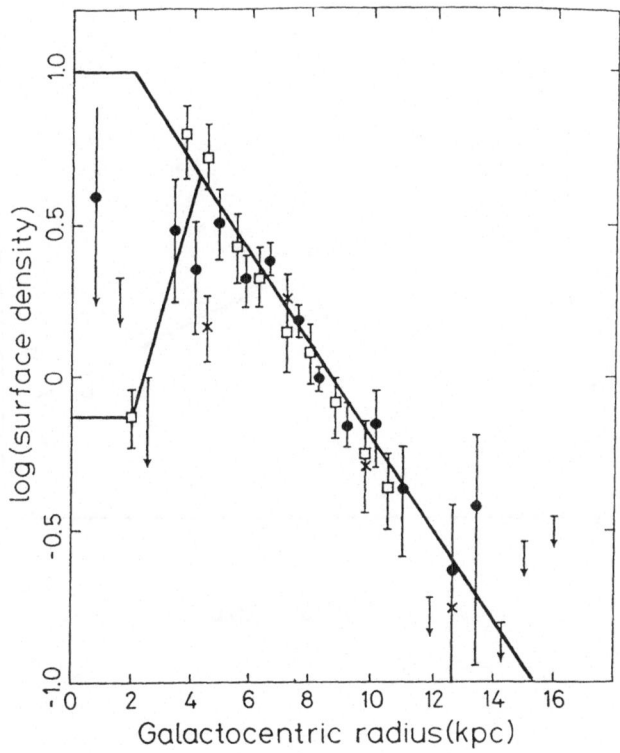

Figure 1.   The Surface density as a function of galactocentric radius for
radio pulsars and massive stars.   The dashed line is an
exponential of 7.5 kpc scale length.

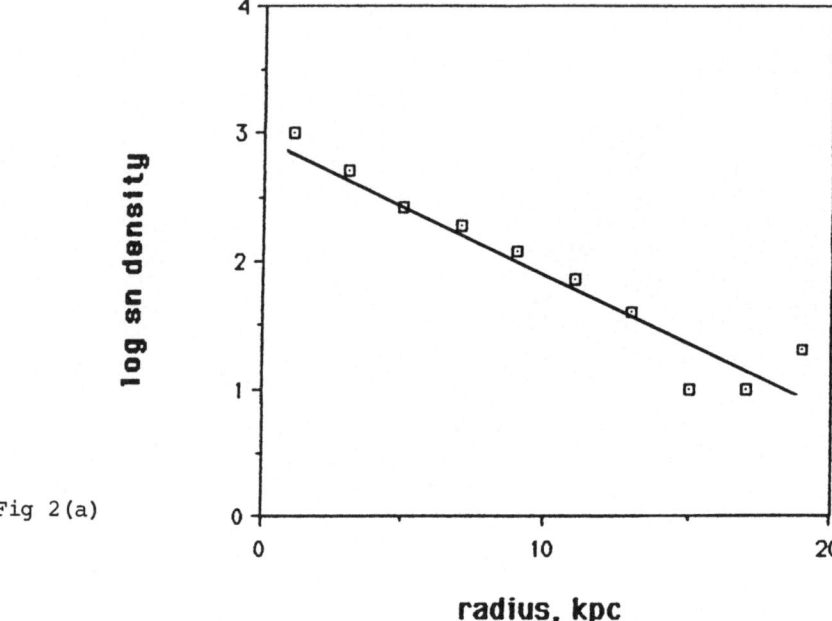

Fig 2(a)

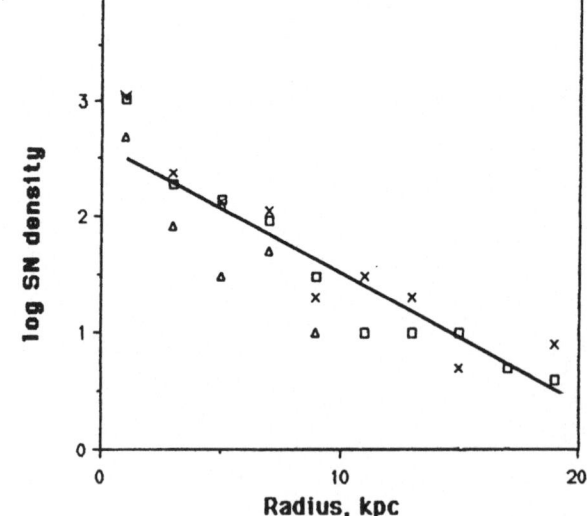

Fig 2(b)

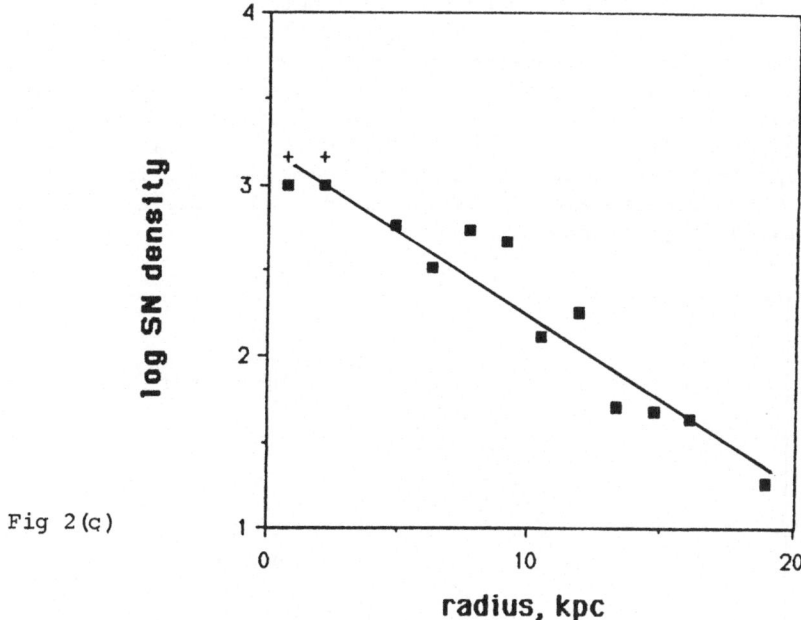

Fig 2(c)

Figure 2. The surface density of Supernovae (arbitrary units) for
external galaxies. a) Surface density of all Supernovae for
galaxies with recession velocity $V > 500$ kms^{-1}. b) Surface
density of supernovae by supernova type and for Sc spirals
(from data given by McCarthy). Squares: SNI, triangles SNII,
crosses: all SN in Sc spirals.
c) The surface density of supernovae in face on Sc spirals
(based on data given by Tammann). The + indicate data
corrected by a 45% core selection effect discussed in the
text.

optical radius of Sc galaxies. It is worth noting that the supernova
surface density scale length does in fact provide a means of determining
H, assuming that the Milky Way scale length is accurately known, and that
our galaxy type is also known.

Selection effects should be considered here. Almost all the
supernovae in this data have been photographically discovered, and
McCarthy suggests that the total supernova count in nuclear regions
should be increased by as much as 45% to account for the poor contrast of
supernova against a bright nucleus. In figure 2(c) the central two bins
are corrected by 45%; clearly the effect of the correction on the overall
distribution is rather weak.

Based on the above data we will now assume that the surface density
distribution of supernovae and pulsars are given by an exponential
distribution with a scale length of 7.5 kpc. We then obtain a curve for
the effective areas of the galaxy within a given horizon. This is
important in obtaining the total pulsar population, and in integrating
the observed density of historical supernovae across the galaxy.

We continue to assume that the galaxy is cylindrically symmetric, and numerically integrate the surface density distribution as a function of radius from the assumed solar system radius (8.5 kpc from the galactic centre). The effective area vs horizon curve is given in Figure 3. The area is normalised to the surface density at the solar system. The effective area of the galaxy is 1260 kpc^2. The effective area increases quadratically (approximating a uniform disc) only within the first 3-4 kpc, thereafter the effective area increases roughly linearly to about 12 kpc before saturating beyond about 18 kpc.

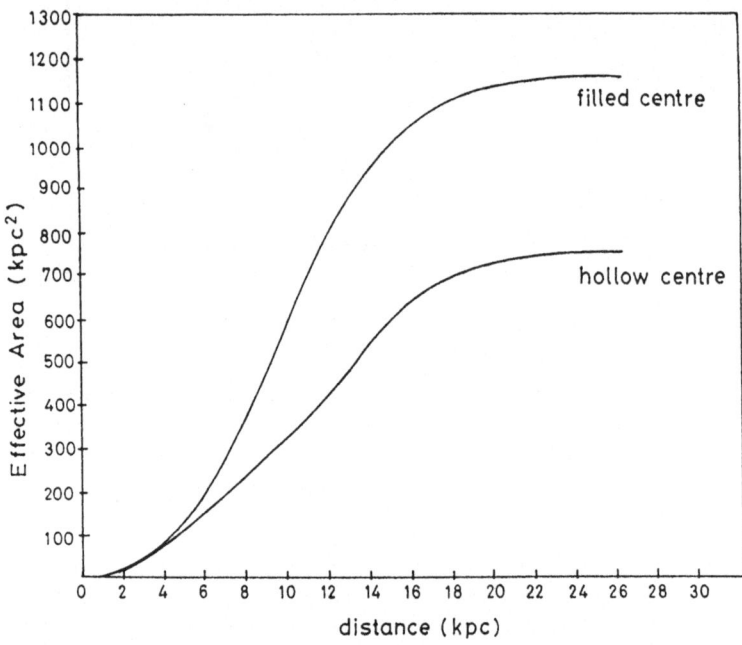

Figure 3. The effective area of the galaxy (within a given horizon, normalised to the surface density at the solar system) as a function of horizon.

3. HISTORICAL SUPERNOVAE

The historical supernovae have been discussed by Clarke and Stephenson[12] and analysed in more detail by Tammann[7]. Tammann shows . that the radial distribution of historical supernovae agrees reasonably well with that for external supernovae. Seven supernovae are considered to have occured in the last 1,000 years, as shown in Figure 4. Of these, only 5 are historically recorded. In addition Tammann considers that Cas A should be included as it is local to the solar system and accurately dated (see Table I which is taken from reference (7)). In addition one of the two young Southern supernova remnants MSH-11-54 and CTB80 is included, to compensate for the lack of Southern historical records. (See reference 7 for more details). All but one of this historical supernova set has a distance ≤ 4 kpc, while the mean distance is 3.7 kpc.

Previous estimates of the supernova rate have assumed a uniform disc model for the galaxy, and have assumed that all supernova were observed historically in the galactic segment 50° wide in which the historical data set occurs.

Strong extinction towards the galactic centre, and the expected increasing surface density at low galactic radii leads to a higher expected rate. If we take a horizon of 4 kpc we have an effective area of 70 kpc^2, compared with a total effective area of the galaxy of 1260 kpc^2. (See Figure 3). Thus we extrapolate that from 7 historical supernovae, 7 x 1260/70 = 125 supernovae actually occurred.

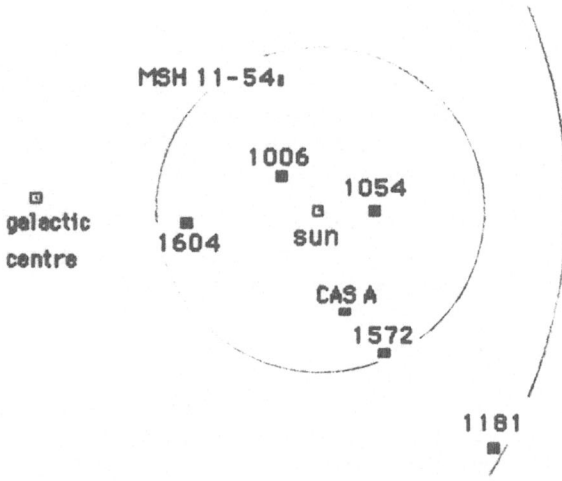

Figure 4. The distribution of historical supernovae on the galactic plane.

It is difficult to identify the correct horizon for this calculation due to the non-uniformity of the data set. It may be more appropriate to consider only the 6 supernovae that occur within the horizon of 4 kpc, thus obtaining ∼ 110 supernovae in the last 1,000 years. Thus we obtain a birthrate of historical supernova of 1 per 8-9 years. The statistical error is ±40%.

4. SUPERNOVAE IN EXTERNAL GALAXIES

In considering supernovae in external galaxies we begin with Tammann's result that the mean time interval between supernovae in Sb-Sbc spirals is $\tau \leq 24.4$ yrs. We consider three sources of bias which may all act to reduce the mean interval. The first problem is that of estimating our own galaxy type. Tammann states[7]: "τ could be 30-40% lower because external supernova rates are based on *total* galactic luminosity whereas our local surface brightness is dominated by the disc component". Thus absorption within our galaxy has given us inadequate knowledge of the galactic bulge and may have caused us to significantly underestimate the total galactic luminosity. It is suggested here that a more appropriate normalisation of galaxy types may be to use the galactic x-ray luminosity since this is less affected by absorption, and to a large extent is due to collapsed objects which can probably be associated with supernov. The second correction considered here is selection against discovery of supernova near galactic centres, due to the poor dynamic range of phtographic plates. Over exposure in the bulge region is estimated by McCarthy to cause up to 45% of supernova in the bulge region to be missed, which lead to an overall deficit of 22% of supernova.

Finally we consider the effect of low luminosity supernova exemplified by SN 1987 A ($M_v \sim -15$) and Cas A ($M_v > -16.5$). Of the historical data set one out of 6 is of low luminosity. The majority of such low luminosity supernova will be missed in external galaxies, leading to a correction factor of 16% on the external rate. If the corrections based on the above arguments are all applied to Tammann's 1983 result, we obtain a new estimate of $\tau \sim 9.5$ yrs. This result is presented as a plausible rate allowed by the data, but as opposed to Tammann's upper limits no rigorous errors are estimated.

5. THE BIRTH RATE OF RADIO PULSARS

It is generally believed that radio pulsars are born in supernova. This is born out by the classic example of the Crab Nebula, but the rigour of this assertion will be further discussed in section 5. The radio pulsar birthrate can be determined from the population statistics of the ~ 400 known pulsars, but the conclusion is dependent on knowledge of the pulsar evolution process. Previous estimates[9] have assumed that the entire population of pulsars have a mean beaming factor F of about 5. That is, the fraction of observed pulsars in the vicinity of the solar system is about 20%. The pulsars which are not observed have emission beams which do not cross the observers position in the sky. The static models which have been used previously assume that the decay of the pulsar radio luminosity with time (due to a decaying magnetic field) accounts for the observed pulsar age distribution. Models of this type do not lead to an explicit solution to the beaming factor, and it has been shown[4,5] that they fail to account for the skewness of the observed number pulsewidth

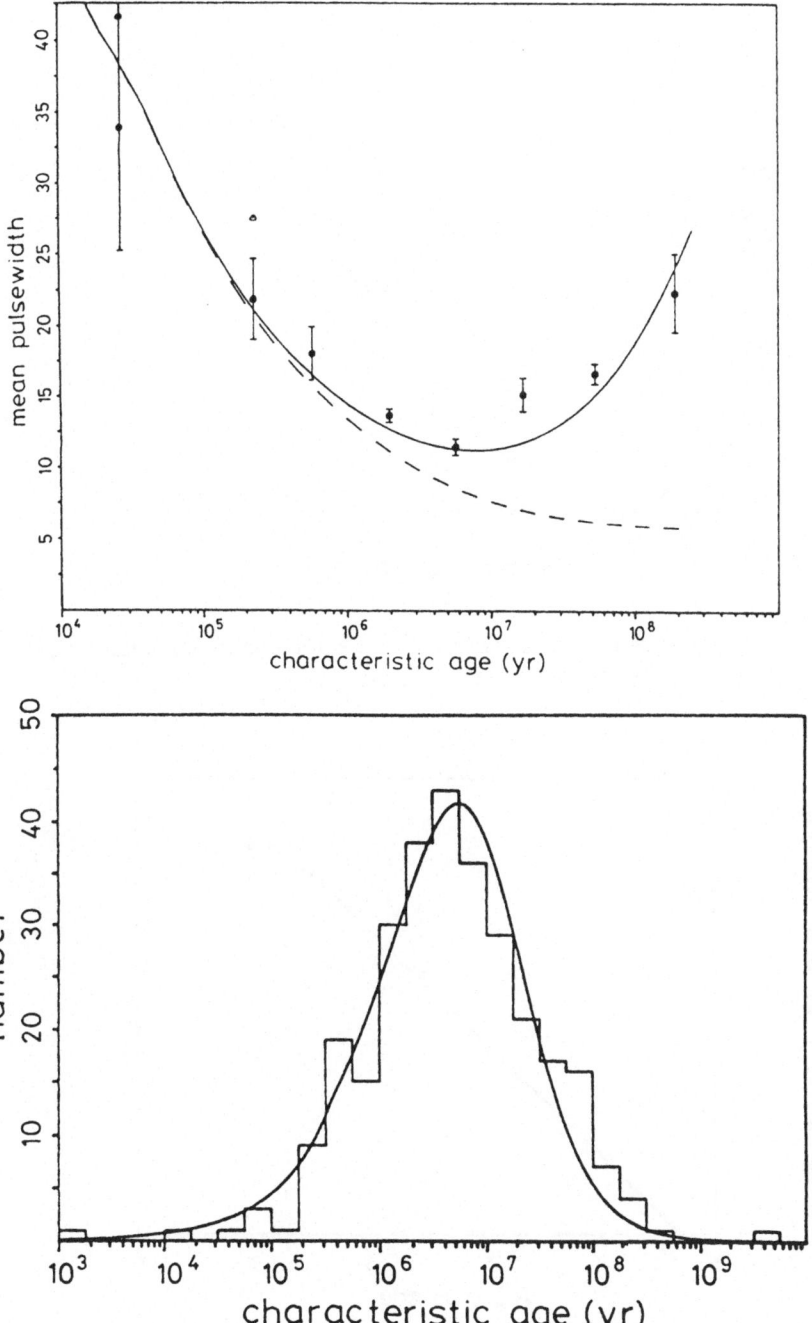

Figure 5.   (a)   The observed pulse width characteristic age distribution,
            and (b) The number age distributions for radio pulsars,
            compared with the prediction of   the CB cone-narrowing
            alignment model.

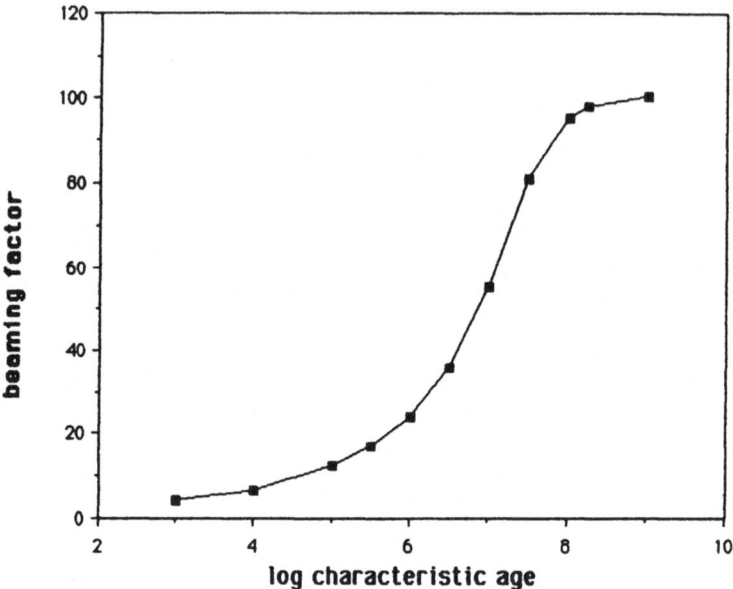

Figure 6. The dependence of beaming factor on characteristic age for radio pulsars.

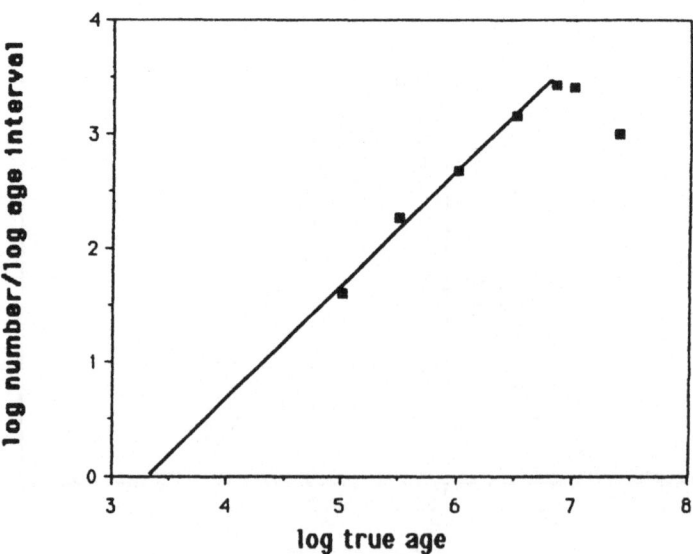

Figure 7. The pulsar number per unit log age interval plotted against age.

distribution, the minimum in the pulsewidth-age distribution, and the peak in the polarisation gradient-age distribution. Moreover on fundamental grounds one would not expect pulsar emission geometry to be age-independent. Electromagnetic torques must bring about evolution of the angle θ between the pulsar emission cone and the spin axis.[14] The sign of the torques have not been theoretically determined; however the data strongly indicates that pulsars align as they age. This gives rise to an increase in the pulsar beaming factor with age.

Candy and Blair have shown that pulsar evolution can be described by alignment:

$$\sin\theta \sim e^{-t/\tau_a} \tag{1}$$

where τ_a is the alignment time scale.

The pulsar emission cone is described by a power low dependence on period P:

$$\beta \sim P^{-\gamma}. \tag{2}$$

That is, the emission cone narrows as the pulsar spins down. The alignment time scale τ_a is 2×10^7 years, and the coefficient γ has a value between 0.3 and 0.7. The cone narrowing-alignment model accurately predicts all the statistical properties of the population. Some examples are given in Figure 5. Independent studies have confirmed the basic form of equation (1) and (2).[15,16]

The above model leads to a definite conclusion for the evolution of the pulsar beaming factor F. F is simply the reciprocal of the fraction of the sky which is swept by the pulsar beam. (See reference (14) for full details.). The dependence of F on pulsar characteristic age is given in figure 6. F rises from 2-3 for young pulsars, to ~ 100 for pulsars with characteristic age $t_c > 10^8$ yrs. Thus we can only observe $\sim 1\%$ of such old pulsars.

With precise knowledge of $F(t_c)$ we can go on to obtain a rigorous birth rate as follows:

If the pulsar population is stable in time it follows that

$$\frac{dN_t}{dt_a} \Delta t_a = K \tag{3}$$

where N_t is the true number of pulsars in some observation volume, and t_a is the true pulsar age. We define $R_t = K/\Delta t_a$, which is the number of pulsars per unit age interval per volume of galaxy sampled. We transform (3) to appropriate logarithmic units

$$\log\left(\frac{dN_t}{d\log t_a}\right) = \log t_a + \log R_t + \log(\ln 10) \tag{4}$$

If the lefthand side of equation (4) is plotted against log t_a, we should obtain a straight line of slope=1. The intercept gives the value of R_t, while the fixed slope is a measure of the internal consistency of the data.

The galactic birthrate B is then given by

$$B = R_t \frac{A_G \rho_s}{N_{obs}} \qquad (5)$$

where A_G is effective surface area of the galaxy, and ρ_s is the local surface density of pulsars. From equation (4) R_t may be determined, and hence B may be determined using knowledge of $A_G \rho_s$.

Lyne, Manchester and Taylor have derived $A_G \rho_s$ from the observed population statistics.[9] They obtain

$$A_G \rho_s = (0.7 \pm 0.17) \times 10^5 .$$

Figure 7 shows the pulsar number per unit log age interval, plotted against age. The data is determined from the fitted number age distribution figure 5(b) using $\tau = 2 \times 10^7$ yrs. The results show that indeed the data is self consistent, the slope is unity and the intercept gives $R_t = 1.1 \times 10^{-3}$. The data in the last two age bins fail to fit the unity slope curve, presumably due to luminosity decay for $t_a > 10^7$ yrs. The derived birthrate is 1 per 4 ± 1 yrs.

The above result is preliminary. A more accurate result may be obtainable by individually correcting pulse profile data using polarisation information[17] and through fine tuning of the fitting parameters. Systematic effects could bring about as much as 50% change in the birthrate.

It is clear, however, that geometrical evolution gives rise to a radical change in the pulsar birthrate. The estimated birthrate of one per 4 years, an order of magnitude larger than previous estimates.

6. DISCUSSION

The above results suggests that there are grounds for believing that the SN rate is more than 1 per ten years in our galaxy. The historical rate and the external galaxy rate are reasonably consistent. The pulsar birth rate is about one per four years. Given the size of the errors this result could be seen as consistent with the SN rate as long as the pulsar – SN association is strong.

In fact the association is not so strong. Only 8% of SN remnants have the filled shells characterising the presence of a central pulsar,[18] and of the 7 historical supernova only two have filled shells. A reduction in he fraction of plerions as supernova remnants age may be expected due to pulsar spin down (reduced energy injection) and alignment (reduced space-filling by the relativistic beams). Amongst

the historical SN direct pulsar association occurs only in the case of
the Crab. We must assume that we are not within the beam geometry of the
pulsar from SN 1181. If we adopt a conservative figure of 30%
association between radio pulsars and supernova we reach the following
conclusion:

    a)   If we accept the pulsar interval of 4 years, we would expect to
        see a supernova interval of 1.3 years.

    b)   If we accept the supernova interval of 10 years, we would expect
        to find a pulsar interval 20-30 years.

These are significant discrepancies. It is difficult to escape from the
conclusion that gravitational collapse must occur through non-optically
luminous events. The discrepancy between the galactic lattitude of
historical supernova ($Z > 200pc$) and progenitor stars ($Z \sim 90$ pc)
suggests that a large fraction of galactic supernovae may be missed due
to strong absorption in the galactic plane. However this argument should
not apply to supernova in external galaxies. Nomoto has advocated silent
collapse of white dwarfs, as a source of unseen supernova.

If the CB pulsar evolutionary model is correct, and if the plerions
truly reflect pulsar content, the galactic gravitational collapse rate
could be ~ 1 per year. This compares with stellar evolutionary studies
by Bahcall and Piran[19] leading to a stellar death rate of one per
years.

A more likely reconciliation of the SN-PSR discrepancy may be to
suppose that pulsars are born in supernovae at 30% of 1 per 7-10 years,
and they are also born in 100% of silent collapses. The implication is
then that the dominant source of pulsars is silent collapse, at a rate of
1 per 4 years.

Since a white dwarf collapse cannot exceed the pulsar critical mass,
this is an attractive possibility. Supernovae also produce pulsars, at a
rate of one per 20-30 years, and produce black holes once per 14-20
years.

Questions remain such as: *why since the advent of the telescope and
the radio telescope has no supernova been seen in our galaxy? Why have
neutrino bursts not been detected prior to SN 1987A, due to much nearer,
but otherwise unseen supernova?*

7. CONCLUSION

This work has emphasised that supernova and pulsars occur deep into
galactic bulges. Surface density data is seen to be consistent for both
supernova in external galaxies, and pulsars and massive stars in our own.
Similarity of thse distributions does not imply a unique relationship
however, and pulsars appear to be associated with only a fraction of the
observed supernova. Surface density data is used to revise the
historical supernova rate, leading to an event rate more frequent than 1
per 10 years, a rate which is not inconsistent with that for external
galaxies. It is suggested that galactic X-ray luminosities should be
used to calibrate our own galaxy against external galaxies.

32

The CB model for the geometrical evolution of pulsars enables the pulsar beaming factor to be explicitly determined. This gives a self consitent birthrate estimate of one per 4 years. It is suggested then that pulsar birth is dominantly associated with silent collapse of white dwarves, with a small contribution from those SN which give rise to pulsars (one per 20-30 years). The remaining supernovae (one per 14-20 years) give rise to black holes.

I wish to thank Bernard Candy, Bernard Schutz, Andrew Lyne, Dick Manchester, Graham Smith and Andy Fabian for useful discussions which contributed to this work. I am grateful to the Institute of Astronomy, Cambridge, for their hospitality when much of this work was undertaken. The work was supported in part by the Australian Research Grants Scheme.

REFERENCES
(1) D. Blair, Contemp.Phys., 28, 457 (1987).
(2) J. Hough. This proceedings.
(3) This statement assumes the traditional cross-section analysis for resonant bars, and not the enhanced cross section model. (J. Weber, this poceedings).
(4) B.N. Candy and D.G. Blair. Mon. Not. Roy. Ast. Soc 205, 281 (1983)
(5) B.N. Candy and D.G. Blair. Ap J. 307, 535 (1986)
(6) B.N. Candy and D.G. Blair. Astron and Astrophys (1986). In press.
(7) G.A. Tammann in Supernovae: A Survey of Current Research. p371-403, ed. M.J. Rees and R.J. Stoneham. Reidel Dordrecht (1982).
(8) D. Sugimoto and K. Nomoto. Space Sci Rev. 25, 155 (1980).
(9) A.G. Lyne, R.N. Manchester and J.H. Taylor. Mon. Not. Roy. Ast. Soc. 213, 613 (1985)
(10) A.G. Lyne (Private Com.).
(11) C.G. Lacey and S.M. Fall. Ap. J. 290, 154 (1985)
(12) M.F. McCarthy in Supernovae and Supernova Remnants Ed. C.B. Cosmovici, Reidel Dordrecht (1974) pp 195-202.
(13) D.H. Clarke and F.R. Stephenson in Supernovae and Supernova Remnants op cit. p355-370.
(14) D.G. Blair and B.N. Candy. The Evolution of Radio Pulsars (in preparation).
(15) A.D. Kuz'min, I.M. Dogkesamanskaya, V.D. Pugachëv. Sov. Astron. Lett. 10, 357 (1984)
(16) A.G. Lyne and R.N. Manchester. (Private Comm.)
(17) J.L. Caswell. Mon. Not. Roy Ast Soc. 187, 431-439 (1987)
(18) J.N. Bahcall and T. Piran. 267, L77 (1983)

Gravitational Radiation from Rotating Stellar Core Collapse*

L. S. Finn
Theoretical Astrophysics
California Institute of Technology
Pasadena, California 91125

ABSTRACT. The gravitational collapse of a rotating stellar core has long been considered a promising burst source of gravitational radiation. In order to determine accurately the gravitational waveform from the collapse and the efficiency of the collapse in converting binding energy into gravitational radiation, it is necessary that a simulation accurately model certain features of the collapse. One purpose of this paper is to examine the scenario for supernova core collapse, and from it determine what must be simulated to achieve accuracy in the waveform and efficiency.

In addition to the hydrodynamic aspects of the collapse, a robust means of extracting the gravitational waveform from the simulation must be provided. In a Newtonian calculation, the standard quadrupole equation is the usual means. A second purpose of this paper is to point out some difficulties in the application of the quadrupole law to finite-difference calculations of gravitational waveforms, and to suggest new, more robust (but mathematically equivalent) means of extracting the gravitational waveforms from Newtonian finite-difference calculations.

1. Introduction

The collapse and bounce of a rotating stellar core in a supernova explosion has long been regarded as a possibly significant burst source of gravitational radiation. There have been numerous attempts to determine the gravitational waveforms of the collapse and its efficiency in converting binding energy into gravitational radiation. The earliest attempts were those of Thuan & Ostriker (1974), who studied the collapse of axisymmetric dust clouds, and Shapiro (1977) and Saenz & Shapiro (1978, 1979,

---

* Research supported by National Science Foundation Grants AST 85-14911, AST 84-51725, and PHY 83-08826.

B. F. Schutz (ed.), Gravitational Wave Data Analysis, 33–53.
© 1989 by Kluwer Academic Publishers.

1981), who studied the collapse of homogeneous Maclaurin ellipsoids in Newtonian gravity. As time has passed, new calculations of waveforms and efficiencies have focussed on increasingly sophisticated treatments of the gravity or of the physical details of the collapse. The most sophisticated of the former calculations are those of Stark & Piran (1985, 1986): they considered the collapse and bounce (or collapse to a black hole) of pressure-undercut rotating polytropes in a fully general relativistic code; while the most sophisticated of the latter calculations are those of Müller (1982): he extracted the gravitational waveforms and efficiencies from detailed axisymmetric core collapse calculations with realistic equations of state, but calculated with Newtonian gravity.

The results of Müller (1982) show that the details of the pre-collapse stellar model are important to the structure of the gravitational waveform, while the calculations of Evans (1986) show that the differences between Newtonian and general relativistic gravity, and details of the super-nuclear equation of state can be important in determining the depth of the bounce and the timescale of the collapse. Since the total energy radiated in gravitational waves is roughly proportional to the negative four thirds power of the maximum central density and the fifth power of the timescale of the bounce, calculations of the efficiency of core collapse in converting binding energy to gravitational radiation are expected to be sensitive to the details of the high density equation of state and strongly dependent on whether the calculation is made with Newtonian or general relativistic gravity. An accurate appraisal of the waveforms resulting from and the efficiency of core collapse will require a fully general relativistic treatment of core collapse utilizing a "realistic" equation of state and realistic pre-collapse initial stellar models.

This paper reports on work in progress toward the goal of a realistic calculation of the gravitational radiation from axisymmetric rotating core collapse. In §2, we review the gravitational collapse scenario, and in §3 we consider what features of the collapse scenario an accurate simulation must model. As a prelude to a fully general relativistic simulation of rotating stellar core collapse, Charles Evans and the author are developing a computer code to simulate core collapse in Newtonian gravity: §4 reports on the present state of that code, comparing its sophistication with the criteria established in §3. In the past, extraction of gravitational radiation from a Newtonian calculation has always been done via the quadrupole law: in §5, we examine some of the difficulties inherent in an application of the quadrupole law to a finite-difference Newtonian gravitational collapse calculation, and question whether the quadrupole law is the best way of determining the gravitational waveforms and efficiencies from a numerical Newtonian hydrodynamical simulation, or whether one of a wide class of mathematically equivalent expressions is more appropriate. Our conclusions are presented in §6.

2. The Gravitational Collapse Scenario

Supernovae come in two varieties, referred to as type I and type II. The two types are distinguished observationally by their light curves: one always reliable discriminant is the lack of hydrogen spectral lines in the emission from type I events, and its presence in type II events. Type I supernovae are usually thought to result from the *thermonuclear explosion* of white dwarfs, while type II supernovae are believed to arise from the *gravitational collapse* of high metallicity (population I) stars.

Present thinking is that type I supernovae only rarely leave a compact remanent: the progenitor star is disrupted in the explosion. More specifically, a type I supernova is *usually* presumed to be an accreting white dwarf in a binary system that has ignited carbon or helium under extremely degenerate conditions. Upon ignition, a spherical detonation or deflagration front passes through the star, burning a large portion (possibly all) of the mass of the star to nuclear statistical equilibrium. The energy released in the thermonuclear burning is sufficient to disrupt the entire star. It is possible that the carbon or helium ignition may take place off-center and the subsequent explosion be aspherical (owing to, *e.g.*, ignition taking place at the base of a shell as opposed to the center of the star). In this case, it is possible that some portion of the white dwarf may be left behind by the explosion. It has been suggested that the millisecond pulsar in the globular cluster M28 is the result of a type I supernova (Michel 1987), although, in this case, it is believed that the event was not due to a thermonuclear explosion, but rather to accretion-induced gravitational collapse.

While type I events are typically believed to result from a thermonuclear explosion, type II supernovae are thought to involve the gravitational collapse of a stellar core to a compact object (*i.e.*, a neutron star or a black hole). High densities and short timescales are a hallmark of gravitational collapse; so, if the collapse proceeds aspherically, we can expect the production of possibly significant amounts of gravitational radiation.

The rates of occurrence of type I and type II supernovae in our galaxy are both approximately equal to 1 in 40 years (Tammann 1982). The energy released in the thermonuclear explosion of a type I supernova is $\sim 10^{51}$ ergs, while the binding energy released in the collapse of a $1.3 M_\odot$ stellar core is $\sim 10^{53}$ ergs. The coupling of this released energy to gravitational radiation is a function of the asphericity, timescale, and matter densities involved in the energy release. All of these factors suggest that a type II supernova will be a more efficient source of gravitational radiation than a type I supernova. This, together with the larger energy available in the explosion of a type II supernova and the nearly equal rates of type I and type II explosions, makes it clear that type II supernova are more important as sources of gravitational radiation. The remainder of this paper is concerned with the gravitational collapse that takes place in a type II supernova and the gravitational waves it emits.

Much of the discussion below is abstracted from Woosley & Weaver's (1986) excellent review of the physics of supernovae explosions. This review is highly recommended for further reading and references on the physics (as opposed to the observations) of type I and type II supernovae.

2.1. THE TYPE II PROGENITOR

A star destined to become a type II supernova begins its life as a main sequence star with an initial mass of anywhere from approximately $8M_\odot$ to $\gtrsim 40M_\odot$. The lower bound of $8M_\odot$ is set by a rough upper limit on the mass of a star that can evolve to a white dwarf (Iben & Renzini 1983), while the upper limit of $\gtrsim 40M_\odot$ is set by the upper limit on the mass of a star that will retain its hydrogen envelope until core collapse (Woosley & Weaver 1986). This upper limit is in some sense a morphological one: type II supernovae are observationally *defined* to be supernovae with hydrogen spectral lines in their emission spectra. Larger stars exist and may explode as supernovae via gravitational core collapse; however, owing to the absence of hydrogen lines in their spectra, they would not be classified as type II supernovae (Chevalier 1976; Filippenko & Sargent 1985).

For the purposes of understanding the evolution of progenitor cores, it is useful to divide progenitors into three classes based on their masses: $8-11M_\odot$, $11-20M_\odot$, and $20-40M_\odot$ (Barkat, Reiss, & Rakavy 1974; Woosley & Weaver 1986). The first class of stars (mass range $8-11M_\odot$) form a transition region between stars that collapse before having begun silicon burning and thus have not formed an iron core, and those that proceed through all six stages of nuclear burning (hydrogen, helium, carbon, neon, oxygen, and silicon) and do form an iron core which then collapses. Stars in the range $11-20M_\odot$ proceed through all six nuclear burning stages non-degenerately, and it is with these stars that we are principally concerned in the discussion below. The heavier stars, those in the mass range $20-40M_\odot$, bypass both carbon and neon burning, proceeding directly from helium to oxygen and silicon burning (Woosley & Weaver 1985). The important consequences this has for the mass of the iron core are discussed below.

2.2. THE PRE-COLLAPSE STAR

The mass of the collapsing core and the properties of the collapse are largely determined by the entropy per baryon S (Bethe, Brown, Applegate, & Lattimer 1979; Bethe 1986; Woosley & Weaver 1986). For a star of mass $\sim 15M_\odot$, the initial entropy distribution in the progenitor star is roughly a constant $25\ k_B$ throughout the star (here and henceforth, entropy is given in units of k_B). As the star evolves, radiation losses from the core (both photons and neutrinos) transport entropy from the core to the envelope (or out of the star entirely), while convection maintains a homogeneous entropy distribution in the core. For stars that burn carbon ($M \sim 11-20M_\odot$), at the time of carbon ignition the core entropy is ~ 3 while the envelope entropy may be as high as ~ 40. As a result of this large entropy, the envelope will have expanded tremendously and the star become a red giant. Excepting the more massive stars ($M \sim 20-40M_\odot$), at the time of core collapse the core entropy is approximately 1 while the entropy in the envelope is approximately 40 (Woosley & Weaver 1986).

In the more massive stars ($M \sim 20-40M_\odot$), nuclear burning proceeds directly from helium to oxygen and silicon burning without carbon or neon burning. Oxygen

ınd silicon burning proceed very rapidly, and as a result the core does not have time
o shed much of its entropy. The cores of these massive stars thus have much higher
entropy than do those of less massive $(8-20M_\odot)$ stars. With a higher entropy, there
s more energy for pressure support and the cores can be significantly more massive
han the cores of the lower mass stars (Woosley & Weaver 1986).

For the low mass stars $(M \sim 8-20M_\odot)$ degenerate electrons provide essentially
ıll of the pressure support of the pre-collapse core. The entropy in the electrons
ıt the time of collapse is approximately half the total entropy of the core. The
Chandrasekhar limit gives the maximum mass that the core can achieve for a given
ilectron entropy per baryon S_e and a given electron fraction Y_e (electrons per baryon);
or a semi-degenerate electron gas, that mass is

$$M_{\text{Ch}} = M_{\text{Ch}}^0 \left[1 + (S_e/\pi Y_e)^2\right],\tag{2.1}$$

where

$$M_{\text{Ch}}^0 = 1.255(Y_e/0.464)^2 M_\odot \tag{2.2}$$

s the Chandrasekhar mass for a cold iron white dwarf ($Y_e = 0.464$ for iron). The
corrections to the Chandrasekhar mass due to general relativity have been ignored
ıere, since for iron white dwarfs other instabilities set in long before the general
relativistic instability does (cf. Shapiro & Teukolsky 1983, pg. 160).

Regardless of the mass of the progenitor, nuclear burning increases the size of
the core until it becomes unstable owing to one (or both) of two processes: electron
capture onto nuclei or photo dissociation of nuclei.

Electron capture onto nuclei induces an instability when the core density reaches a
few times $10^9 \, \text{g cm}^{-3}$. At this density, relativistic electrons provide all of the pressure
support of the stellar core, so the adiabatic index characterizing the equation of state is
very nearly $4/3$ and the mass is very nearly M_{Ch}. With further compression, electrons
are captured onto iron group nuclei in the core of the star, removing pressure support
and softening the equation of state. As the adiabatic index falls below $\Gamma_{\text{Crit}} \sim 4/3$, the
core undergoes gravitational collapse (Bethe, Brown, Applegate, & Lattimer 1979).

The core becomes unstable owing to photo-dissociation of nuclei at core temperatures approaching $T \sim 10^9 \, ^\circ\text{K}$ (Bethe, Brown, Applegate, & Lattimer 1979). As
the core contracts with the accretion of ash from nuclear burning, the temperature of
the core increases. The increasing number of trapped thermal photons dissociate an
increasing number of nuclei into α's and neutrons. In the final analysis, the energy
for the dissociation comes from the thermal energy of the electrons, reducing the
electron entropy and thus the Chandrasekhar mass for the core. When the Chandrasekhar mass falls below the core mass, the core becomes gravitationally unstable
and collapses.

For stars with mass less than about $15M_\odot$, electron capture is the immediate
cause of the core's instability, while for stars more massive than about $20M_\odot$, nuclear
photo dissociation is more important in determining the point of instability. Both
are significant, however, and contribute to the collapse (Woosley & Weaver 1986).

Thus, for stars in the mass range $11 - 20M_\odot$ the stellar core at the time of collapse is approximately the Chandrasekhar mass appropriate for an electron entropy of 1/2 and an electron fraction appropriate to iron group nuclei (~ 0.42): $M_{Ch} \simeq 1.2M_\odot$. For more massive stars, the core mass at the onset of the instability is higher (greater than or on order $2.2M_\odot$, cf. Woosley & Weaver 1985) and is far from the low-entropy Chandrasekhar mass (the core entropy is much higher than in the lighter stars and thus so is M_{Ch}).

2.3. CORE COLLAPSE AND BOUNCE

Once the core has become gravitationally unstable, it collapses rapidly on a timescale of ~ 0.1 seconds. The collapse is nearly adiabatic with the entropy S remaining roughly constant throughout. Since nuclear excited states have a high heat capacity, the temperature does not rise as rapidly as one would expect (Mazurek, Lattimer, & Brown 1979; Bethe, Brown, Applegate, & Lattimer 1979) and thermal pressure does not halt the collapse before nuclear densities are reached.

As the collapse proceeds, the presence of a finite sound speed and the development of an associated sonic point causes the core to segregate into an inner core and an outer core. The inner core collapses subsonically and homologously, with an inward radial velocity that is linear in r. The boundary of the inner core is the sonic point, where the inward radial velocity is equal to the sound speed of the fluid. Late in the collapse, the inner core encloses a mass of between 0.6 and $0.8M_\odot$. In contrast, the outer core collapses supersonically and, because it is unable to maintain sonic communication with itself, does not collapse homologously. The radially inward fluid velocity of the outer core is proportional to $r^{-1/2}$, and the velocities are roughly half the free-fall velocities (the velocities in the outer core never reach free-fall because of the small but non-trivial remaining pressure support).

Core collapse begins when the star's central density reaches a few 10^9 g cm^{-3}. Regardless of whether the collapse was initially induced by nuclear photo-dissociation or electron capture, as the collapse proceeds, electrons are captured onto iron group nuclei releasing neutrinos that initially stream freely out of the core. Once the core reaches a density of a few 10^{11} g cm^{-3}, the neutrino diffusion timescale exceeds the collapse timescale, and the neutrinos are effectively trapped in the core for the remainder of the collapse (Bethe, Brown, Applegate, & Lattimer 1979).

Neutrino trapping is a significant event in the collapse of the core. Before neutrino trapping takes place, electron capture is a non-equilibrium process: the energy and entropy associated with the capture are released in neutrinos which immediately leave the core (Zel'dovich & Novikov 1971, pg. 204). This acts to further destabilize the core. After neutrinos are trapped, however, i) electron capture behaves as an equilibrium process so the proportions of electrons, neutrinos, and nucleons are governed by the Saha equation; ii) the entropy does not decrease (in fact, initially it will increase slightly — cf. Bethe 1986, pg. 53); iii) the total Lepton fraction remains constant; and iv) the degenerate sea of neutrinos becomes available as a weak source of pressure support (approximately 10% of the electron pressure). The trapping of neutrinos is the first event that acts toward stabilizing the core and halting the collapse.

The collapse of the star continues until the core reaches nuclear densities, $\rho_{nuc} \simeq 2 \times 10^{14}$ g cm^{-3}. At nuclear densities and above, the equation of state becomes very stiff and the inner core halts its collapse very quickly. Since it remains in sonic communication throughout, the inner core responds to the stiffening of the equation of state in a coherent fashion: as the equation of state stiffens at the center of the star and then at larger and larger radii, sound waves travel outward from the center and accumulate at the sonic point, building an accretion shock at the boundary between the inner and outer cores. As the inner core rebounds, when the central density is approximately $3 - 4\rho_{nuc}$, the accretion shock begins to propagate outward through the (still infalling) outer core.

Ultimately, the shock is believed responsible for the supernova display; however, whether the shock initially has enough energy to reverse the infall and eject the remainder of the star (Baron, Cooperstein, & Kahana 1985), or whether it stalls (Burrows & Lattimer 1985) and must be recharged by diffusing neutrinos (Bethe & Wilson 1985) or some other mechanism remains to be understood. That question is unimportant to us, since essentially all of the energy radiated as gravitational waves is produced in the aspherical collapse bounce of the core.[1]

3. Elements of a Successful Simulation

The interplay between analytical and numerical work in theoretical astrophysics is a close one. It must be so if we are to use numerical simulations at all because the phenomena we wish to simulate are both too complicated for present and future computers (and programmers), and insufficiently well understood. The interplay between analytic methods and numerical methods is closely akin to the interplay between experimental (or observational) science and theoretical science.

In the present circumstance, we wish to understand quantitatively the gravitational radiation arising from the explosion of a type II supernova. We use our "theoretical" understanding of type II supernovae to pinpoint the physics responsible for the detailed gravitational radiation waveform that emerges from the collapse. This theoretical picture becomes the foundation for a numerical "experiment:" the simulation of the collapse of a stellar core, and the observation of the resulting gravitational radiation waveform for a range of initial parameters (e.g., initial core mass and angular momentum). In this section, we explore what the gravitational core collapse scenario described above tells us about the elements of the core collapse that a successful simulation must model.

---

[1] If the shock stalls and is not revived, then subsequent accretion may cause the neutron star remanent to undergo further collapse to a black hole and emit another burst of gravitational radiation; however, we do not concern ourselves with this possibility here.

Of course, some of our confidence in the above picture of core collapse arises from the feedback of numerical experiments on our physical intuition, and this is evidence of the close interplay between analytical and numerical techniques in the study of supernovae; however, while the scenario described above may depend on numerical experiments to validate some of its *quantitative* predictions, the *qualitative* validity of the scenario does not.

3.1. THE EQUATION OF STATE

The stellar core that undergoes gravitational collapse in a supernova explosion begins life as an equilibrium solution to the equations of stellar structure. That equilibrium is initially a stable equilibrium; however, as the core grows owing to nuclear burning, microphysical effects (*i.e.*, electron capture and nuclear photo-dissociation) softens the equation of state until the core becomes first neutrally stable (with adiabatic index $\Gamma = \Gamma_{Crit} \sim 4/3$) and then unstable ($\Gamma < \Gamma_{Crit}$). As the collapse proceeds, first neutrino trapping moderates the electron capture instability, and finally the equation of state stiffens dramatically as nuclear densities are reached and surpassed. It is the stiffening of the nuclear equation of state that is responsible for the turn-around of the core; without that stiffening, the collapse would proceed to a black hole.

The sudden stiffening of the equation of state and core's transition from gravitational instability to stability are two features of the equation state that a simulation must model accurately. The rapid change in the adiabatic index and the resulting change in the pressure forces on the core are qualitatively different in a supernova collapse than the adiabatic indices and forces in the core of a polytrope that has been induced to collapse by a cut in the pressure support. These differences can be expected to produce a difference in the details (*e.g.*, sharpness) of the hydrodynamic bounce and the resulting gravitational radiation waveform.

The equation of state at the low entropies and high densities involved in core collapse is not well known. Evans (1986) has used one parametrization of the nuclear equation of state (Baron, Cooperstein, & Kahana 1985) and found that experimental uncertainties in two parameters of that model (the nuclear compressibility and nuclear effective adiabatic index) can account for factors of three or more in the maximum central density of the bounce. Correspondingly, uncertainties in the super-nuclear equation of state introduce non-trivial uncertainties into the gravitational waveforms and efficiencies of the collapse.

Also important to the simulation is an accurate modeling of the microphysics that induces the core to collapse. The core begins to collapse when *either* electron capture or nuclear photo-dissociation drives the effective adiabatic index of the equation of state below Γ_{Crit}. As the core collapses, *both* electron capture and nuclear photo-dissociation play significant roles in determining the speed of the collapse (and hence the timescale and depth); hence, it is important that the equation of state accurately model both of these microphysical effects. Since the electron capture instability depends on the phase-space available for neutrino emission, this requires that some form of neutrino transport or leakage be incorporated into the simulation.

3.2. INITIAL DATA

In order that there be gravitational radiation from the supernova, the collapse must proceed aspherically. A spherically symmetric, non-rotating star will collapse spherically symmetrically; so, to produce gravitational radiation we must have an initial stellar model that is non-spherical (owing to, *e.g.*, rotation). We will consider the axisymmetric collapse of an initially axisymmetric rotating star.[1]

In order that the gravitational waveform from the calculation represent the collapse and not be just a transient feature of the initial data, it is important that the initial stellar model be in a state of hydrostatic equilibrium. Non-equilibrium initial data may affect the collapse and the waveform in various ways. For example, one might consider imposing the the angular momentum required for a non-spherical collapse as rigid rotation on a spherical stellar model; however, such a model is not in equilibrium: a rotating stellar core in equilibrium is oblate, not spherical. In the absence of the instabilities that lead to collapse, such a star would pulsate with a period of approximately $\rho_c^{-1/2}$ (in gravitational units). The initial amplitude of the pulsations would be proportional to the square of the eccentricity, which is in turn proportional to the square of the angular momentum for small values of the angular momentum.

In the collapse problem, when a rigidly rotating spherical star is released it begins to pulsate and collapse. The period of the pulsations is approximately $\rho_c^{1/2}$ (where rho_c is the central density). When the star is released, this is also the timescale of the collapse to maximum central density, and when the star has reached nuclear densities, this is also the timescale of the hydrodynamic bounce. During the collapse, as the central density increases the pulsation period decreases, and the pulsation amplitude is adiabatically amplified (Goldreich & Weber 1980). In the case of moderate to high angular momentum collapse (which radiate gravitational waves the most strongly), the initial amplitude of the pulsations will be rather large, and these amplified pulsations can be expected to significantly alter both the outcome of the hydrodynamic collapse and the gravitational waveform that results.

The issue of initial data raises its head in another context. It is not simply enough to specify that the pre-collapse star is rotating: we must specify how much angular momentum is present and also specify the *distribution* of both angular momentum and entropy. Müeller (1982) found that the initial distribution of angular momentum could significantly affect the gravitational radiation waveform and efficiency (they

---

[1] An axisymmetric star can collapse non-axisymmetrically if there are nonaxisymmetric instabilities; however, owing to the lack of computational resources, we are restricted to modeling strictly axisymmetric collapse: the r and θ momentum are dynamical while the ϕ momentum is purely kinematical.

found the efficiency could increase by two orders of magnitude if the angular velocity was not uniform in the pre-collapse star); so the question of distribution is far from trivial. Our knowledge of stellar evolution is not up to this task: to date, there have been no evolutionary calculations of rotating stellar models to the point of core collapse; hence, the question of angular momentum, angular momentum distribution, entropy, and entropy distribution will remain a great uncertainty in any simulation.

3.3. RESOLUTION OF SIMULATION

The simulations we are contemplating are finite-difference, hydrodynamical ones. The core of the stellar model initially has a radius of $\gtrsim 10^3$km, corresponding to the core of a high mass white dwarf, while at the moment of maximum central density, the core has a radius of ~ 10km. In order to be accurate, the numerical simulation must be capable of maintaining adequate radial resolution in the core throughout the entire collapse. This means that the radial grid must move to follow the collapse of the core.

The issue of angular resolution is also an important one, and here we can find a concrete example of just how serious a problem truncation error can be. The majority of the gravitational radiation emitted in the core collapse and bounce is quadrupole radiation, and the collapse will have reflection symmetry across the equatorial plane. Consequently, the density may be represented by a multipole expansion in the even-l Legendre polynomials $P_l(\cos\theta)$:

$$\rho = \sum_{l=0}^{\infty} \rho_l(r) P_l(\cos\theta). \tag{3.1}$$

The finite angular resolution and order of a finite-difference numerical calculation mix multipole moments independently of any *physical* effects: while analytically it is true that

$$\int_1^{-1} P_{l'}(\cos\theta) P_l(\cos\theta) d\cos\theta = \frac{2}{(2l+1)}\delta_{l'l}, \tag{3.2}$$

in a finite-difference numerical calculation we can neither represent nor extract any multipole moment with such perfect resolution. To illustrate what effect this has, define the "numerical overlap integrals" $I_N(l',l)$ by

$$I_N(l',l) = \frac{2l+1}{2}\sum_{k=0}^{N-1} P_{l'}\left(\cos\theta_{k+1/2}\right) P_l\left(\cos\theta_{k+1/2}\right)(\cos\theta_k - \cos\theta_{k+1}), \tag{3.3}$$

where

$$\theta_k \equiv \frac{k\pi}{N}, \tag{3.4}$$

Then, on a finite difference grid with N angular zones in π radians a numerical evaluation of the multipole moment $\rho_{l,N}$ yields

$$\rho_{l,N} = \sum_{l'=0}^{\infty} I_N(l',l)\rho_{l'}. \tag{3.5}$$

Several of the "numerical overlap integrals" $I_N(l',l)$ are given in table 1 for various values of N. In order to represent the first three non-zero moments ($l = 0, 2, 4$) of a scalar quantity (*e.g.*, density) to at least 1% precision requires that the angular resolution be at least $\pi/32$ radians; improvements in the resolution scale as the square of the number of angular grid zones. As a rough rule of thumb, to represent moments of order l requires a resolution of very much better than π/l radians.

Table 1.

N	$I_N(0,0)$	$I_N(0,2)$	$I_N(0,4)$	$I_N(2,2)$	$I_N(2,4)$	$I_N(4,4)$
2	1.000	1.250	-3.656	0.3125	-0.9141	1.485
4	1.000	0.1565	0.4419	1.170	0.8871	0.3858
8	1.000	3.341×10^{-2}	6.643×10^{-2}	1.029	7.015×10^{-2}	1.074
16	1.000	8.110×10^{-3}	1.495×10^{-2}	1.007	1.511×10^{-2}	1.014
32	1.000	2.013×10^{-3}	3.644×10^{-3}	1.002	3.654×10^{-3}	1.003
64	1.000	5.024×10^{-4}	9.054×10^{-4}	1.000	9.061×10^{-4}	1.001
128	1.000	1.258×10^{-4}	2.261×10^{-4}	1.000	2.261×10^{-4}	1.000

4. An Axisymmetric Collapse Code

As a first step toward the ultimate goal of a fully general relativistic hydrodynamic axisymmetric collapse code, Charles Evans and the author are developing a non-relativistic, Newtonian self-gravitating, hydrodynamical, axisymmetric collapse computer code, which is described in this section.

4.1. NUMERICAL TECHNIQUES

We are developing a code that solves the coupled Newton-Euler equations for a self-gravitating Newtonian fluid. The code uses spherical coordinates (r, θ, ϕ), simulating the dynamical freedom of the r and θ coordinates, while the momentum in the ϕ coordinate is carried along as a kinematical variable. In this sense the code is said to be a 2 1/2 dimensional simulation.

The state variables (mass density, internal energy density, momentum density, and gravitational potential) are maintained on a staggered mesh for proper centering of spatial derivatives. The Euler-Newton equations are operator-split (*cf.* Wilson

1979) and time-updates of the state variables are done sequentially (as opposed to simultaneously).

The technique of operator splitting allows us to identify certain combinations of terms in the Newton-Euler equations and use special techniques appropriate to these terms to update the state variables. For example, among the operator split Newton-Euler equations are equations for the conservative transport of mass density, internal energy density, momentum density, and specific angular momentum. These transport equations are handled by second-order accurate conservative flux techniques. To maintain locally accurate angular momentum conservation, consistent advection (Norman & Winkler 1986) is used to transport specific angular momentum.

Time evolution is explicit and first order.

Computer finite difference codes that use spherical coordinates often have difficulties with truncation error at and near coordinate singularities ($r = 0$ and $\theta = 0$ and π). Our code uses numerical regularization of difference operators (a technique developed by Evans 1984, 1986) to control truncation error near coordinate singularities.

In order to follow the collapse of the core through several orders of magnitude in radius, an "adaptive" radial mesh is used. The average motion of the inner few zones, which reside deep in the inner core, is monitored at each time step and this motion is used to homologously scale the entire grid throughout the majority of the collapse. As a result, the grid shrinks as the core collapses and expands as the core bounces. The use of an adaptive mesh allows us to maintain adequate radial resolution of the star throughout the collapse without having an excess of zones at early or late times.

The code is being developed on a Cray XMP-48 at the National Center for Supercomputing Applications. We are making every effort to write the code in standard fortran, using as few Cray extensions as possible while still maintaining a high degree of vectorization for speed.

4.2. ARTIFICIAL VISCOSITY

The Euler equations have no viscous terms, and so our code simulates an inviscid flow. In a truly inviscid flow, shocks are present as jump discontinuities in the state variables and cannot be represented in the finite resolution of the mesh. Additionally, the Euler equations alone do not provide a mechanism for turning the energy in the bulk motion of the pre-shock fluid into internal energy in the post-shock fluid. In an effort to solve this problem, Von Neumann & Richtmyer (1950) introduced "artificial viscosity" terms into the finite difference Euler equations. The purpose of these viscous terms is to allow the code to stably form a shock, to smooth the shock to the resolution of the grid, and to convert the bulk motion of the pre-shock fluid into internal energy of the post-shock fluid.

Artificial viscosity is typically implemented by adding *ad hoc* viscous heating and force terms to the Euler equations: a viscosity that is non-zero for compressional motion in each of the coordinate directions is defined, a viscous acceleration of the fluid proportional to the coordinate derivative of the appropriately defined viscosity is added to the Euler equations, and a viscous heating term proportional to the sum of the squares of the coordinate viscosities is added to the energy equation.

We are in the process of creating a covariant formulation of artificial viscosity for use in our code (Evans & Finn 1987). Like a real viscosity, our artificial viscosity arises from a viscous stress tensor, and its contributions to the Euler equations and the energy equations are derived from the divergence of that stress tensor. Consequently, our artificial viscosity will properly conserve energy in and around shocks (something that present implementations of artificial viscosity cannot guarantee). The form of the viscous stress tensor chosen insures that the viscous forces so derived have the properties required to smooth the shock to the resolution of the grid.

4.3. GRAVITATIONAL WAVE EXTRACTION

Our present code treats only Newtonian self-gravitating fluids. For the reasons discussed previously, the waveforms we extract from this code cannot be assumed represent quantitatively the waveforms that will emerge from a fully relativistic simulation of core collapse. Nonetheless, the extraction of gravitational radiation from a Newtonian code is a useful exercise: not only to obtain a qualitative picture of the collapse waveform that will guide the calibration of a fully relativistic code, but also as a means of exploring the difference between general relativistic and Newtonian gravity in the non-linear regime.

Like electromagnetic radiation, gravitational radiation can be characterized by multipole moments. The moments that characterize the gravitational radiation field are *mass* moments (analogous to electric moments) and *mass-current* moments (analogous to magnetic moments). These moments have simple and evocative Newtonian limits (for an in depth exposition of a multipole formalism for general relativistic gravitational radiation and its Newtonian limit, see Thorne 1980).

Assuming axisymmetry and equatorial plane symmetry (appropriate for our treatment of gravitational core collapse), the first two non-vanishing multipole moments of the gravitational radiation field in a collapsing star are the mass quadrupole and the current octupole (Turner & Wagoner 1979). If the angular velocity Ω in the stellar core is significantly non-uniform, then at lowest order in $\Omega^2 R^3/M$ the current octupole is responsible for the gravitational radiation; otherwise, the mass quadrupole gives the radiation to lowest order in $\Omega^2 R^3/M$. The mass quadrupole always gives the radiation to lowest order in R/T, where T is the timescale and R the core radius at bounce.

Two techniques have been used to extract the gravitational radiation from a Newtonian code. The technique most often used is the quadrupole law (*e.g.,* Müller 1982 for gravitational stellar core collapse; other examples include the gravitational waveform from the head-on collision of two neutron stars, *cf.* Gilden & Shapiro 1984, and also Evans 1987). A more sophisticated approach was developed by Turner & Wagoner (1979) who investigated gravitational radiation from rotating core collapse using a post-Newtonian approximation scheme where the rotation was treated as a small perturbation on a spherical collapse. In addition to the familiar (mass) quadrupole radiation, Turner & Wagoner (1979) explored the octupole radiation from the collapse.

While the use of the quadrupole law in a Newtonian calculation is convenient because all the information required to specify the quadrupole radiation field is folded into the second moment of the density distribution, there are subtle issues involved in the use of the quadrupole law in a finite difference calculation on a finite-extent grid. At present, we extract gravitational radiation waveforms from our code in several mathematically equivalent but numerically different ways (the quadrupole law being one among them), and we are evaluating these different extraction techniques in order to determine the most appropriate techniques for a numerical code (see §5 below; also Evans & Finn 1987).

4.4. EQUATION OF STATE

Our ultimate goal is to simulate stellar core collapse in a fully general relativistic code. One purpose behind our development of a Newtonian code is to provide a workbench for the development of an equation of state that meets all the criteria set forth above. While the equation of state incorporated in our code is still under development, it does have many of the features that we discussed above as being necessary for an accurate simulation of core collapse: *i)* the sudden stiffening of nuclear matter at super-nuclear densities, *ii)* neutrino trapping, and *iii)* a simple modeling of the microphysics that leads to the gravitational instability (*i.e.,* electron capture and nuclear photo-dissociation).

4.5. INITIAL DATA

The initial data for our simulations consist of a stellar core in rotational equilibrium with uniform angular velocity. The equilibrium models created all have a mass just beyond the point of instability and thus are slightly unstable to gravitational collapse. The entropy is constant throughout the core and appropriate to that of a pre-collapse star; the electron fraction is function of the density in the star.

The equilibrium models are all created from an initially spherical equilibrium model through a relaxation technique which imposes a rigid angular velocity to the spherical model, and then deforms it until it is in equilibrium at the given velocity. This procedure is repeated until the model has the desired angular momentum.

5. Gravitational Radiation Waveforms

There are subtle issues involved in the extraction of gravitational waveforms from Newtonian codes, and some of these issues seriously compromise the utility of the quadrupole law. In this section, we briefly examine some of those issues and propose a remedy in the form of a new technique for gravitational radiation extraction.

The quadrupole equation as formulated by Landau & Lifshitz is applicable to isolated slow-motion sources with Newtonian internal gravity (Damour 1987, pg. 21). Subject to the *caveats* discussed by Damour (1987, pg. 21), the Landau & Lifshitz quadrupole equation reduces to the the standard quadrupole formula (*cf.* Misner, Thorne, & Wheeler 1973):

$$h_{ij}^{\mathrm{TT}}(t, \mathbf{x}) = \frac{2}{r} \frac{d^2}{dt^2} \mathbf{I}_{ij}^{\mathrm{TT}}(t - r), \qquad (5.1)$$

where $\mathbf{I}_{ij}$ is defined by

$$\mathbf{I}_{ij} = \int T^{00} \left(x^i x^j - \frac{1}{3} \delta_{ij} r^2 \right) d^3 x, \qquad (5.2)$$

Here, T^{00} is the "time-time" component of the stress-energy tensor, and the super-script ^{TT} indicates that first the Transverse (to the "line of sight" $\mathbf{x}$) and then the Trace-free parts of the indicated expression are to be taken.

Several features of the quadrupole law (5.1) complicate its use in a numerical finite-difference code. The evaluation of the reduced quadrupole moment $\mathbf{I}_{ij}$ itself is a simple matter; however, the calculation of the two numerical time derivatives required to turn $\mathbf{I}_{ij}$ into the metric perturbation introduces significant numerical "noise." The noise arises from the fact that $\mathbf{I}_{ij}$ is largely constant from timestep to timestep, and so the differences of $\mathbf{I}_{ij}$ are very small.

Figure 1 shows the gravitational radiation waveform derived from the quadrupole equation (5.2) for one of our simulations. The presence of the high frequency noise in the waveform is obvious.

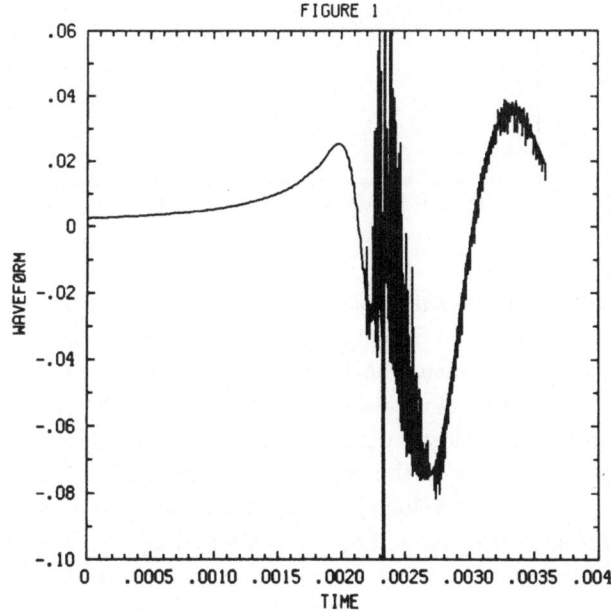

FIGURE 1

Figure 1. In the quadrupole law, the gravitational waveform is the second time derivative of the the quadrupole moment. The numerical evaluation of these two time derivatives introduces a large amount of high-frequency numerical "noise" into the waveform. Time is given in seconds; for the purposes of this discussion, the scale of the waveform is arbitrary.

48

As a test of the proposition that it is the time derivatives that lead to the noise, we can use the continuity equation to eliminate a single time derivative in favor of a spatial divergence:

$$
\begin{aligned}
\frac{d^2}{dt^2}\,I_{ij} &= \frac{d^2}{dt^2}\int d^3x\left(x_i x_j - \frac{1}{3}\delta_{ij}r^2\right)\rho \\
&= -\frac{d}{dt}\int d^3x\left(x_i x_j - \frac{1}{3}\delta_{ij}r^2\right)\nabla\cdot(\mathbf{v}\rho).
\end{aligned}
\tag{5.3}
$$

Figure 2 shows the waveform computed from this second moment of the divergence of the momentum density. The elimination of a single time derivative has eliminated virtually all of the high frequency noise from the waveform.

Figure 2. The continuity equation can be used to evaluate one of the time derivatives in the quadrupole equation, so that only one numerical time derivative need be taken to evaluate the gravitational waveform. The waveform calculated by this technique shows very little of the high-frequency numerical noise found in the the standard quadrupole equation waveform. Time is given in seconds; for the purposes of this discussion, the scale of the waveform is arbitrary.

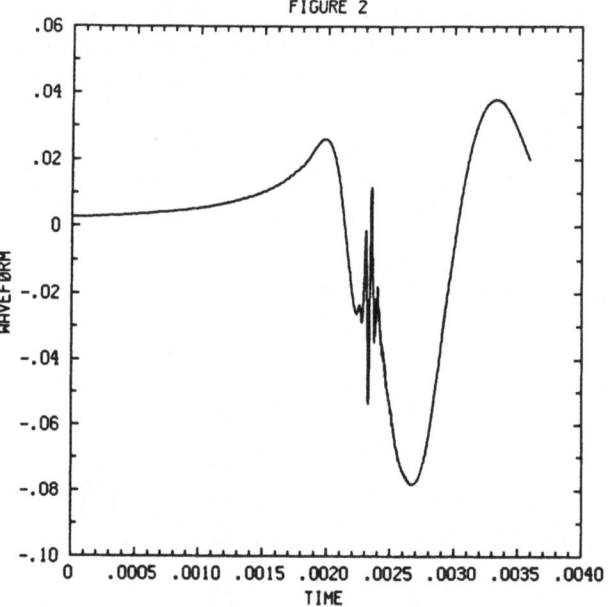

FIGURE 2

A second feature of the quadrupole formula that makes its implementation difficult in a numerical calculation has to do with the finite extent of the finite-difference grid. In a numerical calculation, it is not an unusual occurrence for a shock to propagate off the grid, carrying with it some mass. As a result of the shock propagating of the grid, the quadrupole moment of the enclosed density distribution changes on the timescale of $\Delta r/(v_{\text{shock}} - v_{\text{grid}})$, where Δr is the resolution of the grid, v_{shock} is the shock speed, and v_{grid} is the grid's radial velocity at the outermost grid zone. Even though the mass carried off the grid may be small, it is weighted by the large moment arm of the quadrupole moment and the short time it takes to propagate off the grid, and so can make a large contribution to the second time derivative of the quadrupole moment. In the same way, mass outflow owing to the motion of the outer boundary of the grid (see the discussion of the adaptive mesh in §3 above) changes the quadrupole moment of the configuration enclosed in the finite-difference mesh, and this change is interpreted as gravitational radiation by either of equations 5.1 or 5.3.

Actually, any mass outflow from the finite-difference grid violates some of the assumptions leading to the quadrupole formula as presented above. In deriving the quadrupole formula, there are two surface integrals that have been removed to infinite radius, where (owing to the assumption that the source is isolated) they vanish. The volume integral that defines the quadrupole moment is similarly assumed to be evaluated over all space. The net result is that the quadrupole formula, as it is conventionally written and used (eq. [5.2]), does not allow for mass outflow, nor does the improved version (5.3). This defect could be corrected by the inclusion of the relevant surface terms; however, the surface integrals added would numerically cancel the volume integral of the mass outflow to the order of truncation error, leading to errors similar to those already discussed.

One remedy for the ills of numerical time derivatives and mass outflow is the direct use of the integrated stresses to determine the gravitational radiation waveform. The full Einstein field equations can be written as

$$\bar{h}^{\alpha\beta}{}_{,\mu\nu}\eta^{\mu\nu} = -16\pi\tau^{\alpha\beta}, \tag{5.4}$$

where

$$h_{\alpha\beta} \equiv g_{\alpha\beta} - \eta_{\alpha\beta}, \tag{5.5}$$

$$\bar{h}_{\alpha\beta} \equiv h_{\alpha\beta} - \eta_{\alpha\beta}h_{\mu\nu}\eta^{\mu\nu}, \tag{5.6}$$

$$\bar{h}^{\mu}_{\alpha,\mu} = 0, \tag{5.7}$$

$$\tau^{\alpha\beta} \equiv T^{\alpha\beta} + t^{\alpha\beta}_{LL}. \tag{5.8}$$

Here, $T^{\alpha\beta}$ is the matter stress-energy tensor and $t^{\alpha\beta}_{LL}$ is the Landau-Lifshitz pseudo-tensor (Landau & Lifshitz 1975, §96). Far from the (isolated) source, $\bar{h}^{\alpha\beta}$ becomes the "TT" gauge radiation field of linearized theory; so, presuming an outgoing-wave boundary condition the quadrupole radiation field of a slow-motion source is given

$$h^{TT}_{ij}(t, \mathbf{x}) = \frac{4}{r}\int \tau_{ij}(t - r, \mathbf{x}')d^3x', \tag{5.9}$$

where $\mathbf{x}$ is the radius vector to the observation point and the integral is taken over all space. Equation (5.9) is valid for an arbitrarily strong field, non-singular source (e.g., a neutron star, but not a black hole). For a Newtonian source, τ_{ij} reduces to the stress tensor of a self-gravitating Newtonian fluid:

$$\tau_{ij} \equiv \rho\left(\frac{1 + \epsilon + P}{\rho}\right)v_iv_j + P\gamma_{ij} + \frac{1}{4\pi}\left(\Phi_{,i}\Phi_{,j} - \frac{1}{2}\gamma_{ij}\Phi_{,l}\Phi^{,l}\right), \tag{5.10}$$

where the v_i are the covariant components of the fluid velocity and γ_{ij} is the spatial metric. In principle, equation (5.9) is mathematically equivalent to the quadrupole equation (5.2); the quadrupole formula is obtained from (5.10) expression via two parts integrations and judicious substitution using the equations of motion (*cf.* Misner, Thorne, & Wheeler 1973 §36.10). Why in principle? Up to now, the introduction of artificial viscosity into the Euler and energy equations has been *ad hoc*. In order that the calculation of the waveform via the stresses be mathematically equivalent to the calculation via the quadrupole formula, the viscous terms in the Euler and energy equations must be derivable from a common viscous stress tensor that is part of the overall stress tensor. While the viscous stresses (and forces) should be negligible, consistent handling of the stresses is necessary so that we can monitor their contributions directly and insure that they are in fact negligible.

No time derivatives appear in equation (5.9) for the gravitational radiation field, thus avoiding the major source of numerical noise in the use of the quadrupole formula. When this formula is used in a finite difference code, the stress that leaves the grid can be monitored and compared to the stress remaining, thereby providing an estimate of the errors introduced into the waveform. We expect that the stress leaving the grid will be small, and so without the time derivatives and the large moment arm associated with the quadrupole formula, mass loss during the simulation should introduce only a negligible errors. For further discussion of the utility of equation (5.9) for calculating the quadrupole radiation field, see Evans & Finn (1987).

6. Conclusions

The gravitational collapse of a rotating stellar core in a type II supernova explosion is a possibly significant burst source of gravitational radiation. Numerical simulations are the only techniques available to us that can yield detailed and accurate gravitational waveforms and efficiencies from the core collapse. Our knowledge of the collapse scenario allows us to identify important features of the collapse that a simulation must accurately model:

i) The sudden stiffening of the equation of state at and above nuclear density is responsible for the rapid deceleration of the core and consequent release of gravitational radiation, and thus a detailed model of the nuclear equation of state of the stellar matter is crucial to the accurate determination of the waveform of the collapse and its efficiency in producing gravitational radiation.

ii) The size of the homologous inner core, whose coherent bounce is responsible for the gravitational radiation, is determined by the velocity of the infall and is thus sensitive to the degree of instability in the collapse. Consequently, the two instabilities responsible for the collapse of the stellar core, electron capture and nuclear photo-dissociation, must be modeled accurately.

iii) Since the timescale of the stellar pulsations and the collapse are approximately the same, the initial data from which simulations of rotating core collapse are begun must represent a star in (unstable) equilibrium. Additionally, the initial stellar model must be close to stability: realistically, the processes that bring the star to the point of instability occur on a timescale long compared to the collapse timescale of the star.

Certain standards are required of the numerical resolution employed in the simulation: the radial resolution of the simulation must be capable of resolving the core of the star as its radius changes by two to three orders of magnitude during the collapse, and the angular resolution should be capable of resolving the state variables to a level to a level of precision consistent with that desired of the results of the calculation. More precisely, to separate the quadrupole and hecadecapole moments of the density distribution to 1% precision requires an angular resolution of at least 16 zones in $\pi/2$ radians.

As a test bed for developing equilibrium stellar models, equations of state, and artificial viscosity, and to provide guidance in the development and calibration of a fully general relativistic relativistic code, a Newtonian code has been developed. The use of the quadrupole formula for extracting gravitational radiation from a Newtonian collapse code is fraught with difficulties. It appears that calculation of the gravitational radiation in a Newtonian code may be best handled by a direct integration of the gravitational and fluid stresses, thereby avoiding the noise-producing numerical time derivatives and the large moment arms of multipole moment formulae.

Acknowledgments

The work reported on in this paper was carried out in collaboration with Charles Evans. I thank both Chuck and Kip Thorne for a critical reading of this manuscript.

I also thank both Kip and Bernard Schutz for their assistance in making my participation in the NATO ARW on Gravitational Radiation Data Analysis possible, and I especially thank both Bernard and Andrej Krolak for their efforts in organizing a very valuable meeting.

This work was supported by National Science Foundation Grants AST 85-14911, AST 84-51725, and PHY 83-08826. Computer work in support of the research reported on here was carried out at the National Center for Supercomputing Applications.

References

Barkat, Z., Reiss, Y., Rakavy, G. 1974. *Astrophys. J. (Letters)* , **193**, L21.

Baron, E., Cooperstein, J., & Kahana, S. 1985. *Phys. Rev. Lett.* , **55**, 1.

Bethe, H. A. 1986. "Supernova Theory," in **Highlights of Modern Astrophysics,** Shapiro, S. L. & Teukolsky, S. A. editors, Wiley & Sons, New York, pg. 45.

Bethe, H. A., Brown, G. E., Applegate, J., & Lattimer, J. M. 1979. *Nucl. Phys.,* **A324**, 487.

Bethe, H. A. & Wilson, J. R. 1985. *Astrophys. J.,* **295**, 14-23.

Burrows, A. & Lattimer, J. M. 1985. *Astrophys. J. (Letters)* , **299**, L19-L22.

Chevalier, R. A. 1975. *Astrophys. J.,* **208**, 826-828.

Damour, T. 1987. "An Introduction to the Theory of Gravitational Radiation," in **Gravitation in Astrophysics,** Carter, B. & Hartle, J. B. editors, Plenum Press, New York

Evans, C. R. 1984. "A Method for Numerical Relativity: Simulation of Axisymmetric Gravitational Collapse and Gravitational Radiation," University of Texas (Austin) Thesis, University Microfilms.

Evans, C. R. 1986. "An Approach for Calculating Axisymmetric Gravitational Collapse," in **Dynamical Spacetimes and Numerical Relativity,** Centrella, J. M. editor, Cambridge University Press, Cambridge, pg. 3.

Evans, C. R. 1987. "Gravitational Radiation from Collisions of Compact Stars," to appear in **Proceedings of the Thirteenth Texas Symposium on Relativistic Astrophysics.**

Evans, C. R., Smarr, L. L., & Wilson, J. R. 1986. "Numerical Relativistic Gravitational Collapse with Spatial Time Slices," in **Astrophysical Radiation Hydrodynamics,** Winkler, K-H. & Norman, M. L. editors, Reidel, Dordrecht (Holland), pg. 491.

Evans, C. R. & Finn, L. S. 1987. *In preparation.*

Filippenko, A. V. & Sargent, W. L. W. 1985. *Nature,* **316**, 407-412.

Gilden, D. L. & Shapiro, S. L. 1984. *Astrophys. J.,* **287**, 728.

Goldreich, P. & Weber, S. V. 1980. *Astrophys. J.,* **238**, 991-7.

Iben, I. Jr. & Renzini, A. 1983. *Ann. Rev. Astr. Astrophys.,* **21**, 271.

Landau, L. D. & Lifshitz, E. M. 1975. **Classical Theory of Fields, Fourth Revised English Edition** Pergammon Press, Oxford.

Mazurek, T. J., Lattimer, J. M., & Brown, G. E. 1979. *Astrophys. J.,* **229**, 713.

Michel, F. C. 1987. *California Institute of Technology preprint.*

Misner, C. W., Thorne, K. S., & Wheeler, J. A. 1973. **Gravitation,** W. H. Freeman and Company, San Francisco.

Müller, E. 1982. *Astr. Astrophys.,* **114**, 53-59.

von Neumann, J. & Richtmyer, R. D. 1950. *J. Appl. Physics,* **21**, 232.

Norman, M. L. & Winkler, K. A. 1986. "2-D Eulerian Hydrodynamics with Fluid Interfaces, Self-Gravity and Rotation," in **Astrophysical Radiation Hydrodynamics,** Winkler, K. A. & Norman, M. L. editors, Reidel, Dordrecht (Holland).

Piran, T. & Stark, R. F. 1986. "Numerical Relativity, Rotating Gravitational Collapse and Gravitational Radiation," in **Dynamical Spacetimes and Numerical Relativity,** Centrella, J. M. editor, Cambridge University Press, Cambridge, pg. 40.

Shapiro, S. L. 1977. *Astrophys. J.,* **214,** 566-575.

Saenz, R. A. & Shapiro, S. L. 1978. *Astrophys. J.,* **229,** 286-303.

Saenz, R. A. & Shapiro, S. L. 1979. *Astrophys. J.,* **229,** 1107-1125.

Saenz, R. A. & Shapiro, S. L. 1981. *Astrophys. J.,* **244,** 1033-1038.

Shapiro, S. L. & Teukolsky, S. A. 1983. **Black Holes, White Dwarfs, and Neutron Stars,** John Wiley & Sons, New York.

Stark, R. F. & Piran, T. 1985. *Phys. Rev. Lett. ,* **55,** 891.

Stark, R. F. & Piran, T. 1986. "A Numerical Computation of the Gravitational Radiation From Rotating Gravitational Collapse," in **Proceedings of the Fourth Marcel Grossman Meeting on General Relativity,** Ruffini, R. editor, Elseveir Science Publishers.

Tammann, G. A. 1982. "Supernova Statistics and Related Problems," in **Super-novae: A Survey of Current Research,** Rees, M. J. & Stoneham, R. J. editors, Reidel, Dordrecht (Holland), pg. 371.

Thorne, K. S. 1980. *Rev. Mod. Phys.,* **52,** 299.

Thuan, T. X. & Ostriker, J. P. 1974. *Astrophys. J. (Letters) ,* **204,** L1.

Turner, M. S. & Wagoner, R. V. 1979. "Gravitational Radiation from Slowly-Rotating 'Supernovae': Preliminary Results," in **Sources of Gravitational Radiation,** Smarr, L. L. editor, Cambridge University Press, Cambridge.

Wilson, J. R. 1979. "A Numerical Method for Astrophysics," in **Sources of Gravitational Radiation,** Smarr, L. L. editor, Cambridge University Press, Cambridge.

Woosley, S. E. & Weaver, T. A. 1985. In **Nucleosynthesis and Its Implications On Nuclear and Particle Physics, Proc. Moriand Astrophys. Conf. 5th,** Audouze, J. & van Thuan, T., Reidel, Dordrecht (Holland).

Woosley, S. E. & Weaver, T. A. 1986. *Ann. Rev. Astr. Astrophys.,* **24,** 205-33.

Zel'dovich, Ya. B. & Novikov, I. D. 1971. **Relativistic Astrophysics, Volume 1,** University of Chicago Press, Chicago.

REMARKS ON SN 1987A

Peter Kafka
Max Planck Institute for Physics and Astrophysics
Karl Schwarzschild Strasse 1,
D-8046 Garching, Germany.

ABSTRACT.   Brief observations are made on the implications of the
neutrino observations of SN 1987a and of the bar-antenna data for that
period.   Another collapse in our Galaxy is fairly likely to occur before
laser interferometric antennae reach the 10^{-22} sensitivity level; it
would give a clearer neutrino signal and possibly a reliable detection
by cryogenic bar detectors.

None of the more sensitive of the presently-existing GW antennae was
working on February 23rd 1987 when the supernova in the Large Magellanic
Cloud went off [1] - not to mention the more sensitive antennae planned
for the near future.   Otherwise the birth of GW Astronomy might have
been registered.   Neutrino Astronomy was more lucky.   However, its
example also shows how unsatisfactory it is to discover a unique event
of some complexity just barely above noise.   The significance of the
first event, at 2:52 universal time, reported by the Mont-Blanc-Tunnel
group [2] is now widely doubted, whereas the second one, at 7:35 UT,
first announced by the KAMIOKA group cannot be denied [3].   But the
information content seems to be insufficient to allow unique conclusions
about the nature of the collapse or the rest masses of the observed
neutrinos.   At the moment, puzzling observations by speckle-interfero-
metrical methods complicate the picture even further because they may
imply a jet-like feature, i.e. strong deviations from spherical symmetry
[4].   Only future observations will, hopefully, decide whether a pulsar
has been formed.   Since it is now clear that the progenitor star really
was a well-known blue B3-supergiant [5] of about 20 solar masses, the
immediate or delayed formation of a black hole cannot yet be ruled out.
A group at our institute [6] did indeed try to consider the possibility
of two neutrino pulses from the formation of a heavy neutron star which
after the loss of heat and surplus angular momentum would find itself
above the critcal mass for cold neutron stars (≈ 2 $M_\odot$?) and collapse
further to form a black hole.   As long as such a second collapse cannot
be modelled in detail (including rapid rotation and the ensuing violent
circulation and convection phenomena on the collapse time scale) an
accompanying second neutrino burst cannot be excluded.   The fact that
the Mont-Blanc and the KAMIOKA experiments did not mutually see the other

B. F. Schutz (ed.), Gravitational Wave Data Analysis, 55–57.
© *1989 by Kluwer Academic Publishers.*

pulse would be marginally understandable with assumptions abut the energy spectra [7]. However, the energy radiated in the form of neutrinos in the first pulse would have been so high, namely more than 10% of a solar rest mass (unless one assumes anisotropic emission or unconventional cross-sections) that a conservative physicist will rather consider the Mont-Blanc event as simulated by noise.

For similar reasons the somewhat unlikely features around the time of the Mont-Blanc Event, observed by the gravitational wave groups at Rome [8] and Maryland [9] in their bar data, will rather be ascribed to noise by most physicists. Only a different theory of gravity or a serious revision of the ideas about gravitational wave cross-sections of bars would make the energy-requirements compatible with energy-conservation.

The next nearby collapse event must be expected in our own galaxy, that is 5-7 times closer to us than the recent supernova. Thus, even the present neutrino detectors will see 25-50 times more neutrinos. This next event can be expected roughly within a decade [10], but very likely it will not be accompanied by an optical supernova, due to obscuration by dust. If by that time a world-wide gravitational wave detector network exists, with an amplitude sensitivity of 10^{-20}, the event could be discovered even if only 10^{-6} of a solar rest mass were radiated in gravitational waves, i.e. even without much influence of rotation [11]. In addition to many interesting details of the collapse, the gravity network will also provide the direction of the source up to a few diameters of the full moon and thus considerly facilitate a subsequent optical detection if immediate data evaluation and information of the astronomical community can be guaranteed. (With neutrinos, which did not supply a large scale coherent wave, the directional resolution of a detector network will depend on details of the neutrino diffusion in the source and possible neutrino rest masses. On the other hand, with still larger detector masses more direction preserving ν-e-scatterings will be observed, which may also provide an angular resolution of a few degrees of arc).

Gravitational wave enthusiasts may now put bets on whether gravitational wave astronomy will be born when the first weak "chirp" signals from distant galaxies (forever unreachable by neutrino detectors) will be dug out of the noise of extremely sensitive antennae [12] or rather when a collapse event in our own galaxy will, in coincidence with a neutrino pulse, give clear information about a hidden supernova. If the improvement of sensitivity to the 10^{-22} level will take about 10 years, the odds are for the stellar collapse in our galaxy.

REFERENCES

1.   Shelton, I., *IAU Circular* No. 4316, Feb. 24. 1987
2.   Anglietta, M. *et al.*, *Europhysics Letters*, 15 June 1987
3.   Hirata, K. *et al.*, *Phys.Rev.Lett.* 58, 1490, 1987

     Bionta, R. *et al.*, *Phys.Rev.Lett.*, 58, 1494, 1987
4.   cf. *New Scientist*, 2 July 1987, p.26
5.   Hillebrandt, W. *et al.*, *Nature*, 327, 597 (1987)
6.   Hillebrandt, W. *et al.*, *Astron.Astrophys.*, 180, L20 1987
7.   de Rujula, ..., *CERN*-preprint, 1987
8.   Amaldi, E. *et al.*, *Europhysics Letters*, 3, 1325, 1987
9.   Weber, J., in this workshop
10.   see e.g. Blair, D., in this workshop
11.   Müller, E. in: *Problems of Collapse and Numerical Relativity*,
       p. 271 (D. Bancel and M. Signore eds.) Amsterdam: Reidel 1984
12.   see e.g. Schutz, B., in this workshop.

COALESCING BINARIES TO POST-NEWTONIAN ORDER

Andrzej Krolak[a],[b]
Department of Applied Mathematics & Astronomy
University College Cardiff
Cardiff, Wales, UK.

ABSTRACT.   At present, coalescing binary systems containing neutron
stars or black holes are thought to be the most likely sources of
gravitational waves to be detected by the long baseline interferometers
being   currently   designed.   We   present   the   calculation   of   the
characteristics of the signal from a coalescing binary to the first
post-Newtonian order - the amplitudes of the waves averaged over
orientations of both the source and the detector and characteristic time
for the gravitational wave frequency change due to radiation reaction.
We show that at coalescence the eccentricity of the orbit and tidal
effects can be neglected. We consider the effects of the expansion of
the Universe on the signal. We also calculate the signal-to-noise ratio
for the post-Newtonian amplitudes of the waves under the assumption that
the detector is in the light-recycling configuration and that we use the
matched filter technique to dig out the signal from the noise.

1.   INTRODUCTION

For a long time relativists have considered the supernova explosions as
the most important and interesting sources of gravitational waves for
the detectors to see.   Recently a new class of gravitational wave
sources has emerged: the coalescence of binary systems containing
neutron stars or black holes.   They were first discussed as the
realistic sources of gravitational waves in the Universe by Clark and
Eardley (1977), but it was Kip Thorne who first understood their
importance for modern detectors. These sources would be most accurately
monitored by laser interferometric gravitational wave detectors because
such detectors are capable of making wide-band observations.   At the
moment there are five proposals to build large size (at least 1 km arm
lengths) laser-interferometric detectors, at the California Institute

---

[a] S.E.R.C. Postdoctoral Research Assistant
[b] On leave of absence from the Mathematical Institute, Polish Academy
of Sciences, Warsaw, Poland.

B. F. Schutz (ed.), Gravitational Wave Data Analysis, 59–69.
© *1989 by Kluwer Academic Publishers.*

of Technology, the Massachusetts Institute of Technology, Glasgow University, the Max Planck Institute for Quantum Optics at Garching near Munich, and there is a joint proposal by research groups at Orsay near Paris and at Piza. The proposed network of detectors would be capable of detecting coalescing binaries over a large part of the Universe.

We shall calculate the characteristics of the gravitational wave coming from coalescing binaries including the contribution from the first post-Newtonian corrections. Our calculations will be based on work of Wagoner and Will (1976) (hereafter referred to as WW) on the dynamics of a point-mass binary system in the post-Newtonian approximation.

In Section 2 we evaluate two quantities characterizing the wave: the root-mean-square value $\langle h \rangle$ of the amplitude h of the gravitational wave from a point-mass binary system, averaged over orientations of both the source and the detector and the characteristic time τ for the increase of the frequency of the wave due to radiation reaction. We assume that orbit of the binary is circular. In Section 3 we show that the approximation of a circular orbit and point masses is a very accurate one even in the post-Newtonian order for neutron stars with typical masses when they are some 100 km apart. At this stage of the evolution of the binary, the gravitational wave frequency will be optimal for detection (see Section 4). In Section 4 we calculate the signal-to-noise ratio for the coalescing binaries that can be achieved in laser interferometers that are shot-noise limited and operating in light-recycling configuration. We shall assume that the matched-filter technique is used to analyse the data.

2. THE GRAVITATIONAL WAVE FROM TWO POINT MASSES IN A CIRCULAR ORBIT

Following Wagoner and Will we assume that the orbit of the binary system lies in the x-y plane. The unit vector $\vec{n}$ along the line joining the centre of mass of the binary and the observer far from the source is given by

$$\vec{n} = \sin \alpha \, \vec{e}_x + \cos \alpha \, \vec{e}_z \tag{1}$$

In the aysmptotically flat region far from the source we set up orthonormal basis vectors orthogonal to $\vec{n}$

$$\vec{e}_\alpha = \frac{1}{R} \frac{\partial}{\partial \alpha} \tag{2}$$

$$\vec{e}_\beta = \frac{1}{R \sin \alpha} \frac{\partial}{\partial \beta} \tag{3}$$

where R is the distance of the centre-of-mass of the binary system to the observer.

From WW formulae (68), (70) and (78) we get the following expression for the transverse traceless part h_{ij}^{TT} (i,j = α, β) of the metric perturbation for circular orbits. This part of the metric describes the gravitational radiation of the system.

$$h_{TT}^{ij} = \left[- \frac{4\mu}{R} \right]\left\{ \left[\frac{m}{r} \right]\left[1 - \left[\frac{m}{r} \right]\left[\frac{19}{6} - \frac{1}{2}\frac{\mu}{m} \right]\right]\bar{h}_{Q^{TT}}^{i} + \left[\frac{m}{r} \right]^{3/2}\left[\frac{\delta m}{m} \right]\bar{h}_{CQ^{TT}}^{j} + \right.$$

$$\left. + \left[\frac{m}{r} \right]^{2}\left[1 - \frac{3\mu}{m} \right]\bar{h}_{0^{TT}}^{ij} \right\} \tag{4}$$

where

$$\bar{h}_{Q^{TT}}^{ij} = \begin{bmatrix} \dfrac{1 + \cos^2\alpha}{2}\cos 2\omega t & \Bigg| & \cos\alpha\,\sin 2\omega t \\ \hline \cos\alpha\,\sin 2\omega t & \Bigg| & -\dfrac{1 + \cos^2\alpha}{2}\cos 2\omega t \end{bmatrix}$$

$$\bar{h}_{CQ^{TT}}^{ij} = \sin\alpha \begin{bmatrix} \dfrac{1}{16}(5 + \cos^2\alpha)\sin\omega t + & \Bigg| & -\dfrac{3}{8}\cos\alpha\cos\omega t + \\ \dfrac{9}{16}(1 + \cos^2\alpha)\sin 3\omega t & \Bigg| & -\dfrac{9}{8}\cos\alpha\cos 3\omega t \\ \hline -\dfrac{3}{8}\cos\alpha\cos\omega t + & \Bigg| & -\dfrac{1}{16}(5 + \cos^2\alpha)\sin\omega t + \\ -\dfrac{9}{8}\cos\alpha\cos 3\omega t & \Bigg| & -\dfrac{9}{16}(1 + \cos^2\alpha)\sin 3\omega t \end{bmatrix}$$

$$\bar{h}_{0^{TT}}^{ij} = \sin^2\alpha \begin{bmatrix} -\dfrac{1}{2}\dfrac{(3 + \cos^2\alpha)}{3}\cos 2\omega t + & \Bigg| & -\dfrac{2}{3}\cos\alpha\sin 2\omega t + \\ -\dfrac{2(1 + \cos^2\alpha)}{3}\cos 4\omega t & \Bigg| & -\dfrac{4}{3}\cos\alpha\sin 4\omega t \\ \hline -\dfrac{2}{3}\cos\alpha\sin 2\omega t + & \Bigg| & \dfrac{1}{2}\dfrac{(3 + \cos^2\alpha)}{3}\cos 2\omega t + \\ -\dfrac{4}{3}\cos\alpha\sin 4\omega t & \Bigg| & \dfrac{2(1 + \cos^2\alpha)}{3}\cos 4\omega t \end{bmatrix}$$

where μ, m, δm are reduced mass, total mass and difference of masses respectively, r is the radius of the orbit and ω is the angular orbital frequency.

The response of a detector to a gravitational wave depends not only on the orientation of its source (given by angle α in the case above) but also on the orientation at the detector with respect to the source. The latter orientation can be described by three Eulerian angles ϕ, θ and ψ where ϕ and θ describe the incoming direction of the wave and ψ is the angle between one semi-axis of the ellipse of polarisation and the node direction. Using the formalism given by Schutz and Tinto (1987) and adapting their formulae to our real representation of the wave we have the following formula for the response of the detector $\delta\ell/\ell_0$

$$\frac{\delta\ell}{\ell_0} = h_+ \left[\cos2\phi \cos\theta \sin2\psi + \frac{1}{2}\sin2\phi(1+\cos^2\theta)\cos2\psi\right] +$$

$$+ h_\times \left[\cos2\phi \cos\theta \cos2\psi - \frac{1}{2}\sin2\phi(1+\cos^2\theta)\sin2\psi\right] \qquad (5)$$

where

$$h_+ = h_{TT}^{\alpha\alpha}$$

$$h_\times = h_{TT}^{\alpha\beta}$$

where ℓ_0 is length of one arm of the detector, $\delta\ell$ is the difference between changes of ℓ_0 in each arm. Arms are at right angles.

To obtain a concise characteristic of the gravitational-wave signal we evaluate the average of the square of the response of the detector over one orbital period and then over orientation angles of both the source and the detector. The square root of the resulting quantity is called the root-mean-square (r.m.s.) average and it is denoted by $\langle h \rangle$. It is given by the formula

$$\langle h \rangle^2 = \frac{\omega}{32\pi^2} \int_0^\pi \int_0^{2\pi} \int_0^{2\pi} \int_0^\pi \int_0^{\frac{2\pi}{\omega}} \left[\frac{\delta\ell}{\ell_0}\right]^2 dt \, \sin\theta \, d\theta \, d\phi \, d\psi \, \sin\alpha \, d\alpha \qquad (6)$$

After carrying out integration over one orbital period the squared response of the detector splits into four components corresponding to once, twice, three times, and four times the orbital frequency. To distinguish between various frequency components of $\langle h \rangle^2$ we introduce the subscript I where I is the integer denoting the number of orbital frequencies. For example $\langle h_2 \rangle$ is the r.m.s. mean value corresponding to the response of the detector to the gravitational wave with twice the oribtal frequency.

We carry out integration over the angles. Then we express m/r in terms of $m\omega$ using formulae (54) and (61) of WW for circular orbits.

$$\frac{m}{r} = (m\omega)^{2/3}\left[1 + (m\omega)^{2/3}\left(1 - \frac{1}{3}\frac{\mu}{m}\right)\right] \qquad (7)$$

We only keep terms up to 1st post-Newtonian order. Finally we present the r.m.s. amplitudes in the following convenient form

$$\langle h_1 \rangle = 3.94 \times 10^{-25} \mu_\odot \, m_\odot \, f_{N100} \, R_{100}^{-1}(\delta m_\odot/m_\odot) \qquad (8a)$$

$$\langle h_2 \rangle = 1.02 \times 10^{-23} \mu_\odot \, m_\odot^{2/3} \, f_{N100}^{2/3} \, R_{100}^{-1} \qquad \times$$

$$\times \left[1 - 0.034(1 - 0.514\ \mu_\odot/m_\odot)(m_\odot f_{N100})^{2/3}\right] \tag{8b}$$

$$\langle h_3 \rangle = 9.15 \times 10^{-25}\ \mu_\odot m_\odot f_{N100}\ R_{100}^{-1}\ (\delta m_\odot/m_\odot) \tag{8c}$$

$$\langle h_4 \rangle = 3.24 \times 10^{-25} \mu_\odot (m_\odot f_{N100})^{4/3}\ R_{100}^{-1}\ (1 - 3\mu_\odot/m_\odot) \tag{8d}$$

where $\mu_\odot$, $m_\odot$, $\delta m_\odot$ are the masses μ, m, δm expressed in solar masses, R_{100} is the distance in 100 Mpc and f_{N100} is twice the orbital frequency in 100 Hz.

The radiation reaction on the binary system can be characterized by the following quantity

$$\tau = P/\dot{P}$$

where P is the period of the orbit of the binary and dot denotes derivative with respect to time. To calculate $\dot{P}$ we assume, as in the Newtonian case the following energy balance equation

$$L = -\ \mu\ \frac{d\tilde{E}}{dt} \tag{9}$$

where L is the gravitational energy radiated per unit time as measured by the distant observer and $\tilde{E}$ is the orbital binding energy per unit reduced mass. From formula (58) applied to circular orbits and formula (68) in WW we deduce the expression for the period of the orbit

$$P = 2\pi r\ (r/m)^{\frac{1}{2}}\ (1+(m/2r)(3-\mu/m)) \tag{10}$$

By differentiation of the above formula we can evaluate $P/\dot{P}$ in temrs of $r/\dot{r}$. By substitution of expressions for $\tilde{E}$ and L from WW formulae (70) and (81) in the balance equation (8) and use at WW (68) we obtain an equation for $r/\dot{r}$. We combine the two expressions for $r/\dot{r}$. We again use formula (7) to express m/r in terms of $m\omega$. Then we can express the characteristic time τ in the following convenient form

$$\tau = 7.97\ \mu_\odot^{-1}\ m_\odot^{-2/3}\ f_{N100}^{-8/3}\ \times$$

$$\times \left[1 - 0.030(1 + 1.24\ \mu_\odot/m_\odot)(m_\odot f_{N100})^{2/3}\right] \tag{11}$$

The most uncertain factor about the coalescing binaries is their
rate of occurrence in the Universe. The recent estimate suggests three
events per year out to a distance of 100 Mpc (Clark *et al.* 1979). This
factor puts strong demand on the sensitivity of the detectors (see
Section 4). It also means that detectors must be capable of picking up
the coalescence events from hundreds of megaparsecs, i.e. from
cosmological distances. Therefore we should consider the effect of the
expansion of the Universe on the signal. One finds that if the source
is at a cosmological distance with redshift z, then the formulae for
$\langle h_I \rangle$ and τ apply with the interpretation that R is the *luminosity
distance* d_L, f is the observed frequency f_0 (redshifted with respect to
frequency measured by the observer at the source), and any mass in the
formulae (total mass, reduced mass and the mass difference) is replaced
by (1+z)m.

In the Newtonian case we obtain a remarkably simple formula for
the luminosity distance in terms of the quantities that can be measured
by a network of detectors (see Schutz 1986):

$$d_L = 7.8 \, f_{0 \, 100}^{-2} \, (\langle h_{23} \rangle \, \tau_0)^{-1} \tag{12}$$

where $\langle h_{23} \rangle$ is the r.m.s. average of the amplitude case in the Newtonian
approximation multiplied by 10^{23} and τ_0 is the Newtonian contribution to
τ.

3. ECCENTRICITY OF THE ORBIT AND TIDAL EFFECTS

We shall first consider the corrections due to eccentricity to the
quantities evaluted in the previous section.

Even in the future designs of laser interferometers it may not be
possible to eliminate the seismic noise below 100 Hz. Thus the
frequency of the gravitational wave that can be detected must be higher
than 100 Hz. For the binary system consisting of neutron stars this
means that distance between the members of the binary must be typically
less than around 100 km. We shall consider what this implies for for
the only known progenitor of the binary coalescence, the binary pulsar
PSR 1913+16. At present members of this binary system are some 10^6 km
apart. The eccentricity of the system is 0.6. The evolution of the
eccentricity of a point mass binary system due to radiation reaction has
been calcuated in the Newtonian calculation by Peters (1964). He
deduced the following relation between the semi-major axis α of the
orbit and its eccentricity

$$\alpha(e) = c_0 e^{12/19}/(1 - e^2)\left[1 + (121/304)e^2\right]^{870/2299} \tag{13}$$

where c_0 is the constant of integration.

From the above formula it is not difficult to deduce that when α
decreases by four orders of magnitude then e goes from 0.6 to 1.3 ×
10^{-6}. We can estimate what this implies for amplitudes $\langle h_I \rangle$ and time τ.

At the Newtonian order for the eccentric orbits one finds that the
first order contributions $\langle h_I \rangle_e$ to the r.m.s. values due to eccentricity
at Newtonian order have the following form

$$\langle h_2 \rangle_e \sim \langle h_2 \rangle_0 \ e^2 \tag{14a}$$

$$\langle h_1 \rangle_e \sim \langle h_2 \rangle_0 \ e \tag{14b}$$

$$\langle h_3 \rangle_e \sim \langle h_2 \rangle_0 \ e \tag{14c}$$

where $\langle h_2 \rangle_0$ is the Newtonian r.m.s. average for circular orbits. The largest corrections $\langle h_I \rangle_{1e}$ due to eccentricity to first post-Newtonian r.m.s. values can be of the following form

$$\langle h_I \rangle_{1e} \sim \langle h_I \rangle_1 e \tag{15}$$

where $\langle h_I \rangle_1$ is the post-Newtonian correction in the r.m.s. averages $\langle h_I \rangle$ for circular orbits. From the above estimates we conclude that when $e \sim 10^{-6}$ first order corrections due to eccentricity to $\langle h_I \rangle$ are typically six orders of magnitude smaller than the first post-Newtonian contributions for circular orbits to $\langle h_2 \rangle$ and $\langle h_4 \rangle$ and around four orders of magnitude smaller than the first post-Newtonian contributions for circular orbits to $\langle h_1 \rangle$ and $\langle h_3 \rangle$.

By analyzing the formulae for general post-Newtonian orbits given in Section 4 of WW one finds that the first order corrections due to eccentricity to the period of the orbit are of the order of post-Newtonian corrections multiplied by eccentricity e assuming that e is small. Thus for $e = 10^{-6}$ the first order correction to the characteristic time τ are six order of magnitude smaller than the first post-Newtonian contributions for circular orbits.

Let us consider the tidal effects in the binary system. These effects must be small in order that the point-mass model considered in Section 2 is valid. Tidal effects in the binary may be of two types. They may cause either the mass exchange between the components of the binary or the angular momentum exchange. The tidally-induced mass exchange in neutron-star binaries was estimated by Clark and Eardley (1977). They found that it is unimportant in systems emitting gravitational waves at frequencies below 200 Hz if the masses of the components of the binary are not less than 0.3 solar masses. We can estimate the effects of tidally-induced angular momentum exchange using classical formula due to Darwin

$$\dot{L}_T = (171/4)(m_B^2/m_A)(a/r)^6 \ v(n-\omega) \tag{16}$$

where $\dot{L}_T$ is the time derivative of the orbital angular momentum (subscript T means that it refers to tidal effects), m_A, m_B are masses of the system (body of mass m_A is assumed to be extended and deformed by point-mass m_B), a is the radius of the body A and v is its kinematic viscosity, n is the angular velocity of revolution of mass m_A. The above formula has been derived by Darwin (1908, p.385, equ. (3) and pp. 60, 90-01, 386) under the assumption that body A is homogeneous and incompressible. This should be a good approximation for a neutron star. The well-known result of the above effect is the tendency to synchronize orbital angular velocity ω with angular velocity of rotation of the bodies.

The corrections to $\langle h_I \rangle$ and τ by the above effect are determined by the ratio $\dot{L}_T/\dot{L}_R$ ($\dot{L}_R$ is the change of orbital angular momentum due to radiation reaction). From formula (14) and the standard formula for $\dot{L}_R$ derived by Peters (1964) we get

$$c = \dot{L}_T/\dot{L}_R = -(855/128)v(n-\omega)(m_A/m)^{\frac{1}{2}}(r/m_A)^{7/2}(a/r)^6 \qquad (17)$$

For a typical radius of a neutron star of a ~ 10 km, r ~ 100 km and maximum estimated value of viscosity v ~ 10^5 (Friedman 1985), we have:

$$c \lesssim -1.9 \times 10^{-15}(n - \omega)m_{A_\odot}^{-3} m_\odot^{-\frac{1}{2}} \qquad (18)$$

Thus the tidal effects for a neutron star binary with masses distance some 100 km apart are very small indeed and the point mass model is highly accurate.

4. SIGNAL-TO-NOISE RATIO

Signal-to-noise ratio (S/N) is a crucial factor for the possibility of the detection of the signal. It depends on both the amplitude of the signal and the noise in the detector. It can be considerably increased by application at special signal-processing techniques. The framework for our calculations at S/N is given in the recent review article by Kip Thorne (1987) which the reader can consult for more details.

Because we know the signal from the binary system to a very high accuracy we can use the matched-filter technique (Helstrom 1968)) to dig out the signal from noise. This technique consists of taking the cross-correlation of the expected wave-form of the signal with the signal it-self. When we use this technique the S/N ratio is given by the following formula:

$$(S/N)^2 = \int_0^\infty \frac{2|\tilde{h}(f)|^2}{S(f)}\, df \qquad (19)$$

where $\tilde{h}$ is the Fourier transform of the waveform and S(f) is the spectral density of the noise in the detector. This technique gives a relative improvement of S/N equal to the square root of the number of periods N the binary spends around the starting frequency of observations. For a neutron star binary at 100 Mz of typical masses $\sqrt{N}$ ~ 30.

We expect that in future laser interferometric detectors, the dominant noise will be the shot noise above 100 Hz. In these detectors we expect to apply the technique invented by Drever (1983) called light-recyling. It consists of extracting the light from the inter-ferometer after a half period of the signal, when further storage would reduce the response of the detector, and then reinserting it into the interferometer along with and in phase with new laser light. With this technique the shot noise in the detector is given by the formula:

$$S(f) = \begin{cases} \infty & f < f_s \\ \dfrac{h\,\lambda_e}{cI_o \eta}\, f_o f_k \left[1 + \left[f/f_k\right]^2\right] & f > f_s \end{cases} \tag{20}$$

$$f_o = \frac{(1 - R_E)c}{4\pi L} \qquad\qquad f_k = \frac{(1 - R_c)c}{8\pi L}$$

where h is Planck's constant, λ_e is the wavelength of the laser light, I_o is the power of the laser, η is its efficiency, f_s is the seismic cut-off frequency, L is the length of one arm of the interferometer, R_E is the reflectivity of the end-mirrors of the interferometer, R_c - the reflectivity of the corner mirror.

In order that the noise is minimized we also require a special choice of the reflectivity R_R of the recycling mirror which reflects the laser light back into the interferometer

$$1 - R_R = \frac{4(1 - R_E)}{(1 - R_c)} \tag{21}$$

For the light-recycling technique to give a real improvement in the S/N ratio the reflectivity R_E must be consideably greater than R_c. Then the improvement in S/N is given by $2f_o/f_k$ i.e. around 80 for a proposed detector of arm length 4 km and $1 - R_E = 10^{-4}$.

We use formula (8) to evaluate S/N given by formula (19). To obtain $\langle \tilde{h}_I(f)\rangle$ we use an approximate formula for the Fourier transform valid when frequency f(t) is a slowly varying function of time. If

$$h(t) = A(f(t))\, \cos 2\pi \int^t f(t')dt' \tag{22}$$

Then

$$|\tilde{h}(f)| \simeq A(f)/\sqrt{4\dot{f}}$$

where the dot denotes derivative with respect to time. It is convenient to express $\langle \tilde{h}_I\rangle$ in the following form

$$\langle \tilde{h}_I\rangle = \langle h_I\rangle \sqrt{\tau/4f} \tag{23}$$

Using the above formulae, signal-to-noise ratio for the post-Newtonian amplitudes take the following form (subscripts in (S/N) corresponds to subscripts in the average amplitudes $\langle h_I\rangle$).

$$(S/N)_1^2 \simeq \left[0.38\, A_o\, \frac{\delta m_o}{m_o}\, \frac{\mu_o^{1/2} m_o^{2/3}}{R_{100}}\right]^2 \frac{I_1(\gamma)}{f_{s100}^{5/3}} \tag{24a}$$

$$(S/N)_2^2 \cong \left[17.61 \; A_o \; \frac{\mu_\odot^{1/2} \; m_\odot^{1/3}}{R_{100}} \right]^2 \frac{I_2(y)}{f_{s100}^{7/3}}$$

(24b)

$$(S/N)_3^2 \cong \left[2.24 \; A_o \; \frac{Sm_\odot}{m_\odot} \; \frac{\mu_\odot^{1/2} m_\odot^{2/3}}{} \right]^2 \frac{I_2(y)}{f_{s100}^{5/3}}$$

(24c)

$$(S/N)_4^2 \cong \left[0.56 \; A_o \; \frac{\mu_\odot^{1/2} m_\odot}{R_{100}} \right]^2 \frac{I_4(y)}{f_{s100}}$$

(24d)

where $y = f_k/f_s$

$$A_o = \left[\frac{0.0818 \; \mu m}{\lambda_e} \right]^{1/2} \left[\frac{I_o n}{100W} \right]^{1/2} \left[\frac{10^{-4}}{1 - R_E} \right]^{1/2} \left[\frac{L}{4km} \right]^{1/2}$$

$$I_2(y) = \frac{1}{y^{5/3}} \int_0^{y^2} \frac{y^{1/3}}{1+y} \; dy = \frac{3}{y^{5/3}} \left[y^{2/3} - \frac{1}{3} \ln \frac{1+y^{2/3}}{\sqrt{1-y^{2/3}+y^{4/3}}} - \right.$$

$$\left. - \frac{1}{\sqrt{3}} \; arctg \; \frac{y^{2/3}\sqrt{3}}{2-y^{2/3}} \right]$$

$$I_2(y) = \frac{1}{y^{7/3}} \int_0^{y^2} \frac{y^{2/3}}{1+y} \; dy = \frac{3}{y^{7/3}} \left[\frac{y^{4/3}}{2} + \frac{1}{3} \ln \frac{1+y^{2/3}}{\sqrt{1-y^{2/3}+y^{4/3}}} - \right.$$

$$\left. - \frac{1}{\sqrt{3}} \; arctg \; \frac{y^{2/3} \sqrt{3}}{2-y^{2/3}} \right]$$

$$I_4(y) = \frac{1}{y} \int_0^{y^2} \frac{1}{1+y} \, dy = \frac{\ln(1+y^2)}{y}$$

We have not given the post-Newtonian correction for $(S/N)_2$ because we used only an approximate formula for the Fourier transform of $\langle h_I \rangle$ and the errors in Fourier transform may be of the same order as post-Newtonian corrections.

To obtain the maximum possible values for signal-to-noise ratio we estimate the maxima of the integrals $I(y)$ numerically

$$[I_1(y)]_{max} = 0.5376 \text{ for } y = 1.61$$

$$[I_2(y)]_{max} = 0.4012 \text{ for } y = 1.44$$

$$[I_4(y)]_{max} = 0.8047 \text{ for } y = 1.98.$$

From the above results we see that even though the amplitude of the wave increases with frequency the corresponding signal-to-noise ratio decreases. Therefore it is desirable to isolate the detector from seismic noise as much as possible to keep f_s low.

The obtained signal-to-noise ratios indicate that with future laser interferometers operating at full sensitivity it may be possible to detect post-Newtonian effects in the gravitational waves coming from the binary system. This could be achieved by detectors working in coincidence and by reducing the seismic noise by active seismic isolation.

REFERENCES

Clarke, J.P.A. and Eardley, D.M. 1977, *Astrophys.J.*, 215, 311
Clarke, J.P.A., van den Heuvel, E.P.J. and Sutantyo, W. 1979,
      Astron.Astrophys. 72, 120
Darwin, G.H. 1908. *Scientific Papers*, Vol. II, Cambridge University
      Press
Drever, R. 1983 in *Gravitational Radiation*, eds. Deruelle, N. and
      Piran, T. North Holland, Amsterdam
Friedman J. 1983 *Phys.Rev.Lett.* 51, 14
Helstrom, C.W. 1968, *Statistical Theory of Signal Detection*,
      Pergamon Press
Peters, P.C. 1964. *Phys.Rev.* 136, B1224
Schutz, B.F. 1986. *Nature*, 323, 310
Schutz, B.F. and Tinto, M. 1987. *Mon.Not.R.astr.Soc.*, 224, 131
Thorne, K. 1987. in *300 Years of Gravitation*, eds. Hawking, S.W. and
      Israel, W., Cambridge University Press.
Wagoner, R.V. and Will, C.M. 1976. *Astrophys.J.*, 210, 764

Part 2

PRINCIPLES OF SIGNAL PROCESSING

A REVIEW OF THE STATISTICAL THEORY
OF SIGNAL DETECTION

M.H.A. Davis
Department of Electrical Engineering
Imperial College, London SW7 2BT U.K.

ABSTRACT   This paper reviews some of the techniques of signal detection theory and discusses their use in the analysis of gravitational wave data.   Topics covered include the classical theory of hypothesis testing, detection of known signals in white noise, maximum likelihood estimation of unknown parameters, the application of nonlinear filtering in signal detection and the use of pre-whitening filters to handle correlated noise.

1.   INTRODUCTION

Signal processing plays a vital role in the quest for gravity waves since ultimately the performance of any detector is limited by its ability to recognize characteristic gravitational waveforms in a noisy signal. The problem is similar to classical problems of radar detection, but there are important differences.   One of these is that one has *a priori* no idea when the signal is likely to arrive, and it is vital to estimate the time of arrival as accurately as possible, since this provides directional information if the same event is picked up by two or more detectors. Another difference is that there is no need to do the signal processing in real time. Of course, saving the data for processing later introduces a massive storage requirement, but this seems in any case inevitable.

The purpose of this article is to review the available techniques of signal detection, with the emphasis on those likely to be most useful in view of the above remarks.   In §2 the basic formulation of detection or hypothesis testing is outlined in the simplest situation: distinguishing between two possible distributions for an observed random vector.   The bulk of the article, §§3-5, is concerned with detection of signals in continuous-time white noise, while §6 describes how to whiten non-white noise.   §3 considers the case in which the signal to be detected is exactly known; this leads to the "correlation detector" or "matched filter".   The leading candidates for sources of detectable gravity radiation are coalescing binary star systems;   these produce a "chirp" signal which is known up to three parameters: time of arrival, distance and a "mass parameter".   We are therefore particularly interested in detection of classes of signals depending on a (vector) parameter.   This is considered in §§4,5, where two different methods of parameter estimation are described - maximum likelihood estimation in §4 and a method based on nonlinear filtering in §5. The latter is potentially much more accurate but involves a greatly increased amount of computation.

Some concluding remarks are presented in §7.   Our general conclusion is that data analysis will be in (at least) two phases. The first phase will be a pass through the data using

73

B. F. Schutz (ed.), Gravitational Wave Data Analysis, 73–94.
© *1989 by Kluwer Academic Publishers.*

the realtively low computational effort per unit time techniques of correlation and maximum likelihood estimation. Then small parts of the data in which signals have been detected with high probability can be re-examined using the nonlinear filtering techniques in order to sharpen up parameter estimates.

The last section, §8, contains an annotated list of references for the material covered in this article.

Finally, it is a pleasure to thank Bernard Schutz for introducing me to the world of gravitational wave detection and for inviting me to participate in the stimulating workshop on which these proceedings are based. I would also like to thank Alessandro Pasetti for his collaboration.

2. THE CLASSICAL THEORY OF HYPOTHESIS TESTING

2.1   The basic idea behind signal detection is that the presence of a pulse or other waveform in an observed signal affects its statistical characteristics, i.e. probability distribution. Thus the problem that confronts us is to make a decision about the distribution of a signal after having observed a sample of it. This is known as *hypothesis testing* . In order to introduce the ideas we shall consider in this section the simplest possible setting, which is the following: random variables $Y_1, \ldots Y_n$, are observed, or equivalently a random vector $Y = (Y_1, \ldots, Y_n)^T$ (superscript T denotes transpose). Y is known to have one of two (joint) density functions $p_0(y)$ or $p_1(y)$, $y = (y_1, \ldots, y_n)$. On the basis of our observation we have to decide which it is, and we want to formulate some "best" way of doing this. Note that

(a)   there are two possible ways of making an error: we can say p_1 when p_0 is true ("false alarm") or say p_0 when p_1 is true ("false dismissal"). Here we think of $p_1(p_0)$ as the distribution corresponding to signal present (signal absent) - hence the terminology.

(b)   a *test* is a rule which tells us what decision to make at every possible value y of Y, or equivalently a partition of the range of values of Y into two sets R and the complementary set R^c, the decision rule then being "say p_1 if $Y \in R$; otherwise say p_0".

For any test R we can calculate the probabilities of false alarm and false dismissal, which are

$$P_0(R) = \int_R p_0(y)dy \quad \text{and} \quad 1 - P_1(R) = 1 - \int_R p_1(y)dy$$

respectively, but how do we combine these to produce an overall best test (= choice of region R)? This cannot be done without introducing some extra features into the problem representing in some sense the relative seriousness of the two types of error. There are two standard formulations, which we describe next.

2.2. The Bayesian Approach

Here we consider that "nature" selects one of the distributions p_0, p_1 at random before Y is generated, distribution p_1 being selected with (known) probability ρ. To put it another way, ρ represents our *a priori* degree of certainty that p_1 is true. We also introduce penalties ℓ_1 and ℓ_2 for false alarm and false dismisal respectively, respresenting the seriousness of these mistakes. For any test R we can now evaluate the total expected penalty: if p_0 is true we err with probability $P_0(R)$ and and pay ℓ_1, whereas if p_1 is true then we err with probability $1-P_1(R)$ and penalty

$$(1-\rho)\ell_1 P_0(R) + \rho\ell_2(1 - P_1(R)).$$

It is easy to show that the region R which minimizes this is given by

$$R = \left\{ y : \frac{p_1(y)}{p_0(y)} \geq \frac{(1-\rho)\ell_1}{\rho\ell_2} \right\}.$$

The quantity

$$\Lambda(y) = \frac{p_1(y)}{p_0(y)}$$

is known as the *likelihood ratio* (LR). [To avoid triviality we suppose that $p_0(y) > 0$ for all y. Obviously regions where $p_0(y) = 0$ or $p_1(y) = 0$ should be assigned to R, R^c respectively]. Thus in the Bayesian framework the best test is a *likelihood ratio test:*

$$\text{"say } p_1 \text{ if } \Lambda(Y) \geq k\text{"} \tag{2.1}$$

where $k = (1-\rho)\ell_1/\rho\ell_2$ is a *threshold* depending only on the prior probability and error costs. Bayesian analysis is often criticized on the grounds that ρ, ℓ_1 and ℓ_2 are subjective quantities. The point we want to make is that the optimal test is always a LR test and only the threshold value k depends on these parameters.

2.3. The Neyman-Pearson Approach

In the earlier statistical literature on significance testing, the false alarm probability $P_0(R)$ is denoted α_R and called the *significance level* whereas the "correct alarm" probability $P_1(R)$ is denoted β_R and called the *power* of the test. Thus $\beta_R = 1$ - (false dismissal probability) and should ideally be large, as the name suggests. In the Neyman-Pearson approach no prior probability is assigned to the two hypotheses, but rather a test is sought which maximizes the power subject to a constraint on the significance level. A "most powerful test at significance level α" is a region R such that $\alpha_R \leq \alpha$ and $\beta_R \geq \beta_S$ for any region S such that $\alpha_S \leq \alpha$. In this framework there is just one subjective parameter - the significance level α. In the 1930s Neyman and Pearson showed that a most powerful test exists and is a likelihood ratio test, i.e. the most powerful test at significance level α is defined by

$$R = \{y : \Lambda(y) \geq k\}$$

where k is determined by the requirement that

$$P_0[\Lambda(Y) \geq k] = \alpha,$$

giving the correct significance level. This is in fact not hard to prove.

What is striking is that the apparently very different Bayesian and Neyman-Pearson approaches to hypothesis testing lead to exactly the same result, namely that a LR test (2.1) is optimal. Only the interpretation of the threshold constant k depends on which theory is being invoked. In the remainder of the paper we shall concentrate on computation of the likelihood ratio and shall say nothing further about k, which will always be an arbitrary threshold interpretable in either of the two ways described above.

2.4   An Example

Let $X_1...,X_n$ be a sequence of independent of $N(0,1)$ random variables[1], so that $X = (X_1,...,X_n)^T$ is a random vector with distribution $N(0,I_n)$ [I_n = nxn identity matrix]. Let $\mu = (\mu_1,...,\mu_n)^T$ be a given vector. Think of X as a finite sample of discrete-time white noise and μ as a known signal we are trying to detect in noise X. Thus if μ is absent the received signal is

$$Y_i = X_i , \qquad\qquad i = 1,...,n$$

whereas if μ is present

$$Y_i = \mu_i + X_i \qquad\qquad i = 1,...,n$$

In this case therefore, p_0 is the normal density $N(0,I_n)$ and p_1 is $N(\mu,I_n)$. The likelihood ratio is

$$\Lambda(y) = \frac{\exp(-\frac{1}{2}\sum_i (y_i - \mu_i)^2)}{\exp(-\frac{1}{2}\sum_i y_i^2)}$$

$$= \exp(<y,\mu> - \tfrac{1}{2}<\mu,\mu>) \qquad\qquad (2.2)$$

where $<y,\mu>$ denotes the inner product in Euclidean space $\mathbb{R}^n$:

$$<y,\mu> = \sum_{i=1}^{n} y_i \, \mu_i \, .$$

Thus $\Lambda(y) \geq k_1$ if and only if

$$<y,\mu> \geq k_2$$

where $k_2 = \ln k_1 + \frac{1}{2}<\mu,\mu>$. So the best test is a *correlation test*

"say μ present if $<Y,\mu> \geq k$".

The test involves calculating the one-dimensional statistic $<Y,\mu>$ and this has distribution $N(m,m),N(0,m)$ when the signal μ is present or absent respectively, where m = $<\mu,\mu>$. The false alarm and false dismissal probabilities are therefore easily calculated for any given value of k; they are the areas marked FA and FD in figure 1.

---

[1]$N(\mu,v)$ denotes the normal distribution with mean μ and variance v (or mean μ and covariance matrix v in the vector case)

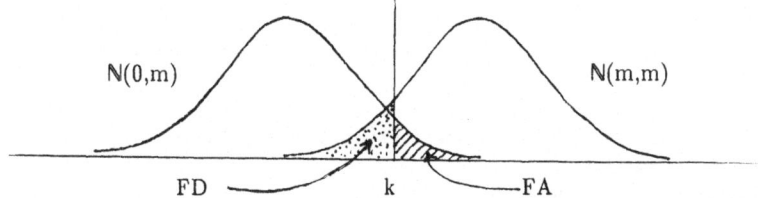

$N(0,m)$ $N(m,m)$

FD   k   FA

Figure 1

3.   DETECTING A KNOWN SIGNAL IN ADDITIVE WHITE NOISE

We now move to a continuous-time setting and suppose that we are trying to detect a known signal $s(t)$ in white noise $n(t)$, observed over a time interval $[0,T]$. Let $y(t)$ denote the observed signal. Then our two hypotheses are

<u>signal absent:</u> $y(t) = n(t)$ $0 \leq t \leq T$

or   (3.1)

<u>signal present:</u> $y(t) = s(t) + n(t)$ $0 \leq t \leq T$

Formally, "white noise" is a stationary process with flat spectral density function $\Phi_n(\omega) = 1$ and corresponding covariance function $r_n(t) = \delta_0(t)$ (the delta function at $t=0$). This is the limiting case of wide-band noise where spectral density is approximately flat over a range of frequencies larger than the bandwidth of the signal to be detected. Of course, white noise is an idealized construct which cannot be formalized mathematically as it stands. The way to make mathematical sense of it is to consider the formal indefinite integral

$$w(t) = \int_0^t n(u)du)$$

and compute the covariance function (here and throughout $\mathbb{E}$ denotes expectation or mean value)

$$\mathbb{E}w(t)w(\tau) = \mathbb{E}\int_0^t n(u)du \int_0^\tau n(v)dv$$

$$= \int_0^t \int_0^\tau \delta_0(u-v)dv\,du$$

$$= \begin{cases} t, & t \leq \tau \\ \tau, & \tau \leq t \end{cases} = \min(t,\tau)$$

This is the covariance function of a well-defined process, namely *Brownian motion* or the *Wiener process* $w(t)$ which can be characterized in the following way:

(i) $w(0) = 0$

(ii)   (independent increments) $(w(t_4) - w(t_3))$ and $(w(t_2) - w(t_1))$ are independent for any t_i, i=1...4 such that $0 \leq t_1 \leq t_2 \leq t_3 \leq t_4$.

(iii)   For any times t_1, t_2, $(w(t_2) - w(t_1))$ has normal distribution $N(0,|t_2-t_1|)$

(iv) $w(t)$ has continuous sample functions.

If we define

$$Y(t) = \int_0^t y(u)du \qquad (3.2)$$

then (3.1) becomes

signal absent   : $Y(t) = w(t), \quad 0 \leq t \leq T$

signal present : $Y(t) = \int_0^t s(u)du + w(t), \ 0 \leq t \leq T$

Think of the sample function $\{Y(t),0\leq t<T\}$ as a "random variable" taking values in the space of continuous functions $C([0,T];\mathbb{R})$ (note point (iv) above) rather than in the space of n-vectors (as in the example in §2.4). Its distribution with signal absent is Wiener measure, the distribution of Brownian motion; with signal present the density with respect to Wiener measure is given by the *Cameron-Martin formula*

$$\Lambda(Y) = \exp\left(\int_0^T s(t)dY(t) - \frac{1}{2}\int_0^T s^2(t)dt \right)$$

where the first integral is a so-called Wiener integral, the precise definition of which need not detain us at this point. We just note from (3.2) that $dY(t) = y(t)dt$, giving the LR in terms of the observed signal $y(\cdot)$ as

$$\Lambda(y) = \exp\left(\int_0^T s(t)y(t)dt - \frac{1}{2}\int_0^T s^2(t)dt \right) \qquad (3.3)$$

Let us denote

$$<y,s> = \int_0^T y(t)s(t)dt$$

(the inner product in $L_2[0,T]$); then (3.3) becomes

$$\Lambda(y) = \exp(<y,s> - \tfrac{1}{2}<s,s>) \qquad (3.4)$$

This is identical in form to the LR (2.2) of Example 2.4 with the L_2 inner product replacing the $\mathbb{R}^n$ inner product. This is not surprising since one can think of the continuous-time case as the limit as $\max_i \Delta t_i \to 0$ of the sampled version

$$y(t_i) = s(t_i) + n(t_i) \tag{3.5}$$

where $n(t_i)$, $i = 1,2,...$ are i.i.d. normal random variables.

The hypothesis testing theory outlined in §2 extends verbatim to function-space-valued random variables, and we conclude that the best test of signal present vs. signal absent is a LR test which takes the form

$$\text{"say signal present if } <y,s> \geq k\text{"}, \tag{3.6}$$

since the term $-<s,s>/2$ in (3.4) is not data-dependent. (3.6) is the well-known *correlation detector* or *matched filter*. The latter term arises from the fact that we can express the test statistic as

$$<y,s> = h*y(T) = \int_0^T h(T\text{-}t)y(t)dt,$$

i.e. the output at time T of a linear filter whose impulse response is $h(t) = s(T\text{-}t)$. Of course, in practice one will be computing it digitally, i.e using the approximation

$$<y,s> \approx \sum_i y(t_i)s(t_i)$$

obtained from (3.5).

The theory outlined in §2 shows that the correlation detector (3.6) provides the optimal test (in at least two different senses.) A subsidiary but intuitively appealing fact is that the matched filter *maximizes the signal to noise ratio* over all linear filters. Let $f(t)$, $0 \leq t \leq T$ be any L_2 function. Then with signal present

$$<y,f> = <s,f> + <n,f>.$$

Now $\mathbb{E}<s,f>^2 = <s,f>^2$ since s and f are deterministic, and $\mathbb{E}<n,f>^2 = <f,f>$, so the signal-to-noise ratio is

$$S/N = \frac{<s,f>^2}{<f.f>}$$

The *Schwarz inequality* states that for any f,

$$<s,f>^2 \leq <s,s><f,f>$$

with equality if and only if $f = \gamma s$ for some constant γ. Thus

$$S/N \leq <s,s> \tag{3.7}$$

with equality when f=s, the matched filter. Put another way, the *loss* in signal-to-noise occasioned by using an imperfectly matched filter f is just 1 - $<s,f>^2$ if one (without loss of generality) normalises the signals so that $<s,s> = <f,f> = 1$. Thinking of s and f as "templates" against which one correlates the incoming signal, this helps to evaluate the effect of using the wrong template - it is just a question of the "angle" between f and s, thought of as infinite-dimensional vectors.

4. SIGNALS WITH UNKNOWN PARAMETERS

4.1. In gravity wave - and almost all other - detectors the signals to be detected are not precisely known but depend on one or more unknown parameters. Our main example of this will be the "chirp" waveform produced by a coalescing binary star system. In this case there are three parameters, the distance, the time of arrival τ and the mass parameter α. This is discussed in detail in §4.4 below, but first we consider a general detection problem in which the signal depends on parameters $\theta_1,...,\theta_d$. The observed signal y(t) is therefore given by

<u>signal absent</u>: $y(t) = n(t)$ $0 \leq t \leq T$

$$(4.1)$$

<u>signal present</u>: $y(t) = s(t,\theta) + n(t)$ $0 \leq t \leq T$

Here n(t) is, as before, white noise with spectral density $\Phi_n(\omega) = 1$, and $s(t,\theta)$ is a class of signal functions depending smoothly on a d-dimensional parameter vector θ, which is unknown to the observer.

If one is prepared to take a Bayesian point of view and specify a prior distribution for the parameter vector θ, then it is possible to formulate a likelihood ratio test which minimizes an average error cost along the lines of §2.2. This procedure is discussed below in §5. It has the disadvantage that the LR is hard to compute, and the may be no very rational basis on which to assign a prior distribution. In this section we consider what can be done in a non-Bayesian context, i.e. regarding θ as a parameter with no probabilistic interpretation. In this case there is generally no clear concept of a *best test*, since the best tests for different values of θ may conflict with each other. Occasionally one comes across examples where no such conflict arrive. Consider for instance a modification of Example 2.4 in which under the "signal present" hypothesis we have

$$Y_i = \theta \mu_i + X_i \qquad i = 1,...,n$$

where μ is a known vector and θ an unknown scalar parameter, but *it is known that* $\theta > 0$. The best test for a fixed value of θ is "say p_1 if $<Y,\mu> \geq k$" where k is determined by $P_0[<Y,\mu> \geq k] = \alpha$ (α pre-specified). Thus k does not depend on θ and the same test is optimal for all θ (though its power depends on θ). This is known as a *uniformly most powerful test*, but basically it only arises in artificial examples in statistics textbooks. In this example the existence of a uniformly best test would fail if we did not know that $\theta > 0$.

4.2   Maximum Likelihood Tests

Faced with this situation, the only sensible procedure is to form some kind of estimate of θ and then to perform a test assuming that the estimated value of θ is the value actually occurring. It will be seen below that this procedure is actually optimal in the Bayesian case. In the non-Bayesian case the form of estimator used is normally the *maximum-likelihood estimator* . Let us return to the signal-in-white-noise problem (4.1). From (3.3), the LR for a specific value of θ is

$$\Lambda(y,\theta) = \exp(\int_0^T s(t,\theta)y(t)dt - \tfrac{1}{2} \int_0^T s^2(t,\theta)dt) \qquad (4.2)$$

The *maximum-likelihood* (ML) *estimate* $\hat{\theta}_{ML}$ of θ is the value of θ that maximizes $\Lambda(y,\theta)$ at the observed sample function y. [It is in principle possible that $\hat{\theta}_{ML}$ might not be unique, but this need not worry us here]. The ML test then consists of the following threshold test:

"say signal present if $\max_\theta \Lambda(y,\theta) \geq k$".

Here the threshold value k is chosen to control the false alarm probability (=significance level, α), i.e. one chooses k so that

$$P_0[\max_\theta \Lambda(y,\theta) \geq k] = \alpha.$$

This probability may be difficult or impossible to evaluate analytically, but is readily estimated by a crude Monte Carlo method (simulate the noise process 5000 times and find the level k which is exceeded by $\max_\theta \Lambda(y,\theta)$ in just 50 runs, for $\alpha = .01$). If the signal is declared to be present then $\hat{\theta}_{ML}$ is the estimated value of θ. This procedure has no overall optimality property, but its power β for specific values of θ can be estimated by the Monte Carlo method in the same way as the significance level.

An important property of the ML estimator is that it is invariant under (possibly nonlinear) changes of parametrization. Suppose we choose a new parameter vector $\eta = \phi(\theta)$ where ϕ is some one-to-one function such that ϕ and the inverse function ϕ^{-1} is continuous. Then since $\theta = \phi^{-1}(\eta)$ the LR written in terms of η is just

$$\tilde{\Lambda}(y,\eta) = \Lambda(y,\phi^{-1}(\eta)) = \Lambda(y,\theta),$$

Clearly $\hat{\eta}$ maximizes $\tilde{\Lambda}(y,\eta)$ if and only if $\hat{\theta} := \phi^{-1}(\hat{\eta})$, maximizes $\Lambda(y,\theta)$; consequently

$$\hat{\eta}_{ML} = \phi(\hat{\theta}_{ML}) \qquad (4.3)$$

Consider for example extimating the frequency of a sinusoidal signal. It seems rather arbitrary whether one chooses to parametrize it by frequency ($\sin 2\pi ft$) or by period ($\sin 2\pi t/\tau$). (4.3) tells us that it *is* arbitrary, in that

$$\hat{f}_{ML} = 1/\hat{\tau}_{ML}.$$

This property is not true of all estimators. It does not hold, for example, in the case of the Bayesian estimators introduced in §5 below.

4.3 Computation of the ML estimate $\hat{\theta}_{ML}$

In some cases this can be done analytically, while in others some form of searching or other numerical maximization technique must be used. Of course, maximizing $\Lambda(y,\theta)$ is equivalant to maximizing $\ell n\Lambda(y,\theta)$. Some examples are

(i)　　Amplitude : 　　　　　　$s(t,\theta) = \theta s(t)$.

In this case

82

$$\ell n \Lambda(y,\theta) = \theta \int_0^T s(t)y(t)dt - \tfrac{1}{2}\theta^2 \int_0^T s^2(t)dt$$

so $\hat\theta_{ML}$ is given in closed form by

$$\hat\theta_{ML} = \frac{<s,y>}{<s,s>}$$

(ii) Time-of-arrival : $s(t,\theta) = s(t-\theta)$ $(s(t) = 0$ for $t < 0)$
Assuming that the duration of the signal is much less than T, $\int s^2(t,\theta)dt$ is essentially independent of θ, so $\hat\theta_{ML}$ maximizes

$$\theta \to \int_0^T s(t-\theta)y(t)dt =: <s_\theta,y>.$$

Thus $\hat\theta_{ML}$ is the delay θ at which the template s_θ has maximum correlation with the observed signal. This is normally found by global search.

(iii) Frequency: $s(t,\theta) = \sin\theta t$

Here it is convenient to normalize by the length T of the observation interval and consider

$$\tfrac{1}{T}\ell n\Lambda(y,\theta) = \tfrac{1}{T}\int_0^T \sin\theta t\, y(t)dt - \tfrac{1}{2T}\int_0^T \sin^2\theta t\, dt$$

$$= \tfrac{1}{T}\int_0^T \sin\theta t\, y(t)dt - \tfrac{1}{4}(1 - \text{sinc } 2\theta T)$$

$\hat\theta_{ML}$ can now be computed numerically using Newton's method or some other hill-climbing technique.

If there is no signal present, then y(t) is white noise, $y(t) = n(t)$. In this case

$$X(T,\theta) := \tfrac{1}{T}\int_0^T \sin\theta t\, n(t)dt$$

is a random variable with distribution $N(0,\sigma^2_{T,\theta})$ where

$$\sigma^2_{T,\theta} = \tfrac{1}{2T}\Big(1 - \text{sinc } 2\theta T\Big).$$

Thus

$$\tfrac{1}{T}\ell n\Lambda(y,\theta) = X(T,\theta) - \tfrac{1}{4}(1 - \text{sinc } 2\theta T).$$

This expression can be used for computing false alarm probabilities. If a signal is present at angular frequency ω then $y(t) = \sin\omega t + n(t)$ and we find that

$$\tfrac{1}{T}\ell n\Lambda(y,\theta) = X(T,\theta) + \tfrac{1}{2}\text{sinc}(\theta-\omega)T - \tfrac{1}{2}\text{sinc}(\theta+\omega)T - \tfrac{1}{4}(1 - \text{sinc } 2\theta T). \quad (4.4)$$

It is clear from this expression that $\hat\theta_{ML}$ is close to the true value ω with high probability if T is sufficiently large, since $\text{sinc}(\theta-\omega)T$ takes its maximum value of 1 at $\theta=\omega$ for any T whereas the other terms are small (note that $\sigma^2_{T,\theta} < 1/T$ for any θ).

4.4.   The "chirp" waveform from coalescing binaries

Consider a binary star system with masses m_1 and m_2. We define $M_T = m_1 + m_2$ and $\mu = m_1 m_2/(m_1+m_2)$. The "Mass parameter" is $\alpha = M_T^{2/3}\mu$ (for example, $\alpha = 1.4$ when $m_1=m_2 = 1.4$). The instantaneous frequency of the system is $f(t)$ which, in units of 100Hz, satisfies

$$\frac{f}{df/dt} = \frac{7.97}{\alpha} f^{-8/3} \text{ secs.}$$

The solution of this equation from an initial time τ is

$$f(t) = (f^{-8/3}(\tau) - .334\alpha(t - \tau))^{-3/8}$$

and the gravitational waveform is then

$$s(t,\theta) = K\alpha f^{2/3}(t) \sin\left(\int_{\tau}^{t} f(u)du\right), \quad t \ge \tau,$$

where the amplitude K is inversely proportional to the distance from the source. Let us assume temporarily that the lower cutoff frequency of the detector is 100Hz and that τ is the time at which this frequency is attained, so that $f(\tau) = 1$ and τ is effectively the time of arrival at the detector. The parameter θ is then the 3-vector $\theta = (K,\alpha,\tau)$ and we find that

$$s(t,\theta) = K\alpha(1-.334\alpha(t-\tau))^{-1/4} \sin \frac{2597}{\alpha}(1 - (1 - .334\alpha(t-\tau))^{5/8})$$

for $\tau \le t \le \tau^*$ where τ^* is the time at which $f(\tau^*)$ is equal to the upper cutoff frequency of the detector. The signal "explodes" at $t-\tau = (.334\alpha)^{-1}$ so $\tau^* \le \tau + (.334\alpha)^{-1}$. ML estimates of the parameters $\theta = (K,\alpha,\tau)$ are found by maximizing the likelihood function (4.2) using global search or other techniques as outlined above.

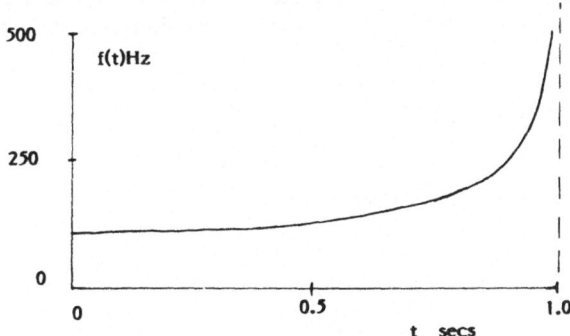

Figure 2

Figure 2 shows a graph of f(t) when $\alpha = 3$, with $\tau = 0$, i.e. f(0) = 100 Hz. In this case the explosion time is 1 sec. The graph is revealing because it shows the importance of keeping the lower cutoff frequency as low as possible. Note from (3.7) that the signal-to-noise ratio with the matched filter is (roughly) proportional to the observation time T of the signal. Figure 2 shows that the signal spends almost all its time at frequencies below 250Hz. Let us fix the upper cutoff frequency arbitrarily at 2.5 kHz, attained at t= .99981. Then the following table shows the duration of the observation interval (δ) as a function of the lower cutoff frequency (ζ)

ζ(Hz)	δ(secs)
100	.99981
150	.33899
200	.15730
250	.08667

These figures speak for themselves,.

5.   DETECTION AND NONLINEAR FILTERING

In this section we will consider the same problem as in §4, i.e. the detection problem of choosing between

signal present: $y(t) = s(t,\theta) + n(t)$ $0 \le t \le T$

signal absent: $y(t) = n(t)$ $0 \le t \le T$

where n(t) is white noise, or in mathematically precise form

signal present: $Y(t) = \int_0^t s(t,\theta)dt + w(t), \quad 0 \le t \le T$

signal absent: $Y(t) = w(t),$ $0 \le t \le T$

where w(t) is a Brownian motion process. But now we take a Bayesian viewpoint and consider θ as a random variable which Nature selects for us before the observed signal y(t) is generated. We suppose that θ has a distribution specified by a known density function $\psi(\theta)$ and that w(t) and n(t) are independent. The signal is present with probability ρ.

In an artificial situation where θ is observed as well as y(t), the signal s(t, θ) is "known" and we are back to the binary testing problem of §3; the likelihood ratio is

$$\Lambda(t,\theta) = \exp\left(\int_0^T s(t,\theta)y(t)dt - \frac{1}{2}\int_0^T s^2(t,\theta)dt \right)$$

and as in §2.2 the best test is to say signal present if

$$\Lambda(t,\theta) \ge \frac{\ell_1\rho}{\ell_2(1-\rho)}$$

where ℓ_1, ℓ_2 are the penalities for false alarm and false dismissal respectively.

In fact, of course, we observe y(t), but not θ, so what we require is the marginal likelihood ratio relating the marginal function space densities of y under the two hypotheses. This is denoted $\hat{\Lambda}$ and is given by the following expression:

$$\hat{\Lambda}(y) = \exp\left(\int_0^T \hat{s}(t)dY(t) - \tfrac{1}{2} \int_0^T \hat{s}^2(t)dt \right) \tag{5.1}$$

In this expression $\hat{s}(t)$ is the causal minimum mean-square error estimate of $s(t,\theta)$ given the observed signal under the "signal present" hypothesis, or equivalently the conditional expectation of $s(t,\theta)$ given the observed signal up to t, i.e.

$$\hat{s}(t) = \mathbb{E}[s(t,\theta)|Y(u), 0 \leq u \leq t]$$

The first integral in (5.1) is an *Ito stochastic integral*; space limitations preclude any detailed description of this integral, but basically it is defined by a forward difference formula:

$$\int_0^t \hat{s}(t)dY(u) = \lim_{\substack{\max|\Delta t_i| \to 0 \\ 0 \leq t_1 \cdots \leq t_n \leq t}} \sum_{i=1}^{n-1} \hat{s}(t_i)(Y(t_{i+1}) - Y(t_i))$$

In ordinary integration theory, any difference formula involving sums of the form

$$\sum_{i=1}^{n-1} \hat{s}(\tilde{t}_i)(Y(t_{i+1}) - Y(t_i)),$$

where $t_i \leq \tilde{t}_i \leq t_{i+1}$, would give the same result, but in the stochastic case, because of the highly erratic nature of the Brownian motion sample path, different choices of $\tilde{t}_i$ give different results, so one has to specify which one is meant. For the Ito integral, $\tilde{t}_i = t_i$.

Formula (5.1) shows that *nonlinear filtering* is an essential part of Bayesian detection theory, since the likelihood ratio involves $\hat{s}(t)$, the computation of which is a filtering problem. There is a general result known as the *Kallianpur-Striebel (KS) formula* which provides - in theory - the answer to this problem. Let $f(\theta)$ be any function of the parameter θ. Then the KS formula states that

$$\mathbb{E}[f(\theta)|Y(u), 0 \leq u \leq t] = \frac{\displaystyle\int_{\mathbb{R}^d} f(\theta)\exp\left(\int_0^t s(u,\theta)y(u)du - \tfrac{1}{2}\int_0^t s^2(u,\theta)du \right)\psi(\theta)d\theta}{\displaystyle\int_{\mathbb{R}^d} \exp\left(\int_0^t s(u,\theta)y(u)du - \tfrac{1}{2}\int_0^t s^2(u,\theta)du \right)\psi(\theta)d\theta} \tag{5.2}$$

The *expected value* of $f(\theta)$ is given by the integral with respect to the density function ψ:

$$\mathbb{E}[f(\theta)] = \int_{\mathbb{R}^d} f(\theta)\psi(\theta)d\theta.$$

Similarly, the *conditional expectation* of θ given $\{Y(u), 0 \le u \le t\}$ should be given by the integral with respect to a *conditional density* $\hat{\psi}_t(\theta, Y)$, i.e. we expect to be able to calculate it in the following way:

$$\mathbb{E}[f(\theta)|Y(u), 0 \le u \le t] = \int_{\mathbb{R}^d} f(\theta)\hat{\psi}_t(\theta, Y)d\theta. \tag{5.3}$$

Comparing (5.2) and (5.3), and denoting by $e_t(\theta, Y)$ the exponential term in (5.2), we conclude that $\hat{\psi}_t(\theta, Y)$ is given by

$$\hat{\psi}_t(\theta, Y) = \frac{e_t(\theta, Y)\psi(\theta)}{\int_{\mathbb{R}^d} e_t(\theta', Y)\psi(\theta')d\theta'},$$

The denominator of this expression does not depend on θ and is just a Y-dependent normalising constant which is there to ensure that

$$\int_{\mathbb{R}^d} \hat{\psi}_t(\theta, Y)d\theta = 1.$$

Thus $e_t(\theta, Y)\psi(\theta)$ is an *unnormalized conditional density* of θ given $\{Y(u), 0 \le u \le t\}$. In (5.2) the stochastic integral appearing in $e_t(\theta, Y)$ has been written as an ordinary integral in the real observation process $y(t)$; there is no ambiguity here since the integrand $s(t, \theta)$ is (apart from the θ-dependence) deterministic, and there is no need to insist on the Ito integral..

The KS formula allows us to calculate the quantity $\hat{s}(t)$ appearing in the likelihood ratio formula (5.1): we just take $f(\theta) = s(t, \theta)$ and obtain

$$\hat{s}(t) = \int_{\mathbb{R}^d} s(t, \theta)\hat{\psi}_t(\theta, Y)d\theta$$

Similarly, minimum mean square error estimates of other quantities can be obtained, for instance the parameter values themselves:

$$\hat{\theta}_j(t) := \mathbb{E}[\theta_j | Y(u), 0 \le u \le t] = \int_{\mathbb{R}^d} \theta_j\, \hat{\psi}_t(\theta, Y)d\theta.$$

The above theory provides the exactly optimal Bayesian test, but, as the reader may already have appreciated, implementing it would be a truly massive computational task, owing to the necessity of computing $\hat{s}(t)$ over the whole range of observation times t. The KS formula has a certain amount of recursive structure to it, in that the exponential term $e_t(\theta, Y)$ can be updated from time t_1 to some later time t_2 in the obvious way:

$$e_{t_2}(\theta, Y) = e_{t_1}(\theta, Y) \exp\left(\int_{t_1}^{t_2} s(u, \theta)y(u)du - \frac{1}{2}\int_{t_1}^{t_2} s^2(u, \theta)du \right);$$

but nevertheless the integrals with respect to $d\theta$ must be evaluate separately at each time the estimate is required. Apart from these questions of computational complexity, the Bayesian test is open to the usual objection that specification of the prior density ψ and the prior probability ρ is a subjective matter. For these reasons it seems unlikely that the Bayesian test in its exact form will ever be used in gravitational wave data analysis, owing to the huge quantities of data involved. However there are two positive points - one general and one specific - to be set against this negative picture:

(a)   It can be argued that the LR formula (5.1) should be regarded as providing "structural information" on how to handle unknown parameters. (5.1) indicates that the exact signal $s(t,\theta)$ should be replaced by its "best estimate" $\hat{s}(t)$. If - as is certainly the case - $\hat{s}(t)$ is hard to compute, then it would seem a reasonable approach to replace $\hat{s}(t)$ by some more easily computable sub-optimal estimate. The one that springs to mind is of course the maximum likelihood estimate. Due to the parametrization invariance property of ML estimators noted in §4.2, we see that

$$\widehat{[s(t,\theta)]}_{ML} = s(t, \hat{\theta}_{ML})$$

and hence replacing $\hat{s}(t)$ by the ML estimate leads directly to the tests of §4. This is an important additional justification for using this kind of test.

(b)   In gravitational wave data analysis involving, say, detection of coalescing binaries, it is very important not only to detect the signal but to obtain the best possible estimate of the parameters. This applies particularly to time-of-arrival estimation in case the same event is picked up at two or more detectors, since the relative times of arrival provide directional information. The KS formula could certainly be used for this purpose without any serious computational difficulty. Suppose for example that a chirp has been detected, using for example a maximum likelihood test. Then ML estimates of the parameters will be known and the chirp will be contained with virtual certainty in some time interval from t_0 to t_1 of at most a few seconds' duration. None of the data outside this interval has any relevance to estimating the parameters; this can be done directly using the KS formula (5.2) which here takes the form

$$\hat{\theta}_i = \frac{\displaystyle\int_{\mathbb{R}^d} \theta_i \exp\left(\int_{t_0}^{t_1} s(u,\theta)y(u)du - \frac{1}{2}\int_{t_0}^{t_1} s^2(u,\theta)d\theta \right) \psi(\theta)d\theta}{\displaystyle\int_{\mathbb{R}^d} \exp\left(\int_{t_0}^{t_1} s(u,\theta)y(u)du - \frac{1}{2}\int_{t_0}^{t_2} s^2(u,\theta)du \right) \psi(\theta)d\theta}$$

Thus estimating each θ_i requires the evaluation of 2 d-fold integrals, and the exponential terms are calculated over only a few seconds of data. This is a perfectly feasible proposition. It is, of course, necessary to specify the prior density ψ. A conservative choice is a "flat" distribution (e.g. a uniform distribution over $[t_0, t_1]$ for the time of arrival parameter), reflecting "ignorance" or lack of prejudice; or some density function centred around the ML estimates could be used.

6.   TESTS WITH NON-WHITE NOISE

All the discussion so far has been under the assumption that the noise in the observed signal is white, i.e. has a flat spectral density $\Phi_n(\omega) = \sigma^2$. The actual spectral density can be estimated accurately given the large amount of data the detectors generate. What happens if the estimated spectral density is significantly non-flat over the signal bandwidth? The answer, conveniently, is that we can reduce the problem to the white noise case by using a *pre-whitening filter,* under one condition, namely that the noise contains *some* white noise component. This is a pretty harmless assumption since it is only saying that there is at least a small amount of wide-band noise in the system. Let us suppose then that the noise is represented as

$$n(t) = n_w(t) + n_c(t)$$

where n_w is white noise with spectral density $\Phi_w(\omega) = 1$ (we choose the units so that $\sigma^2 = 1$) and $n_c(t)$ is "coloured" noise independent of n_c, with spectral density $\Phi_c(\omega)$ which is assumed known from previous measurement. The noise spectral density and covariance function are then $\Phi_n(\omega) = 1 + \Phi_c(\omega)$ and $r_n(t) = \delta(t) + K(t)$ respectively, where

$$K(t) = \frac{1}{2\pi} \int\limits_{-\infty}^{\infty} \Phi_c(\omega)e^{i\omega t}\,d\omega$$

As before, we consider observations over a time interval $[0,T]$. It is convenient to introduce the following "operator" notation: if $f(t)$ is a function defined on $[0,T]$ then Kf is the function whose value at t is

$$Kf(t) = \int\limits_0^T K(t\text{-}u)f(u)du$$

Replacing K by the delta function gives us the "identity operator"

$$If(t) = \int\limits_0^T \delta(t\text{-}u)f(u)du = f(t)$$

Thus the covariance function considered as an operator is just

$$r_n = I + K$$

Note that K is a *non-causal* operator in that for any t the value of $Kf(t)$ depends on the values of $f(u)$ for all u, $0 \le u \le T$, not just on the values of $f(u)$ up to the present time t. We shall also need to consider causal, non-stationary operators of the form

$$kf(t) = \int\limits_0^T k(t,u)f(u)du = \int\limits_0^t k(t,u)f(u)du \qquad (6.1)$$

k is *non stationary* if it depends on twp parameters t,u as shown, not just on the difference t-u. It is *causal* if $k(t,u) = 0$ for $u > t$; then the two expressions given in (6.1) are the same. We denote by k^* the operator defined by $k^*(t,u) = k(u,t)$; thus

$$k^*f(t) = \int\limits_0^T k(u,t)f(u)du = \int\limits_t^T k(u,t)f(u)dt.$$

k^* is *anti-causal* in that $k^*f(t)$ depends on the values of $f(u)$ in the future, $u \geq t$. Of course, the identity operator I is both causal and anti-causal. Finally *composition* of operators is defined in the obvious way, for example $kk^*f(t) = k(k^*f)(t)$.

The first result is that gives any covariance operator K, the operator I+K can be uniquely factored as a product of causal and anti-causal operators in the following way:

$$I + K = (I+k^*)(I+k) \qquad\qquad (6.2)$$

where k is a (non-stationary) causal operator. It is easily checked that if $\tilde{n}(t)$ is a white noise process then $(I+k)\tilde{n}$ is a process whose covariance function is $(I+k^*)(I+k) = I+K$. But this is exactly the covariance of the noise process $n(t)$ in our detector, which can therefore be thought of as having been produced by *causal linear filtering* of white noise $\tilde{n}(t)$. We can thus envisage the detection problem in the way shown in Figure 3.

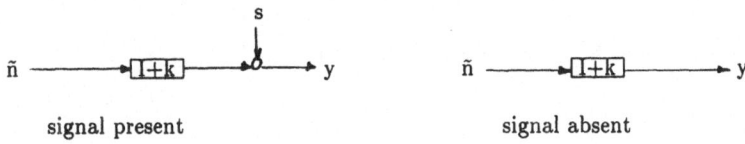

signal present   signal absent

Figure 3

The main result in this subject is that it is possible to invert the filter (I+k) so as to recover the white noise $\tilde{n}$ in a causal way. The key to this is the following Wiener-Hopf integral equation to be solved for a causal operator $h(t,u)$

$$h(t,u) + \int_0^t K(u-t)h(t,\tau)dt = K(u-t) \qquad\qquad 0 \leq u \leq t \leq T \qquad\qquad (6.3)$$

This has a unique solution $h(t,u)$ for any continuous covariance function K. If the spectral density Φ_c is a rational function (as it will be in any simple coloured noise model) then (6.3) can be solved by Wiener's method of spectral factorization; we will not go into the details here.

A nice property of a causal operator such as h is that the operator (I-h) is *causally invertible:* if we write formally

$$(I-h)^{-1} = I + h + h^2 + h^3 + \dots$$

and, as usual, interpret h^2 as the operator $h^2f = h(hf)$, h^3 as $h(h^2f)$ etc then it is not hard to show the sum $(h + h^2 + h^3 + \dots)$ converges to another causal operator. In fact this limit is precisely the operator k appearing in the factorization (6.2). It is straightforward to check this. First, one can easily verify that if h satisfies (6.3) then the non-causal operator H defined by

$$H(t,u) = h(t,u) + h^*(t,u) - \int_0^t h^*(t-\tau)h(\tau,u)d\tau \qquad\qquad (6.4)$$

satisfies

$$H(t,u) + \int_0^T H(t,\tau)K(\tau-u)d\tau = K(t-u) \qquad\qquad 0 \le t, u \le T$$

which we can write as

$$(I + K)H = K \tag{6.5}$$

Using this fact we have

$$
\begin{aligned}
(I+K)(I-h)(I-h^*) &= (I+K)(I-h^*-h+hh^*) \\
&= (I+K)(I-H) && \text{(by 6.4))} \\
&= (I+K) - (I+K)H \\
&= (I+K) - K && \text{(by (6.5))} \\
&= I && (6.6)
\end{aligned}
$$

Now $(I-h)$ is invertible, as is $(I-h^*)$, so (6.6) is equivalent to

$$(I+K) = (I - h^*)^{-1}(I - h)^{-1}.$$

Comparing this with (6.2) we see that

$$(I-h)^{-1} = I + k$$

so that $(h + h^2 + h^3 + ...) = k$, as claimed. $(I-h)$ is the *whitening filter*. Returning to the detection problem we saw above that the "signal present" hypothesis can be represented as

$$y(t) = s(t,\theta) + (I + k)\tilde{n}(t) \tag{6.7}$$

Hence

$$
\begin{aligned}
(I - h)y &= (I - h)s_\theta + (I - h)(I + k)\tilde{n} \\
&= (I - h)s_\theta + \tilde{n}
\end{aligned}
$$

Defining $\tilde{y} = (I-h)y$, $\tilde{s}_\theta = (I-h)s_\theta$, this becomes

$$\tilde{y}(t) = \tilde{s}(t,\theta) + \tilde{n}(t)$$

Since $\tilde{n}$ is white noise, the effect of filtering the observed signal through $(I-h)$ is therefore to remove the "colouring" and reduce the detection problem to the white noise case as studied before. We are using here the fact that the noise appears *additively* in (6.7) so that filtering y is equivalent to filltering the signal and noise separately and can be represented in either of the two ways shown in fig. 4

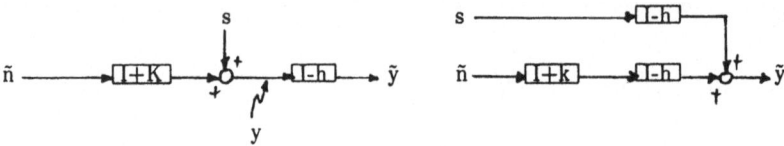

Figure 4

It is an essential feature of this procedure that (I-h) is invertible: this means that we lose no information in passing from y to ỹ (we could in principle recover y from ỹ as $(I-h)^{-1}\tilde{y}$) and this in turn means that there is no loss in testing power. If we filtered y using some non-invertible filter then we might be throwing away some essential characteristic of y, reducing our ability to distinguish the presence or absence of a signal.

To summarize: the procedure for dealing with coloured noise is to estimate its spectral density in the form $\sigma^2 + \Phi_c(\omega)$; solve the Wiener-Hopf equation (6.3) (which can be done by spectral factorization using Φ_c as long as this is rational); filter both the observed data y(t) and the template signals s(t,θ) through (I-h), giving ỹ(t) and s̃(t,θ); and apply the tests for signal in white noise to ỹ and s̃. Note that it is not necessary to compute K or k explicitly at any stage.

7.   CONCLUDING REMARKS

7.1   Use of several detectors

Detection of a gravity wave, particularly at low signal-to-noise ratios, will not be believable unless it is verified independently, i.e. unless the same event is picked up by two or more detectors. Further, all parameters of the signal other than the time of arrival (i.e. the mass parameter and range in the case of coalescing binaries) should be the same (to within statistical error) at each detector. In this sense all detectors have to work independently. However there are advantages to be gained by pooling the data from several detectors in two respect, (a) increasing the signal-to-noise and (b) increasing the accuracy of parameter estimates. The second of these seems to me the more important, sincee for positive detection the signal really ought to be way above threshold in each detector.

Suppose we have two detectors with noise strengths σ_1^2, σ_2^2 and a signal depending on time of arrival τ_i at the i'th detector, i = 1,2, and on another vector parameter θ (the same at each detector). Then the received signal at detector i with signal present is

$$y_i(i) = s(t, \tau_i, \theta) + n_i(t) \qquad\qquad i = 1,2,$$

The LR given both signals $y = (y_1, y_2)$ is

$$\ell n\ \Lambda(y, \tau_1, \tau_2, \theta) = \int_0^T s(t, \tau_1, \theta) y_1(t) dt + \int_0^T s(t, \tau_2, \theta) y_2(t) dt$$

$$- \tfrac{1}{2} \int_0^T s^2(t, \tau_1, \theta) dt - \tfrac{1}{2} \int_0^T s^2(t, \tau_2, \theta) dt$$

$$= \ell n\ \Lambda_1(y_1, \tau_1, \theta) + \ell n\ \Lambda_2(y_2, \tau_2, \theta) \qquad\qquad (7.1)$$

where Λ_i is the LR given y_i only. Evidently the maximum likelihood estimates $\hat{\theta}_{ML}$, $\hat{\tau}_{1,ML}$, $\tau_{2,ML}$ depend on all the data (y_1, y_2) although $\hat{\tau}_{1,ML}$ is only affected by y_2 through θ_{ML} (from (7.1), it is obtained by maximizing $\ell n\ \Lambda_1$). It is very difficult to quantify how much more accurate these estimate will be than estimates based on y_1 and y_2 individually. Of course the signal-to-noise is improved. For example if the same signal s(t) (not depending on any parameters) arrives at both detectors then the averaged observed signal is

$$\bar{y}(t) = \tfrac{1}{2}(y_1(t) + y_2(t) = s(t) + \tfrac{1}{2}(n_1(t) + n_2(t))$$

giving for the mateched filter

$$S/N = \frac{4}{\sigma_1^2 + \sigma_2^2} <s,s>$$

as opposed to $<s,s>/\sigma_1^2$ for detector 1 alone. However some of this advantage is lost if the times of arrival area not the same.

Turning to Bayesian parameter estimation, we can use the two-dimensional analysis of the Kallianpur- Striebel formula (5.2) to give estimates based on the pooled data (y_1, y_2). Here it is preferable to take as parameters $\tau = \tau_1$ and $\delta = \tau_2 - \tau_1$ rather than τ_1 and τ_2, because it is reasonable to assume that τ, δ, θ are independent. The prior density ψ_δ of δ can be inferred by assuming that the direction from which the gravity wave comes is uniformly distributed over space). The joint prior density of τ, δ, θ is then

$$\psi(\tau, \delta, \theta) = \psi_\tau(\tau)\psi_\delta(\delta)\psi_\theta(\theta)$$

where ψ_θ and ψ_τ are chosen following the considerations discussed at the end of §5. If $f(\tau, \delta, \theta)$ is any function of the parameters and $\theta \in \mathbf{R}^d$, the best estimate of f is

$$E[f(\tau, \delta^\theta,)|y_1(u), y_2(u), 0 \le u \le T] = \frac{\displaystyle\int_{\mathbf{R}^{d+2}} f(\tau, \delta, \theta)\Lambda(y, \tau, \delta, \theta)\psi_\tau(\tau)\psi_\delta(\delta)\psi_\theta(\theta)d\tau d\delta d\theta}{\displaystyle\int_{\mathbf{R}^{d+2}} \Lambda(y, \tau, \delta, \theta)\psi_\tau(\tau)\psi_\delta(\delta)\psi_\theta(\theta)d\tau d\delta d\theta}$$

Taking for example $f(\tau, \delta, \theta) = \tau$ and $f(\tau, \delta, \theta) = \delta$ we obtain the Bayesian estimates for τ and δ respectively, and hence for τ_1 and τ_2. These should be much more reliable than the corresponding ML estimates.

7.2   Detecting the "wrong" signal

Suppose we are searching for gravitational waves from coalescing binaries when in fact a signal generated by some completely different event arrives. What happens?

To get a feeling for this kind of question, let us return to the example of §4.3(iii): estimating the frequency of a sine wave, $s(t, \theta) = \sin \theta t$. We saw from (4.3) that when the "true" signal is $\sin \omega t$ then $\hat{\theta}_{ML}$ is likely to be close to ω as long as the observation time T is sufficiently large. But suppose the true signal is actually a constant, $s(t) = a$, while we continue to use the templates $\sin \theta t$. Then with signal present the LR is

$$\ell n\ \Lambda(y, \theta) = X(\tau, \theta) + \frac{a}{T\theta}\Big(1 - \cos \theta T\Big) - \tfrac{1}{4}\Big(1 - \text{sinc } 2\theta T\Big) \tag{7.2}$$

(notation as in (4.4)). In contrast to (4.4), there is no deterministic term in (7.2) which is large compared to the noise term $X(T, \theta)$, even for large T. From this we conclude the following :
(i)   Using templates $s(t, \theta)$ we are unlikely to detect a signal from some completely different source, because of the reduced S/N occasioned by a "mismatched" filter

(ii)   If such a signal *is* detected the corresponding ML estimate $\hat{\theta}_{ML}$ will be completely random.

Of course, these remarks do not apply if the true signal is close to one of the templates, "closeness" being measured by the L_2 inner product, as pointed out in §3.

As a final cautionary remark, let me insist that it is a *completely meaningless question* to ask whether there   is *some* signal buried in the noise without specifying in any way what that signal is supposed to be. In the absence of such a specification, what is wrong with the noise itself as a candidate for the "signal"?   (Answer: nothing).   Thus one is obliged to formulate a hypothesis of some kind.   If this is in fact simply inappropriate then one is much more likely to be rewarded by failure to detect anything than by positive detection of some totally spurious signal.

7.3   Final remarks

Probably the only thing in this article which will be wholly unfamiliar to workers in gravitational wave detection is the Kallianpur-Striebel formula (5.2).   I feel that the use of this has considedrable potential, particularly for providing accurate time-of-arrival extimates, and some preliminary simulations carried out by Alessandro Pasetti indicate that this is so.   We plan to report on this further [19].

8.   REFERENCES

The hypothesis testing theory of §2 is covered in many books on statistics and decision theory, for example Kendall and Stuart [1], Fisz [2], Ferguson [3].   There are several excellent textbooks which cover the detection theory of §§3,4, for example Van Trees [4], Whalen [5] or Helstrom [6]. However there seem to be no more recent books which cover this as well as the newer material of §§5,6.   The best general reference for this is undoubtedly the recent textbook by Wong and Hajek [7].   Wong's introductory book [8] can also be consulted for a readable account of most of the relevant material.   Gallager [9] contains an extensive account of white noise channels and linear filtering.   Nonlinear filtering is a big subject, but the standard textbooks on it, Liptser and Shiryaev [10] and Kallianpur [11], are only accessible to specialists as well as being - as accounts of filtering theory - rather out of date.   An introduction with emphasis on more recent developments is Davis and Marcus [12].   The likelihood ratio formula (5.1) was introduced by Kailath, whose papers [13], [14] are still excellent reading on this subject.   Ito's stochastic calculus is covered in Wong and Hajek [7]; other excellent introductions to it are Arnold [15] and Chung and Williams [16].   Beware of any treatment that claims to cover this subject at a "heuristic" or "engineering" level.   It can't be done.   The whitening filter of §6 is discussed in detail in Kailath's article [17].   Davis [18] can also be consulted for material on linear filtering and the associated integral equations.   Solution of the Wiener-Hopf equation is covered in Wong and Hajek [7].

[1]   M. Kendall and A. Stuart, *The Advanced Theory of Statistics*, vol II, 4th ed., Griffin, London 1979.

[2]   M. Fisz, *Probability Theory and Mathematical Statistics*, (3rd ed) Wiley, New York 1963.

[3]   T.S. Ferguson, *Mathematical Statistics: a Decision Theoretic Approach*, Academic Press, New York 1967.

[4]   H.L. Van Trees, *Detection, Estimation and Modulation Theory*, vol. I, Wiley-

Interscience, New York 1968.

[5]   A.D. Whalen, *Detection of signals in Noise,* Academic Press, New York
        1971.

[6]   C.W. Helstrom, *Statistical Theory of Signal Detection,* 2nd. ed, Pergamon
        Press, London 1968.

[7]   E. Wong and B. Hajek, *Stochastic Processes in Engineering Systems,*
        Springer-Verlag, New York, 1985.

[8]   E. Wong, *Introduction to Random Processes,* Springer-Verlag, New York
        1983.

[9]   R.G. Gallager, *Information Theory and Reliable Communication,* Wiley, New
        York 1968.

[10]   R.S. Liptser and A.N. Shiryaev, *Statistics of Random Processes* I, II,
        Springer-Verlag, New York 1977.

[11]   G. Kallianpur, *Stochastic Filtering Theory,* Springer-Verlag, New York
        1978.

[12]   M.H.A. Davis and S.I. Marcus, An introduction to nonlinear filtering, in
        *Stochastic Systems : the Mathematics of Filtering and Identification
        and Applications,* eds. M. Hazewinkel and J.C. Willems,
        D. Reidel, Dordrecht 1981.

[13]   T. Kailath, A general likelihood ratio formula for random signals in
        Gaussian noise, IEEE Trans. Information Theory IT-15, 1969, 350-361.

[14]   T. Kailath, The structure of Radon-Nikodym derivatives with respect to
        Wiener and related measures, Ann. Math.Stat. 42(1971), 1054-1067.

[15]   L. Arnold, *Stochastic Differential Equations,* Wiley, New York 1974.

[16]   K.L. Chung and R.J. Williams, *Introduction to Stochastic Integration,*
        Birkhäuser, Boston 1983.

[17]   T. Kailath, The innovations approach to detection and estimation theory,
        Proc. IEEE 58(1970)680-695.

[18]   M.H.A. Davis, *Linear Estimation and Stochastic Control,* Chapman & Hall,
        London 1977.

[19]   M.H.A. Davis and A. Pasetti, The use of nonlinear filtering in the
        analysis of gravitational wave data, paper in preparation, Imperial
        College, London 1987.

RADIO PULSAR SEARCH TECHNIQUES

A.G. Lyne
University of Manchester
Nuffield Radio Astronomy Laboratories
Jodrell Bank, Macclesfield,
Cheshire SK11 9DL

ABSTRACT.   Pulsar astronomers have been seeking periodic waveforms in noisy data for nearly 20 years. Throughout this time the sensitivity of these searches have been limited by the available computing resources so that efficient algorithms have been at a premium. The same limitations are likely to prevail in the search for unknown periodic phenomena in the data from gravitational wave detectors. In this paper we discuss the problem as encountered by radio astronomers and indicate the various techniques that are used.

This paper is not concerned so much with the theory of detection, rather, it is more of a guide to the problems and current practical limitations of such searches.

1.   SEARCHES FOR PULSE TRAINS IN NOISE

The fundamental problem is simply described; a time sequence of N equispaced samples ($10^5 < N < 10^8$) contains random noise and a train of pulses having a profile of unknown form and a repetition period of P samples. Individually the pulses are much smaller than the rms noise. By folding the data at the period P, the pulses add up coherently, the noise incoherently and the signal-to-noise ratio of the pulse is increased by $\sqrt{(N/P)}$. Unfortunately we do not know the width, W, of the pulse or its position through the pulse period. Figure 1 shows the profiles of the pulses from the pulsars found in a recent survey (Clifton and Lyne 1986) and indicates that the pulse widths can vary from perhaps 1% of the period to nearly 50%.

The task is to perform with minimum effort a 3-dimensional search of the time sequence for pulses having unknown period, pulse width and pulse phase. Two main classes of technique have been used, firstly, what I call a periodogram analysis and secondly a Fourier transform based analysis.

B. F. Schutz (ed.), Gravitational Wave Data Analysis, 95–103.

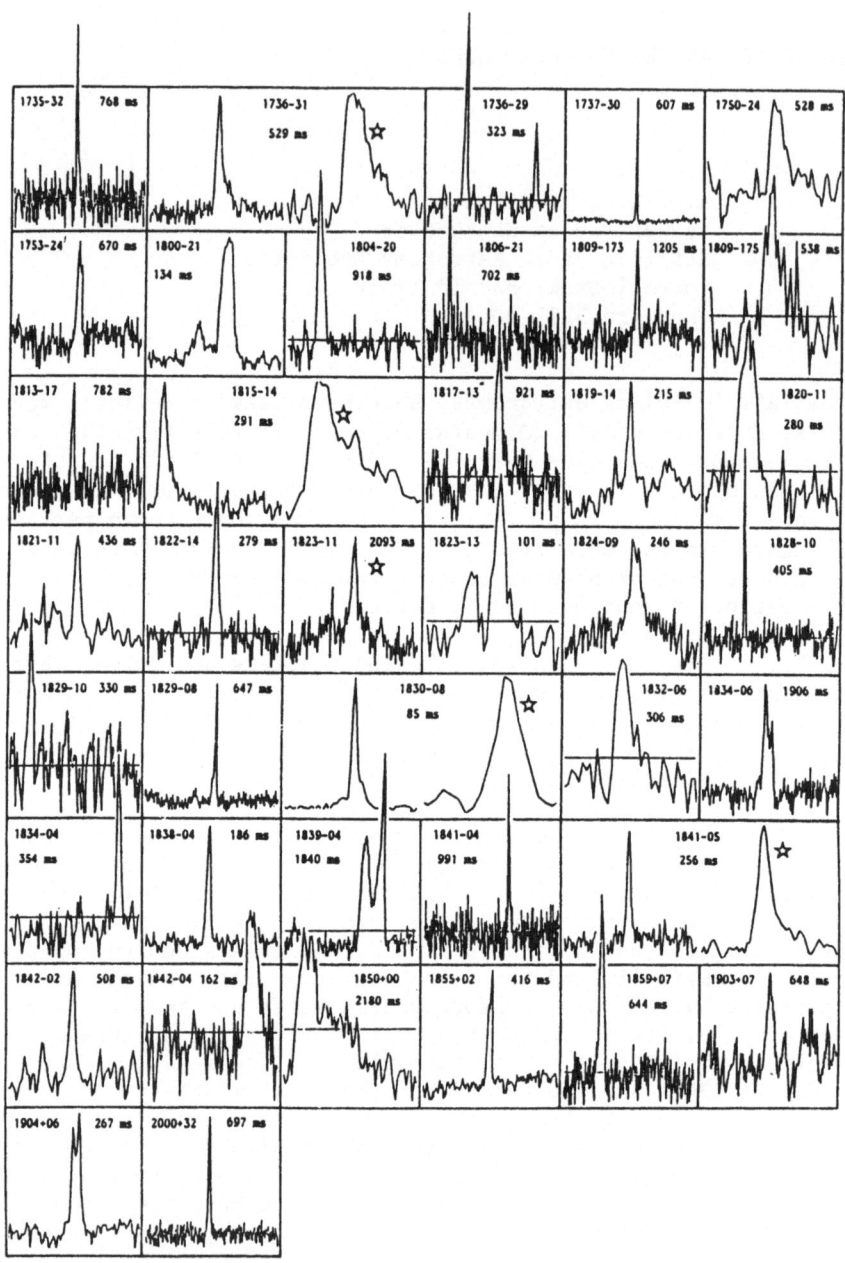

Figure 1 The profiles of 40 new pulsars detected in a recent survey at Jodrell Bank (Clifton, Jones and Lyne 1987). The whole of the pulsar period is shown.

2. PERIODOGRAM ANALYSIS

The periodogram analysis was used in the early days and is conceptually easy to follow, involving folding the data for roughly N different values of P, thus providing a profile of length P samples for each. This profile then has to be searched for pulses of different widths and phases. This is usually carried out by repeated convolution of the profile with 'top-hats' of increasing width, scanning for the peak and comparing its height with the expected rms value of the noise. For a given total pulse power, this process enables pulses of width W to be detected with signal-to-noise ratios which are increased by a factor of about $\sqrt{(P/2W)}$ over a sinusoidal waveform. Thus a pulsar with $W \simeq P/2$ would need to be nearly an order of magnitude more luminous than one with $W = 0.01P$ to have the same detectability. This dependency of sensitivity upon pulse width is a fundamental one for all search processes.
   Although many relatively efficient algorithms were developed for this procedure, it requires much computing power and memory and is not used in modern searches.

3. FOURIER ANALYSIS

This involves taking the fast Fourier transform (FFT) of the time series and inspecting the resultant spectrum for signals together with any associated harmonics. The FFT of the real time sequence is complex and hermitian so that all the information is available in just the N/2 positive frequency elements. Figure 2 shows a sketch of the Fourier relationship between the time and frequency domains for a pulse train. The amplitudes and phases of the individual harmonics are determined

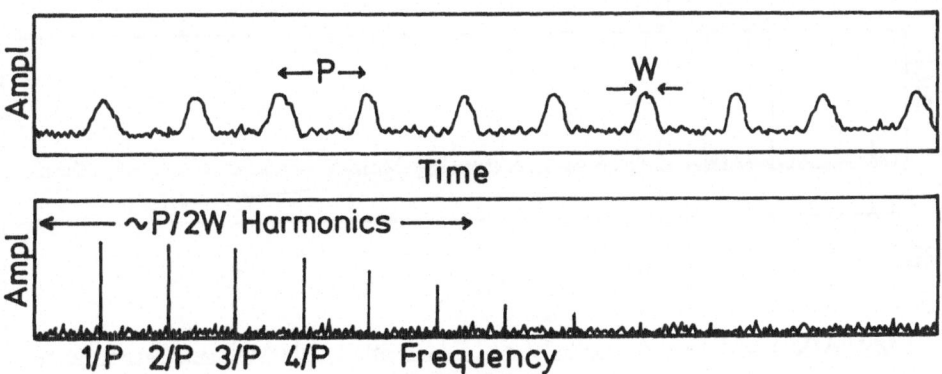

Figure 2 The Fourier relationship between the time and frequency domains for a pulse train. A time series of N points provides a complex spectrum of length N/2.

by the average pulse profile in the time domain. For a nearly sinusoidal pulse there will be a large fundamental spectral feature with small harmonics, while for a very narrow pulse (e.g. a delta function) the harmonics will extend with constant amplitude up to the Nyquist frequency. In general, the first P/2W harmonics have large amplitude. Individually, they may not be statistically significant.

The problem now is how to combine information from the unknown number of harmonics of an unknown fundamental frequency in order to maximise the detectability of the signal.

The solution is to use an incoherent addition of the harmonics, i.e. we just work with the amplitudes. Consider taking the N/4 lowest frequency elements, i.e. the first half of the spectrum, and adding them in turn to those elements which correspond to their first harmonics (Figure 3). If a fundamental and first harmonic have equal amplitude, for large signal-to-noise ratios the amplitude of the signal will double while the noise in the rest of the spectrum, being incoherent will only increase by a factor of $\sqrt{2}$. The signal-to-noise ratio in the resulting spectrum will increase by a factor of about $\sqrt{2}$. Thus a fundamental and its harmonic might have individual signal-to-noise ratios of say 4 and be considered statistically insignificant in a spectrum of say 10^5 elements. However, after the addition described above, they combine to provide a signal-to-noise ratio of 5.7, a value unlikely to occur by chance and statistically very significant. (The probability density functions of noise in the spectra before and after addition are somewhat different and I do not offer a precise analysis of the statistics here). Unfortunately because we are using no phase information this process does not work if the signal-to-noise ratios of the elements before addition are less than about 1.5. However, this is not a practical limitation since, with these large data sets, we are interested only in about 6 sigma events after addition. Such events may be provided for instance, by 16 1.5 sigma harmonics.

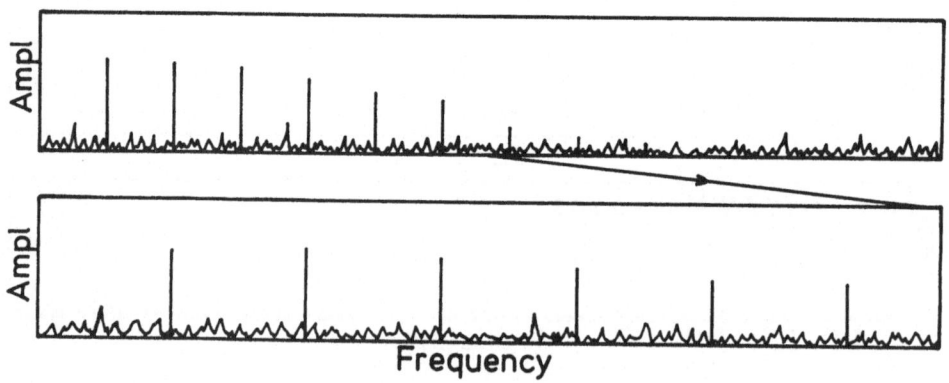

Figure 3 The first half of the spectrum is expanded and added to the unexpanded spectrum so that each fundamental is added onto its first harmonic.

The procedure then is as follows: after performing the FFT, the amplitudes are inspected and any significant peaks remembered. The fundamentals are then added on to their corresponding first harmonics and again any significant peaks remembered. The process is repeated for additions of 4, 8, 16 and 32 harmonics. These of course correspond to searching for successively narrower pulses. Many of these resulting suspects will be the result of the addition of noise spikes in the spectra and, in general, are unlikely to have the correct phase relationships that would result from a true pulse train. It is possible now to eliminate most of these by extracting the complex harmonics of each suspect frequency and performing an inverse FFT in order to recover a pulse profile. For most noise-derived suspects, the harmonics will not add together properly and not produce a large pulse, although of course, by chance, a few will. The signal-to-noise ratio in the profile is then used as the criterion of confidence in a suspect.

Modern computers can perform large FFTs with great efficiency, while the harmonic adding and inspection can be carried out in an even shorter time. The total memory required is little more than that occupied by the input time sequence.

4. THE PULSAR SEARCH PROBLEM

In the search for radio pulsars there is unfortunately a further dimension which concerns us, that of frequency dispersion of the pulses due to the ionised component of the interstellar medium. This causes the pulses to sweep with time from high to low frequency. With a finite receiver bandwidth the pulse may take a substantial time to pass through the band so broadening the pulse. As we have seen, any broadening of the pulse will result in a decrease in sensitivity by an amount proportional to $\sqrt{W}$. However, we do need large total bandwidth, B, since the sensitivity is proportional to $\sqrt{B}$. The solution is to split the receiver band and record the outputs of several contiguous narrow channels. The pulses now appear as a series of sloping periodic stripes as shown in figure 4a, where the slope is determined by the disperion measure.

There are two approaches which have been used to solve this problem. The first is to assume a number of trial slopes and add the data up along lines parallel to the slopes, so producing for each a time series which can be processed using the techniques discussed in the last section.

The second approach and the one which turns out to be most efficient involves taking the two-dimensional FFT of the data array. The resulting spectrum is shown in figure 4b in which the pulses appear as a series of harmonically related features lying along a diagonal which passes through the origin and whose slope is determined by the dispersion measure. The search process then simply involves extracting 1-dimensional diagonal segments from the array and seeking fundamentals and related harmonics in each using the same processes described in the last section.

In the search for millisecond pulsars the amount of data required is enormous because it turns out that the total number of data samples

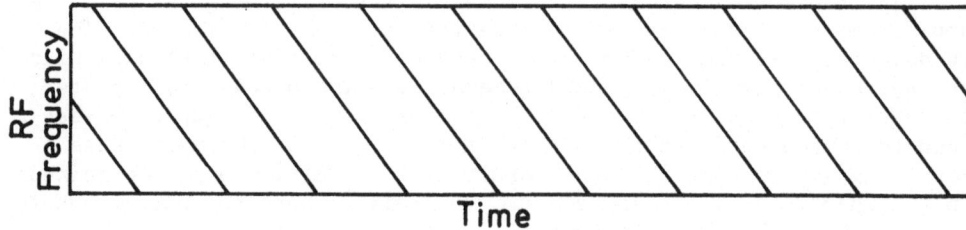

Figure 4a   The appearance of dispersed pulses in the RF frequency-time domain.

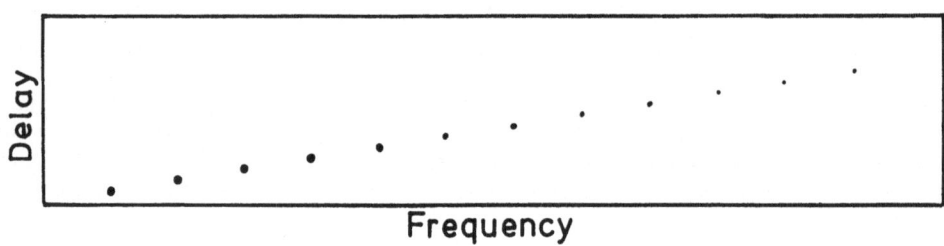

Figure 4b   The 2-D FFT of the RF frequency-time domain in Figure 4a.

depends upon the inverse square of the minimum detectable period. This can be seen to be reasonable in a very simple case: consider doubling the sample rate in the two-dimensional search above. If the dispersion broadening is going to be reduced, also by a factor of two, we will have to reduce the bandwidth of the individual frequency channels and hence double their number to retain the same overall bandwidth. The result is a quadrupling of the total number of samples which have to be recorded.

5.   A SEARCH FOR MILLISECOND PULSARS

These techniques have been applied recently in a search for millisecond pulsars at Jodrell Bank. This survey illustrates a number of practical difficulties encountered and various means of surmounting

them. The survey was planned to cover a region along the galactic
plane, recording 32 frequency channels every 300 μs. The resulting
10^5 samples a second had to be recorded on magnetic tape for the
duration of the survey, about 3 weeks. If stored as one byte/sample
on standard 1600 bpi magnetic tapes, about 5000 tapes would be required,
a quite unmanagable number. The purchase of a tape transport which
operated at 6250 bpi reduced this number to about 1300, still impractical. The solution lay in reducing the number of bits recorded for
each sample. After a number of simulations of long time sequences in
which the signal in a single pulse was only a tiny fraction of the rms
noise, I found that the sensitivity of the search was reduced by only
about 20% by the use of single-bit digitisation. This simply involves
recording either a 0 or 1, depending upon whether the instantaneous
receiver output is less or greater than the long term mean. A simple
theoretical argument confirms this and shows that the sensitivity is
$\sqrt{(2/_\pi)}= 0.80$ of a system using many bits. This was a small penalty to
pay for the reduction in the number of tapes required to only 160.

The processing involved the analysis of 2-dimensional arrays of
data of size 32 x 256,000 samples. This was carried out on a Cyber
205 supercomputer at the University of Manchester Regional Computer
Centre. The 2-D FFT takes only about 5 seconds of CPU time and the
whole algorithm about 15 seconds.

The main problem in running the program was the large size of the
8 million word data set and only 2 million word main memory of the
computer. This gave rise to severe paging problems with data being
swapped between main memory and disk. The solution was to pack the
main data arrays during the processing as bytes. At 20% extra CPU
costs, this solved the problem.

For any larger data sets than this the paging limitations of
present day computers provide severe restrictions. This was illustrated when recently a weak, small diameter, steep spectrum, highly
linearly polarised radio source was detected in the globular cluster
M28 (Hamilton et al. 1985, Erickson et al. 1987). It seemed very
likely that that this was a pulsar. In order to reach a sufficient
sensitivity to detect the source if it were a pulsar, about 90
minutes data from the 76m telescope at Jodrell Bank were required,
sampled at 300 μs intervals, so providing a total data set of over
500 million samples. There was no practical possibility of using
the Cyber 205, even though it would have required only about 20
minutes CPU, because of the paging problem. The solution was to use
a Cray-XMP at the Los Alamos National Laboratory in New Mexico. This
machine had a much larger main memory, and used different code provided by John Middleditch, On that machine the job was at least
possible even though it actually used about 5 hours of CPU time.

This analysis produced a suspect periodicity at about the 6σ level
with a period of 3.054314 milliseconds and had to be confirmed by
further observation (Lyne et al. 1987). Rather than record such large
quantities of data and use the Cray again, bearing in mind that the
pulsar might be a member of a binary system and hence show corresponding Doppler effects, it is possible to use a fairly narrow filter
in real-time, followed by a modest amount of analysis. This was

achieved by performing on-line coherent integrations folding at the suspect period, each of perhaps 1 minute duration for a total of about 2 hours. The pulse was not visible in the individual one minute integrations, but when they were added together with a number of trial drift rates, the pulse appeared for one particular drift rate. This process took only a few minutes on a VAX 780.

6. CONCLUSION

Some of the practical difficulties associated with pulsar searches have been described above. There are a few comments or observations which may be of value in designing a system to seek periodicities in gravitation wave data.

To first order, the amount of CPU time required is independent of the 'shape' of the 2-D data array and takes about 2 seconds for each million data samples on a modern supercomputer. For example, it would take about 20 minutes to process a single time series of 500 million points or one of 32 x 16 million points, if paging is no problem.

In the course of this work we have developed a new FFT algorithm for use on a vector machine like the Cyber 205 (Ashworth and Lyne 1987). This consists of splitting a large FFT into a number of smaller ones which can be executed in parallel. The execution of an N-point FFT where N = m x n is carried out in 4 stages:
1. Perform m n-point FFTs.
2. Transpose the m x n array.
3. Phase rotate each element.
4. Perform n m-point FFTs.
This process takes about N μseconds on a Cyber 205 or Cray supercomputer.

So far, the algorithms I have described take no account of any period derivative, such as would be encountered if the pulsar were very young or a member of a binary system or if gravitational waves were arising from a coalescing binary system. This of course requires a further dimension in the search and probably another order of magnitude increase in computing resource. Two solutions seem possible:
1. 'Rebin' the time series data with a number of parabolic 'stretches' and analyse using the normal techniques described above.
2. In the FFT algorithm above, carry out steps 1 and 2 once only, and repeat steps 3 and 4 with extra quadratic phase rotation to perform the "stretching". The resulting spectra is then searched for fundamentals and associated harmonics. In this way the bulk of the FFT has only to be carried out once.

The two main problems which restrict this kind of work at present are the limited sizes of main memory in the supercomputers and the means of recording the data for transport to the off-line processor. The first of these problems is likely to be eased in the future as the next generation of supercomputers becomes available. Similarly the advent of the optical disk promises to reduce the physical bulk of the recording medium by one of two orders of magnitude.

REFERENCES

Ashworth, M. and Lyne, A.G. (1987). Parallel Computing. (In press).
Clifton, T.R., Jones, A.W. and Lyne, A.G. (1987). IAU Symposium No. 125. The Origin and Evolution of Neutron Stars. Eds. D.J. Helfand and J.H. Huang, Reidel, p.47.
Clifton, T.R. and Lyne, A.G. (1986). Nature 320, 43.
Erickson, W.C., Mahoney, M.J., Becker, R.H. and Helfand, D.J. (1987). Astrophys.J., 314, L45.
Hamilton, T.T., Helfand, D.J. and Becker, R.H. (1985). Astr.J., 90, 606.
Lyne, A.G., Brinklow, A., Middleditch, J., Kulkarni, S.R., Backer, D.C. and Clifton, T.R. (1987). Nature 328, 399.

REFERENCES

SAMPLE COVARIANCE TECHNIQUES IN THE DETECTION OF GRAVITATIONAL WAVES

A D Irving
Rutherford Appleton Laboratory
Chilton
Didcot
Oxon   OX11 0QX

ABSTRACT

Sample covariance estimates for stationary processes, non stationary
events and the effects of filters using linear systems are discussed.
The extraction of the Green's or Impulse response function from the time
delayed movement estimates using the Winer-Volterra expansion is
discussed for linear and non linear systems. The effect of additive
outliers on the power spectral density estimates is illustrated and
recent reviews referred. Finally the need to interpret the observations
in terms of the operator eigenstates or 'natural' eigenfunctions of the
radiating source is discussed.

KEYWORDS

Sample covariance, Power Spectral Density, Linear Systems, Non Linear
Systems, Non Stationary Events, Robust Estimates, Operator Eigenstates.

INTRODUCTION.

During the course of this workshop we have heard about three commonly
used data identification and processing techniques
(i)   Matched Filtering,
(ii)   Fast Fourier Transformation of the time series data, and
(iii)   Maximum Likelihood Methods.
     Although I shall present a fourth technique, which is complementary
to the above, my talk is intended rather to make you think about your
proposed event detection and analysis techniques.
     In our experimental measurements of the output signal $\{y(t)\}$, we
will need to decide if an event input $\{x(t)\}$ is present. This decision
will be based upon our sequence of observations $\{y(t)\}$ which are
contaminated by noise $\{n(t)\}$. Without loss of generality we will assume
that $\{x(t)\}$, $\{y(t)\}$ and $\{n(t)\}$ are zero mean real sequences of data.
The event detection problem may be formulated as a binary hypothesis
testing procedure

105

B. F. Schutz (ed.), Gravitational Wave Data Analysis, 105–122.
© 1989 by Kluwer Academic Publishers.

$$H_0 : y(t) = n(t) \qquad \forall\ t \qquad\qquad \text{the null hypothesis}$$
$$\text{there is only noise}$$

versus

$$H_1 : y(t) = n(t) + x(t) \quad \forall\ t \qquad \text{the alternative hypothesis}$$
$$\text{there is a signal and noise}$$

Once the event signal $\{x(t)\}$ has been identified, we will need to estimate or extract the relevant information or parameters $\hat{\alpha}(t)$ from the data sequence $\{y(t)\}$. The parameterisation of the signal $\hat{\alpha}(t)$ will then need to be tested against the equivalent parameters predicted by theory using the binary hypothesis,

$$H_0 : \hat{\alpha}(t) = \alpha(t) \qquad\qquad \text{the theory is not rejected}$$

versus

$$H_1 : \hat{\alpha}(t) \neq \alpha(t) \qquad\qquad \text{the theory is rejected}$$

The information extracted ranges from individual events such as star formation or neutron star collapse to global phenomena such as the isotropy and deceleration rate of the universe. This information will need to be extracted from noisy signals.

"Statistical analyses are built upon mathematical models which link reality with the mechanism generating the observations" (MO84). The mathematical models are usually functionally simple and the parameterisation of the model chosen such that each parameter may be identified with some aspect of the observations (IR87). Simple, in this context, must be considered a relative term and any model building or parameterisation process must be part of a chain of progress. Typical links in such a chain will be (PA86) 'incomplete experimental guidance → pitfalls of simplicity → paradox → progress → incomplete experimental guidance → ...'. Such chains repeat themselves endlessly although the cycle time will vary from months to decades and even centuries, depending on the resource available. Experimental measurements form the first link and the theorist interprets the data, usually using the simplest assumptions which are compatible with the experimental data. Improved experimental measurements may indicate that some of the simplifying assumptions made were in fact a pitfall. Such pitfalls are unavoidable and the theorist now resolves the paradox with a revised set of assumptions (PA86). The experimentalist further refines the observational techniques and the next cycle begins.

When we are developing our data analysis techniques we may be guided (or misled) by the properties of the theoretical model against which we will ultimately test our experimental data. There will obviously be a trade-off between flexibility in the data analysis and the ease with which we can compare the experimental and theoretical parameters.

As an example the square of a length interval ds in a cartesian frame of reference is given by

$$ds^2 = c^2dt^2 - dx^2 - dy^2 - dz^2.$$

The quadratic measure of length or mean square metric means that we are in a Hilbert space (AN71).

The need to retain the same measure of length, $ds = \sqrt{length^2}$, may be used as an argument to restrict our Maximum Likelihood Estimator to being the Last Squares Estimator.

The properties of Hilbert spaces have been used to clarify the structure of time series and in solving problems of statistical inference (PA61,LO63). Loéve (LO63) shows the connection between second order stationary time series and the reproducing kernel of Hilbert spaces. In particular a Hilbert space may represent a time series if there exists a one-to-one linear mapping $\{\psi(f(t)) = X(t)\}$ where $\{X(t)\}$ is the time series and ψ a linear operator, and that the inner product exists and is $(f(s), f(t)) = E[X(s), X(t)]$ where E is the sample expectation operator.

The fact that operator theory may be related to statistical prediction theory allows a geometric interpretation to be made of the statistical quantities obtained. One obvious example (LA77) is that in quantum mechanics the scalar product of a wave function is its probability density and that the basis set formed by the wave functions constitutes a Hilbert space which remains a Hilbert space when the basis set of vectors is transformed into other representations.

When we consider a perturbed system we try to represent the eigenvector solution in terms of the eigenfunctions of the unperturbed system. The eigenfunctions form the basis set for the Hilbert space, and in order to discover the symmetries of the system one would identify those unitrary and anti unitrary operators, which preserve the properties of the system, and display the symmetries of the problem.

STATIONARY TIME SERIES

Stationary time series are the simplest and the most well understood. Weak or second order stationary processes are ones for which the covariance C is invariant to time translations. The covariance is defined as the expectation

$$C(s,t) = E[x(t)\ x(s)]$$

and for weak stationary processes the covariance is simply a function of the time difference or delay (t-s) and

$$C(s,t) = E[x(t)\ x(s)] = C(t-s)$$

If the signals are completely deterministic, for example, a pure CHIRP, then to be completely rigorous, deterministic signals, with finite duration cannot be band limited. See Dupaz (DU86) for details.

A bit harsh but even when we apply a perfect sinusoidal input to a real system and wait for transient effects to disappear, the system output will not be a perfect sine wave. The distortions present in real systems are due to harmonics at multiples of the excitation frequency, and random noise. Real systems are always somewhat nonlinear and when one excites a nonlinear system with for example a sinusoidal input, the output is periodic but nonsinusoidal and a Fourier series analysis reveals a discrete spectrum of harmonics. Random components in the output signal can arise from random noise

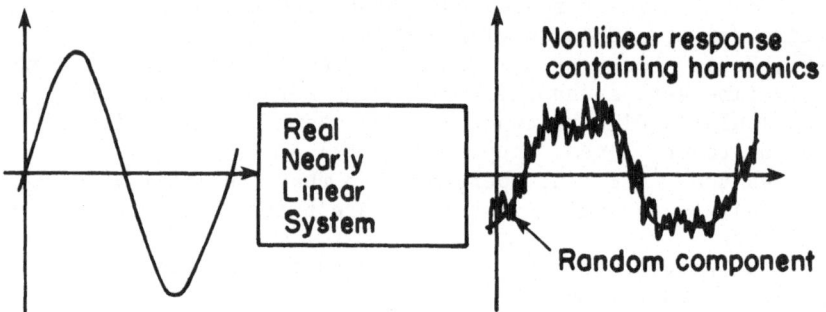

Figure 1:   Real System Response to Sinusoidal Input.

contributed by the experimental apparatus including the amplifiers and transducers, earth loop and aerial affects, time-varying environmental effects such as temperature, and so on.

When the effects become large relative to the basic signal, as in Figure 1, then we must use more sophisticated data analysis techniques such as spectral analysis.

"It is natural to feel that the measurement of power spectra is simple, and that no problems deserving extended discussion arise. After all, are there not commercial "wave analyzers" of many sorts; have not Fourier series served for many years to analyze the frequencies of many signals, (musical instruments, human voices, etc). Why should there be a serious problem?" (BL57)

Parzen (PA60) summed up the main objection to the application of the Fourier transform directly to the time series data by saying

"It has long been traditional among physical scientists to regard time series as arising from a superposition of sinusoidal waves of various amplitudes, frequencies, and phases. In the theory of time series analysis and statistical communications theory, a central role is played by the notion of the spectrum of a time series. For the time series analyst, the spectrum represents a basic tool for determining the mechanism generating an observed time series. For the communication theorist, the spectrum provides the major concept in terms of which to analyze the effects of passing stochastic processes (representing either signal or noise) through linear (and, to some extent, non-linear) devices. Spectral (or harmonic) analysis is concerned with the theory of the decomposition of a time series into sinusoidal components. For many time functions $\{X(t)\}$, such a decomposition is provided by the two-sided Fourier transform

$$F(\omega) = \int_{-\infty}^{\infty} e^{-it\omega} X(t) \, dt.$$

Unfortunately, no meaning can be attached to this integral for many stochastic processes $\{X(t)\}$ since their sample functions are nonperiodic or undamped functions and therefore do not belong to the class of functions dealt within the usual theories of Fourier series and Fourier integrals. Nevertheless, it is possible to define a notion of harmonic analysis of stochastic processes (that is, a method of assigning to each frequency ω a measure of its contribution to the "content" of the process)" as shown by N Wiener (WI30) and A Khintchine (KH34).

The Fourier Series

$$y(t) = A_o/2 + \sum_n A_n \cos(n\pi t/a) + \sum_n B_n \sin(n\pi t/a),$$

allows the periodic structure of a stationary time series to be identified.

It must be emphasised that the conditions for such a spectral analysis are the assumptions of stationarity, ie that at least the time shifted mean $E[x(t + k)] = E[x(t)]$, and covariance $E[x(t + k) \times (t + \tau + k)] = E[x(t) \times (t + \tau)]$ elements remain the same for all time shifts or translations k and E represents the expectation or averaging operator.

The assumption of stationarity is not valid if we are detecting events.

The Fourier Series is expressed in terms of the power spectral density coefficients A_n and B_n, with the variance, $C_{xx}(0) = \sum_n \sqrt{A_n^2 + B_n^2}$, being equal to the area under the power spectral density.

The frequency components of the variance were identified by Rice (RI44) as

$$\sigma_j^2 = \sqrt{A_j^2 + B_j^2} = C_{xx}(0)$$

and these components form the basis set for the second order statistical moments in a Hilbert space. This basis set forms a diagonal matrix where the eigenvalues are equal to the variance frequency components, i.e. $\lambda_i = \sigma_i^2$. This matrix remains in diagonal form after the operation of linear time invariant shifts (TU61).

This analysis of the variance interpretation of the power spectral density of a linear stationary system is also true for the time delayed covariances (TU61) because the Wiener-Khintchine theorem relates the power spectral density $H(\omega)$ to the autocovariance $C(\tau)$.

$$C_{xx}(\tau) = \int H_{xx}(\omega) e^{i\omega\tau} \, d\omega$$

We can interpret the spectrum of a stationary process as an energy-frequency distribution if $\{X(t)\}$ is stationary. Then the process itself has a spectral representation of the form

$$X(t) = \int e^{i\omega t} \, dF(\omega)$$

where $F(\omega)$ is the Fourier transform of the time function $x(t)$ (PA62) and that evolutionary spectral estimates have physical meaning only when the process possesses the appropriate spectral representation (PR81). Traditionally the power spectral density has been associated with an energy interpretation, however the entropy, probability density and analysis of the variance interpretations are also used.

The inverse Fourier-Stieltjes transform of a time series $\{X(t)\}$ is (WI30)

$$X(t) = \int_{-\infty}^{\infty} e^{i\omega t} dF(\omega)$$

or in terms of a unitary linear operator this is

$$X(t) = U^t X(0) = \int e^{i\omega\tau} dF(\omega)$$

and includes both the Fourier series and the Fourier integrals as special cases. The Fourier-Stieltjes transform forms the basis of Wiener's (WI30) 'Generalised Harmonic Analysis' for stationary stochastic processes in terms of steady state functions $F(\omega)$, which are not necessarily differentiable, although the nondifferentiability is a mathematical property not a physical property (MC74).

The Wiener-Khintchine theorem allows the power spectral density function to be estimated when the autocovariance of the time series data $\{X(t)\}$ exists and converges in the mean square sense.

The sample power spectral density $H_{xx}(\omega)$ may be estimated using the linear operator

$$H_{xx}(\omega) = \int C_{xx}(\tau) \, e^{-i\omega\tau} d\tau$$

and its inverse the autocovariance

$$C_{xx}(\tau) = \int H_{xx}(\omega) \, e^{+i\omega\tau} d\omega$$

form a Fourier transform pair (WI30,KH34).

Similarly the crosscovariance function between two signals x and y is given by

$$C_{xy}(\tau) = \int H_{xy}(\omega) \, e^{+i\omega\tau} d\omega$$

and its Fourier Transform pair the cross spectrum function is

$$H_{xy}(\omega) = \int C_{xy}(\tau) \, e^{-i\omega\tau} d\tau$$

The sample autocovariance may be estimated (AN71) using.

$$C_{xx}(\tau) = \frac{1}{n-|\tau|} \sum_{i=1}^{n-|\tau|} (x_i - \mu_x)(x_{i+|\tau|} - \mu_x)$$

$$C_{xy}(\tau) = \frac{1}{n-|\tau|} \sum_{i=1}^{n-|\tau|} (x_i - \mu_x)(y_{i+|\tau|} - \mu_y)$$

As the Fourier Transform F is a linear operator then we know that

$$F\left[\sum_i a_i f_i(x)\right] = \sum_i a_i \, F[f_i(x)]$$

similarly for the covariance function we know for n independent sequences (PA62)

$$C_{xx}(\tau) = \sum_{i=1}^{n} C_{xxi}(\tau)$$

which may be thought of as the sum of the covariances of superposed series $\{x_i t\}$.
The Fourier Transform pair property allow the covariance function $C_{xx}(\tau)$ to be defined as the characteristic function of the time series. The characteristic function of a random variable $\phi_x(\tau)$ may be defined as (PA60) the expectation of the complex exponential function $e^{i\tau x}$ with respect to the probability law of $\{X\}$, ie.

$$\phi_{xx}(\tau) = \int e^{i\omega\tau} \, dF_{xx}(\omega)$$

which may be seen to be simply Fourier-Stieltjes transform given above.
    The power spectral density $H_{xx}(\omega)$ has the properties of a probability density function of the form (PA62)

$$H_{xx}(\omega) = \frac{1}{2\pi} \int e^{-i\omega\tau} \phi_{xx}(\tau) \, d\tau$$

and the autocovariance function $C_{xx}(\tau)$ has the properties of the characteristic function of the random variate $\{X(t)\}$.

NONSTATIONARY TIME SERIES AND WINDOW FUNCTIONS OR FILTERS

There has been a considerable amount of work on time series which are assumed to be stationary or weakly stationary. However little work has been done on nonstationary or nonlinear time series and work has only recently started on ensuring that the time series estimation techniques are robust. For nonstationary time series the covariance function

$$C(s,t) = E[x(s) \ x(t)]$$

depends upon both s and t. If the process is harmonizable we may write by analogy with the stationary case the Wiener-Khintchine theorem

$$C(s,t) = \iint e^{is\omega_s - it\omega_t} \ dF(\omega_s) \ dF(\omega_t)$$

For recent summaries the reader is directed to Rao's review on Harmonizable processes (HA85) for nonstationary series, to Tjostheim's review of recent developments in nonlinear time series analysis (TJ86) and Saleem's review on robust techniques (SA86). I shall now briefly mention the problems with linear filters or windows when the time series is nonstationary then in the following section indicate how to detect and analyse nonstationary signals when the system is nonlinear.

One technique we could use for nonstationary events could be "physical spectrum" analysis but as Priestly points out (PR81), if we introduce "a second type of "spectrum", called the "physical spectrum", $S_x(f,t,W)$, defined as follows. Let $W(t)$ be an appropriate real-valued function s.t. $W(0) > 0$, $|W(t)|$ is "small" outside the neighbourhood of t=0 and

$$\int_{-\infty}^{\infty} W(t)dt = 1$$

Define

$$S_x(f,t,W) = \left| \int_{-\infty}^{\infty} W(t-u)X(u)e^{-2\pi ifu}du \right|^2 .$$

The quantity, $S_x(f,t,W)$, as defined is of little use as a definition of a time-dependent spectrum since it involves an arbitrary weight or window function $W(t)$. Thus, each possible choice of $W(t)$ leads to a different expression for $S_x(f,t,W)$, and no particular $S_x(f,t,W)$ is characteristic purely of the process $\{X(t)\}$. However, suppose we consider an "idealized" form of $S_x(f,t,W)$ in which the function $W(t)$ becomes a δ-function in the limit, so that in this situation $S_x(f,t,W)$ would describe the behaviour of $X(t)$ at the single time point t and would not be "contaminated" by the behaviour of $X(t)$ at other time points. Unfortunately, it is impossible to achieve this type of "idealized" spectrum by considering a limiting form of the expression for S_x. For, as $W(t) \to \delta(t)$, the "filter" $\{W(t-u)e^{-2\pi i f u}\}$ looses all form of frequency domain selectivity, and, in fact

$$S_x(f,t,W) \to |X(t)|^2.$$

Note now that $S_x(f,t,W)$ has lost all dependence on the frequency variable f.

On the other hand, if we attempt to design a filter which has perfect frequency selectivity, then (ideally) its transfer function must have zero bandwidth, and consequently its impulse response function will never die out. This, in turn, means that $S_x(f,t,W)$ will now depend on the behaviour of X(t) over all time, and will not, therefore, describe the spectral properties of the process in the neighbourhood of the specific time instant t. In other words, in the latter case $S_x(f,t,W)$ will have lost all form of time domain selectivity.

This feature is a consequence of the fundamental uncertainty principle for time-dependent spectra, which states, roughly speaking, that in using the linear filtering approach one cannot obtain simultaneously an arbitrarily high degree of resolution in both the time and frequency domains".

Whenevery we use a window or filter function W, with us a priori reason for choosing any particular window, we may think of the weighted output series, $y(t) = \int W(t-u) X(u) e^{-i2\pi f u} du$, as viewing the input series X(t) through the window W(t-u). I prefer to think of this as a window with distorting glass in it as each window function will produce a different output series. Further we may be able to compare different y(t) outputs corresponding to different window functions for linear systems but we will not in general be able to for nonlinear systems.

NONLINEARITIES

Intuitively, we may think of a nonlinear system as possessing a set of autofrequencies or natural frequencies which may be located by exciting the system by noise or by signals with specific information.

We will also have to take into account the non-linearities in our system. This may be achieved in a variety of ways and I shall briefly outline just one. First consider a linear system which is stimulated with the input sequence $\{x(t)\}$ and we obtain the output sequency $\{y(t)\}$. The most general formulation of the linear time invariant system may be written as the convolution equation (SI86)

$$y(t) = \int h(\tau) x(t-\tau) d\tau$$

where h(t) is the impulse response or Green's function of the system and τ is a time delay. If our input sequence $\{x(t)\}$ has an autocovariance function $C_{xx}(\tau)$ and we measure the input-output covariance $C_{xy}(\tau)$ we may determine the Green's function using the Wiener-Hopf equation (CO86)

114

$$C_{xy}(\tau) = \int h(\tau)\, C_{xx}\,(t - \tau)\, d\tau$$

and similarly the output sequence autocovariance will be

$$C_{yy}(\tau) = \int h(\tau)\, C_{xy}\,(t - \tau)\, d\tau$$

If we now add in a linear filter which has a Green's function $g(t)$ the output autocovariance will be modified as

$$C_{zz}(\tau) = \int g(\tau)\, C_{yy}\,(t - \tau)\, d\tau.$$

Generally speaking in linear operator theory the system will be time dependent so that we may expect linear operators to be of the form (BI84)

$$L\, y(t) = x(t) \text{ and } y(t) = L^{-1}\, x(t)$$

It is reasonable to assume (BE84) that the inverse operator L^{-1} is time dependent integral operator involving a kernel $G(t,\tau)$ called the Green's function (named after the founder of the method) so that the inverse operator may be written as

$$y(t) = L^{-1}\, x(t) = \int G(t,\tau)\, x(\tau) d\tau$$

A subclass of such operators for linear time invariant systems is the convolution equation.

$$y(t) = \int h(t - \tau)\, x(\tau)\, d\tau = \int h(\tau)\, x(t - \tau) d\tau$$

where $h(\tau)$ is the impulse response function for the linear time invariant system.

We can now move on to non linear systems. The trick is to retain the linear operator L in a new operator Λ which includes a nonlinear operator N so that

$$\Lambda\, y(t) = \{L + N\}\, y(t) = x(t) + m$$

where m is a constant mean level, usually on the input of the system. Usually we wish to solve for the output $y(t)$ given a particular input series $\{x(t)\}$ to the system, this is done using the inverse operator Λ^{-1} as (BI84)

$$y(t) = \Lambda^{-1}\, (x(t) + m) = \{L^{-1} + N^{-1}\}\, (x(t) + m)$$

or

$$y(t) = L^{-1} (x(t) + m) \qquad \text{as before}$$

$$+ N^{-1} (x(t) + m) \quad \text{the nonlinear term}$$

As will be shown in the next section this should allow us to identify if the system under study is linear or not.

If the system is nonlinear we may describe the nonlinearities in many ways, the perturbation expansion above being one method frequently employed. For stochastic nonlinear systems we may describe the nonlinearities in terms of the higher order time delayed moments or correlations between the input $\{x(t)\}$ data series and the output $\{y(t)\}$ time series (BI84,TJ86).

NONLINEAR NONSTATIONARY TIME SERIES ANALYSIS

When the input time series has variations in amplitude which cause the response function of the system to change. We need to be able to describe
(i)   the input time series variations
(ii)   the changes induced in the impulse function due to the boundary conditions
(iii)   the effect on the output of the system.
If possible we should relate all of the above to the sample time delayed moments.

That is we need to describe the nonlinear, a time varying, impulse response function $h(x(t),t)$.

For deterministic functions we may expand the Green's function as a Taylors series

$$h(x(t + \tau), \, t + \tau) = h(x(t),t) + \{x(t + \tau)-x(t)\} \frac{\partial h(x(t),t)}{\partial x(t)} + \dots.$$

For linear systems we know that

$$y(t) = \int h(\tau) \, x(t - \tau) \, d\tau$$

and $C_{xy}(\tau) = \int h(\tau) \, C_{xx}(t - \tau) \, d\tau$

which when $C_{xx}(t - \tau) = \delta(t - \tau)$ becomes

$$C_{xy}(\tau) = h(\tau) = E[x(t) \, y(t + \tau)] \quad \text{the impulse response function.}$$

For non linear systems may write by analogy

$$h(t,\tau) = e^{D} h = h(t) + \frac{\partial h(t)}{\partial x(t)} + \dots.$$

$$e^D = 1 + \frac{\partial}{\partial x^2} + \ldots\ldots, \quad D = \frac{\partial}{\partial x(t)}$$

or in vector form as

$$e^D\underline{h} = \underline{h}(t) + \frac{\partial \underline{h}(t)}{\partial \underline{x}(s)} + \frac{1}{2!} \frac{\partial^2 \underline{h}(t)}{\partial \underline{x}(s)^2} + \ldots\ldots$$

For linear systems the operator e^D yields the convolution equation

$$y(t) = \int h(\tau) \, x(t - \tau) d\tau = \int h(t - \tau) \, x(t) dt$$

and

$$C_{xy}(t) = \int h(\tau) \, C_{xx}(t - \tau) d\tau = \int h(t - \tau) C_{xx}(\tau) dt$$

Now put $h(t) \to e^D h(t)$ using the differential operator so for non-linear systems we may write the Volterra Series of convolution as

$$y(t) = \int e^D h(\tau) \, x(t - \tau) d\tau = \int h_1(\tau) \, x(t - \tau_1) d\tau_1 +$$

$$\iint h_2(\tau_1, \tau_2) \, x(t - \tau_1) x(t - \tau_2) d\tau_1 d\tau_2 + \ldots\ldots$$

similarly for the crosscovariance we will have

$$C_{xy}(\tau) = \int e^D h(t - \tau) \, C_{xx}(\tau) \, d\tau$$

So when the system is nonlinear and/or nonstationary we may expand the impulse response or Green's function of the system as a Taylors series and modify the convolution equation $y(t) = \int h(\tau) \, x(t - \tau) d\tau$ to yield the Volterra series (VO87) which was first applied by Wiener (WI30) to the study of non-linear stationary systems, see for example Schretzen (SC80). The Green's functions may be determined in terms of the time delayed moment expansion, the second moment yields the linear impulse response for example.

When we are dealing with independent events on a stationary background we could detect the event using the difference between the time dependent covariance and the stationary background covariance, ie

$$\Delta C_{xy}(\tau, t) = C_{xy}(\tau, t) - E[C_{xy}(\tau, t)]$$

then process $\Delta C_{xy}(\tau, t)$ for the information of the event.

ROBUST ESTIMATORS

I am most grateful to Dr Dewey of MIT for providing me with his data containing an outlier or contaminent due to an electrical surge when his tape drive turned on, see figure 2. Any data set could have outliers in it. Outliers may manifest themselves as simple delta functions through to complete distributions contaminating the data. They may be in the time domain as shown, or in the frequency domain for example as in sinusoidal earth loop pickup or some other domain.

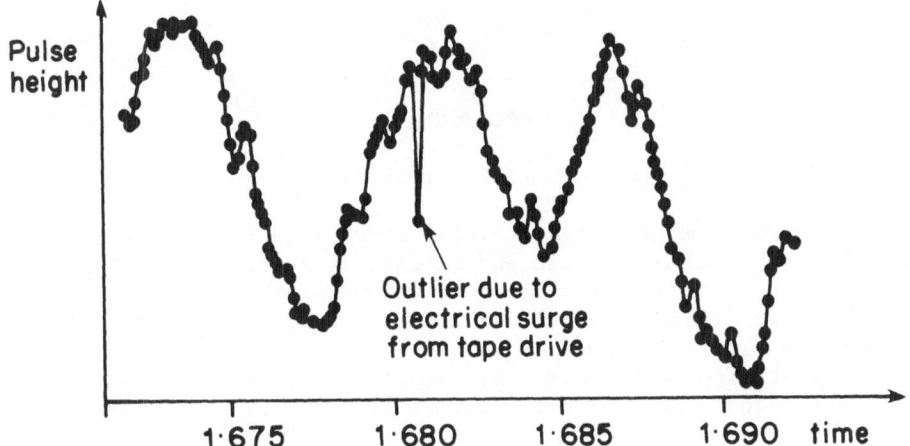

Figure 2:   Time Domain Outlier

It must be said at this point that there are very few robust statistical techniques available for location of outliers or the estimation of parameters but the interested reader is directed to the references cited in "Outliers in Statistical Data" by Barnett and Lewis (BA84).

The non robust nature of statistical measures has been known for a long time and was summarised well for stationary time series analysis by Tukey (TU60). In 1979 Thomson (TO79) at the Bell Labs performed the quality control of curvature of waveguides using the power spectral density. The measuring machine malfunctioned at two points in about 10^9 and the qualitative effect of these two outliers (one outlier would produce the same effect as shown in IR87) was as shown in figure 3.

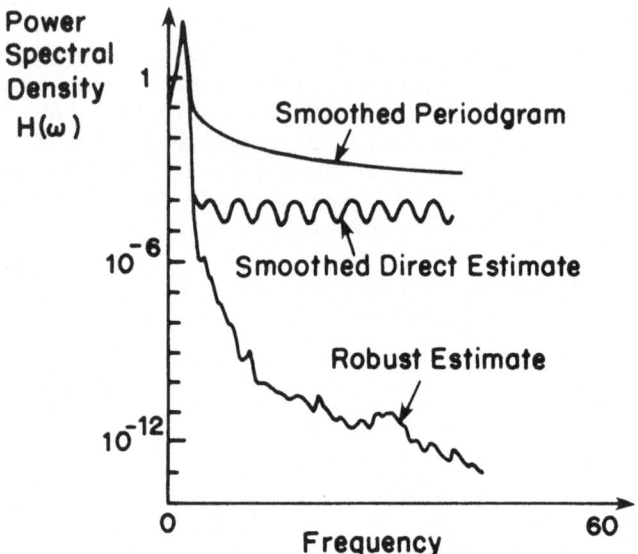

Figure 3: Effect of Outliers on the Power Spectral Density

This spectacular example has recently encouraged the development of more robust statistical estimators and outlier identification techniques. Although there is a long way to go before we can say that there are generally robust maximum likelihood, covariance, power spectral density estimators available.

As you are trying to identify very small signals embedded within a feedback error signal you will need all the help you can get using a variety of estimators to extract the signal information from the noise.

MATHEMATICAL AND PHYSICAL EQUIVILENCE

When we solve a problem we can usually choose between a variety of techniques and representations. For example if we wish to obtain the impulse response function of a system we could solve the integral convolution equation

$$y(t) = \int h(\tau)\, x(t - \tau) d\tau$$

or obtain a best estimate for h(t) using least squares (SP70)

$$\underline{h}(t) = (\underline{\underline{X}}\, \underline{\underline{X}}^T)^{-1}\, \underline{\underline{X}}^T\, \underline{y}$$

where T denotes the transpose of the data matrix $\underline{\underline{X}}$.

If we wish to solve a particular problem with a set of known initial and boundary conditions using one technique we can still formulate the problem in different ways. For example if we wish to solve for the electron eigenvalues of the unperturbed Hydrogen atom could choose to solve the two body Hamiltonian in cartesian, cylindrical polar or spherical polar coordinates. All the solutions would be mathematically equivalent. However the ease by which we can extract the relevant information about the system will be very dependent on our choice of basis set eigenfunctions. In this example we would be led to the spherical polar coordinates by the boundary conditions which indicates that the natural eigenfunctions for this problem would be the spherical harmonics.

Before we impose any initial or boundary conditions we can often identify the natural eigenfunctions of the problem from the form of linear operator being used. For example

Linear Operator	"Natural" Series Expansion (Eigenstates of the Operator)
$L = \dfrac{d^2}{dx^2} - \theta^2$	Fourier
$L = (1 - x^2)\dfrac{d^2}{dx^2} - x\dfrac{d}{dx}$	Tchebycheff
$L = \dfrac{d}{dx}(1 - x^2)\dfrac{d}{dx}$	Legendre
$L = \nabla^2 - \xi^2$	Bessel x Spherical Harmonic or Hankel x Spherical Harmonic

So we should be able to at least restrict the types of expansions, likely to be appropriate, simply by looking at our differential or integral equations.

In the inverse scattering problem we know that asympotically the Born approximation applies and that the solution should be of the form

$$\sim f(r)\, Y_{\ell m}(\theta, \phi)$$

which may be analysed in terms of a multipole expansion of radiation eminating from the distant (point like) source. Even if you prefer to describe the incoming wave as 'plane' you will still need to extract the spherical harmonic admixture if you wish to determine the stellar mass ratio. A more complete treatment of the analysis of 2-D and 3-D nonstationary waves is given in Hannan (HA70) or Yaglom (YA62).

General relativity predicts that the lowest source multipole which contributes to gravitational radiation is the quadrupole (TH77). Most alternative relativity theories predict the presence of all multipoles (HA79), in particular systems containing neutron stars may radiate predominantly a dipole field (HA79). To resolve such differences you must be able to extract the multipole nature of the incoming radiation and compare your experimental results with the current theories. It may be that this is very difficult with an individual event and that you need to build up ensemble averages from many events as a function of, for example, solid angle and source orientation.

SUMMARY

My talk is intended to make you think of your data analysis rather than simply present the covariance analysis. You have several years breathing space before the next generation of gravitational wave detectors are built and you should use this time to develop a variety data analysis systems which have a significant investment of resources. Be guided by the theoretical predictions, for example the eigenfunctions expected, but your signals are just as likely to be unintelligable 'scruff' which takes years to decode in terms of the underlying physical processes. Talk to all your friends in Geophysics to Econometrics and make best use of the expertise available to solve your nonlinear nonstationary multivariate signals processing in a robust manner.

This work has been supported by the United Kingdom Science and Engineering Research Council.

REFERENCES

AN71   Anderson T W, The Statistical Analysis of Time Series, John Wiley
        Sons, New York, 1971.

BA84   Barnett V and Lewis T, Outliers in Statistical Data, John Wiley
        and Sons, Chichester, 1984.

BE84   Bellman R and Adomian G, Partial Differential Equations; New
        Methods for their Treatment and Solution, D Reidel Publishing
        Company, Dordrecht, 1984.

BI84   Billings S A, Gray J O and Owens D H, Nonlinear Systems Design,
        Peter Peregrinus Ltd, London, 1984.

BL58   Blackman R B and Tukey J W, The Measurement of Power Spectra from
        the Point of View of Communications Engineering, Part I, 37, Bell
        System Tech. J., 1958, 185-282.

CO86   Cooper G R, and McGillem C D, Probabilistic Methods of Signal and
        System Analysis, Holt Rinhart and Winston, New York, 1986.

DU86   Duparz J, Probability, Signals Noise, North Oxford Academic,
        London, 198.

HA70   Hannon E J, Multiple Time Series, John Wiley & Sons, New York,
        1970.

HA79   Hawking S W and Israel W, General Relativity: An Einstein
        Centenary Survey, Cambridge University Press, Cambridge 1979.

IR87   Irving A D, Application of Statistical Techniques to the Analysis
        and Validation of Multivariate Time Series Simulations, In
        Preparation, 1987.

KH34   Kintchine A. MAth Ann, 109, p 604, 1934.

LA77   Landu L D and Lifshitz E M, Quantum Mechanics Non-Relativistic
        Theory, Pergamon Press, Oxford, 1977.

LO63   Loeve M, Probability Theory, D Van Nostrad Co Limited, Canada,
        1963.

MC74   McShane E J, Stochastic Calculus and Stochastic Models, Academic
        Press, New York, 1974.

MO84   Morrison D F, Multivariate Statistical Methods, McGraw Hill,
        London, 1984.

PA60   Parzen E, Modern Probability Theory and Its Applications, John
        Wiley, New York, 1960.

PA61   Parzen E, An Approach to Time Series Analysis, Annals of
        Mathematical Statistics, Vol 32, 1961.

PA62   Parzen E, Stochastic Processes, Holden Day, San Francisco, 1962.

PA86   Pais A, Inward Bound, Oxford University Press, 1986.

PR81   Priestly M B, Spectral Analysis and Time Series Academic Press,
        London, 1981.

RI44   Rice O, Mathematical Analysis of Random Noise, Bell System
        Technical Journal, Vol 23 and 24, 1944.

RO67   Robinson E A, Multichannel Time Series Analysis with Digital
        Computer Programs, Good Pond Press, Houston, 1967.

SA86   Saleem A K and Poor H V, Robust Techniques for Signal Processing:
        A Survey, Proc. IEEE, Vol 73, No 3, March 1985.

SC80   Schetzen M, The Volterra and Wiener Theories of Nonlinear
        Systems, John Wiley, New York, 1980.

SI86   Siebert W, Circuits Signals and Systems, The MIT Press,
        Cambridge, Massachusetts, 198.

SP70   Speedey C B, Brown R F and Goodwin G C, Control Theory:
        Identification and Optimal Control, Oliver Boyd, Edinburgh, 1970.

TO79   Thomson D J, Spectrum Estimation Techniques for Characteristics
        and Development of WT4 Waveguide, The Bell System Technical
        Journal, Vol 56, No 9, 1977, 1769-1823 and 1983-2005.

TJ86   Tjostheim D, Recent Developments in Nonlinear Time Series
        Modelling, In Communications and Networks, A Survey of Recent
        Advances, (Ed Blake I and Poor H), Springer-Verlag, New York,
        1986.

TH77   Thorne K S, Multipole Expansions of Gravitational Radiation,
       Reviews of Modern Physics, Vol 52, N2 Part 1, p 299-340.

TU61   Tukey J W, Discussion Emphasizing the Connection between Analysis
       of the Variance and Spectrum Analysis, Technometrics, 3, p 191,
       1961.

VO87   Volterra V, Sopra le Funjioni che Dipendono de Altre Funzioni,
       Rend. R. Accademia dei Lincei 2° Sem, 1887, 97-105 and 141-146.

WI30   Wiener N, Generalised Harmonic Analysis, Acta Mathematica,
       Uppsala, Vol 55, 1930, 117-258.

YA62   Yaglom A M, An Introduction to the Theory of Stationary Random
       Functions, Englewood Cliffs, Prentice Hall, New Jersey, 1962.

Part 3

QUANTUM LIMITS ON DETECTORS

PARAMETRIC TRANSDUCERS AND QUANTUM NONDEMOLITION IN BAR DETECTORS

Mark F. Bocko, Department of Electrical Engineering
Univ. of Rochester, Rochester, New York 14627   USA

Warren W. Johnson, Department of Physics and
Astronomy, Louisiana State University, Baton Rouge,
Louisiana 70803   USA

ABSTRACT.   We describe a RF parametric transducer which is
being developed for gravitational radiation bar antennae.
With this transducer it is possible to measure a quantum
nondemolition observable and thus potentially exceed the
"quantum limit" for bar detectors.   A more immediate benefit
of this type of transducer to the development of
gravitational wave antennae is the reduction of the
requirements on the electrical Q and the pump phase noise.
We summarize our progress and plans for the further
development of this transducer.

1.   Introduction

The existence of a quantum limit for the sensitivity of
bar gravitational radiation detectors and quantum
nondemolition (QND) measurement techniques to circumvent this
limit have been recognized for some time.   For a thorough
discussion of the issues of principle the reader is referred
to the extensive review article by Caves et al.[1]   The purpose
of the present contribution is to discuss the practical
realization of one quantum nondemolition measurement method
with a parametric transducer.   An important point we make in
this paper is that, in addition to the ability to perform a
QND measurement, there are advantages of parametric
transducers which can contribute to the immediate improvement
of gravitational radiation detectors.

In the first section of this paper we describe the
parametric transducer we are developing and and how it may be

B. F. Schutz (ed.), Gravitational Wave Data Analysis, 125–134.

used to perform a QND measurement. The most recent
experimental results are then reported. In a following
section we explain the major advantage of the parametric
transducer - the reduction of the effect of the transducer
electrical read-out fluctuations on the antenna noise. In a
final section we indicate our plans for further development
and the achievable sensitivities.

2. Parametric Transducers

All gravitational radiation antenna transducers are
parametric in the sense that an impedance which is part of a
low noise electrical circuit is modulated by the motion of
the gravitational radiation antenna. The distinction between
the parametric transducer described here and the conventional
"DC" transducers is the frequency of operation. The
inductively modulated transducer developed at Stanford[2] and
the capacitively modulated transducer developed by the Rome
group[3] work by allowing the antenna motion to modulate an
impedance element in which there is stored a static magnetic
or electric field. Thus an electrical signal appears at the
same frequency as the gravitational wave antenna. In another
class of transducers an impedance in a radio frequency
circuit is modulated by the antenna motion.[4-7] A high
frequency alternating field is present in the modulated
impedance element and the signal appears as side-bands
imposed upon the "pump" field. The modulated impedance is
usually part of a tuned circuit or a microwave cavity which
makes it easier to impedance match to available low noise
amplifiers.

The parametric transducer which we have developed
operates at an electrical read-out frequency near 4 MHz. It
is a balanced radio frequency bridge circuit in which a pair
of capacitors, which make up half of the bridge, are
modulated 180° out of phase by the motion of a mechanically
resonant diaphragm. Here, as in almost all transducer
schemes, one senses the motion of a second, low mass,
resonator which is attached to the end face of the
gravitational radiation antenna to allow one to achieve
stronger electromechanical coupling. In operation, the
bridge circuit is balanced by externally controllable means
and a large RF electric field is imposed on the capacitors by
exciting the bridge through a center tapped transformer which
constitutes the other half of the bridge circuit. An output
signal in the form of a current through the "resonating
inductor" will appear if the bridge is unbalanced by the
motion of the mechanical resonator. This current is sensed
by a low noise gallium arsenide transistor (GAT) amplifier.

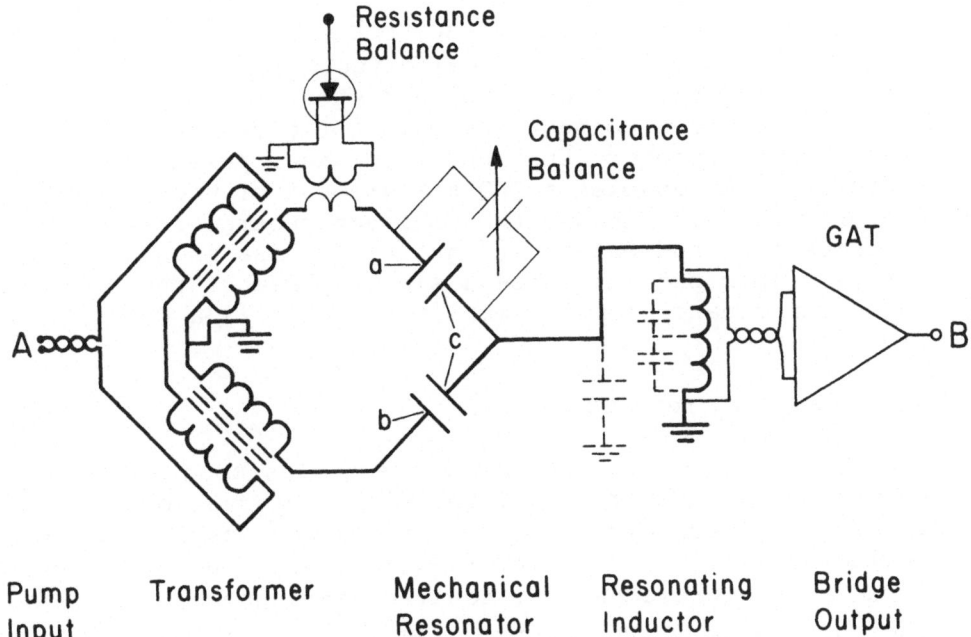

Figure 1. The electromechancial schematic of the RF
parametric transducer is shown. The three plate capacitor
forms one half of a radio frequency bridge circuit. The
bridge is excited through the transformer. The read-out arm
of the bridge is the "resonating inductor" which resonates at
4.06 MHz with the parallel combination of the capacitors.
The balancing capacitor and resistor are shown. The
capacitors drawn in broken lines represent the unwanted
parasitic capacitances which shift the frequency of the read-
out resonance.

The excitation source is referred to as the "pump" and is
chosen to have the following time dependence;

$$V_p(t) = \frac{V_0}{2} \left\{ (1-f)\cos[(\omega_2 + \omega_1)t] + (1+f)\cos[(\omega_2 - \omega_1)t] \right\} \qquad (1)$$

where ω_2 and ω_1 are respectively the angular frequencies of
the electrical read-out and mechanical resonator and f is a

parameter which takes values -1 < f < 1. It has been shown
that this transducer is equivalent to a familiar parametric
convertor.[8] When f = +1, the device functions as a phase
preserving parametric upconvertor and when f = -1 it
functions as a phase conjugating parametric upconvertor. It
has been shown elsewhere that the equal superposition of the
two pumps, f = 0, enables a QND measurement of the mechanical
resonator. This is because the coherent superposition of a
phase preserving and phase conjugating process is phase
sensitive. Thus one may sense a single phase of the
mechanical resonator complex amplitude which is a QND
observable.

3. Experimental results

We have tested at cryogenic temperatures a parametric
transducer fabricated from a large single crystal of
sapphire. The mechanical resonant frequency was 2.3 kHz and
the bridge electrical resonant frequency was 4.06 MHz. We
achieved an electromechanical coupling strength of 2×10^{-4},
defined by $\beta = E_0^2 (\omega_2/\omega_1)(C_0/m\omega_1^2)$, where E_0 is the peak
electric field in the capacitor, C_0 is the capacitance of one
capacitor and m is the effective mass of the mechanical
resonator. We also measured a wide band electronic noise
equivalent to a displacement of 10^{-15} m/$\sqrt{\text{Hz}}$. The details of
the experiments will be published elsewhere.[9]

The most significant result of our tests is that we have
gained a quantitative understanding of the dominant source of
noise in parametric transducers, the phase and amplitude
fluctuations of the pump oscillator, and demonstrated that
this noise source can be dramatically reduced. If the bridge
circuit is not perfectly balanced, the fluctuations of the
pump at frequencies separated from the carrier by $\pm\omega_1$ feed
through to the amplifier and appear as an additional source
of noise at the amplifier input. In Figure 2 we show the
equivalent displacement noise of the transducer as a function
of the coupling strength for a number of cases which differ
by the degree of bridge imbalance. At weak coupling levels
the effective displacement noise decreases as the coupling is
increased because the dominant source of noise is the
additive noise of the amplifier. We find that the
displacement noise reaches a lower limit for a value of
electromechanical coupling which depends on the degree of
balance. This is the effect of the pump noise. In this
limit, although an increase of the pump amplitude gives a
larger signal, it also increases the noise at the same rate
because the phase and amplitude noise of the pump are

proportional to the pump amplitude.

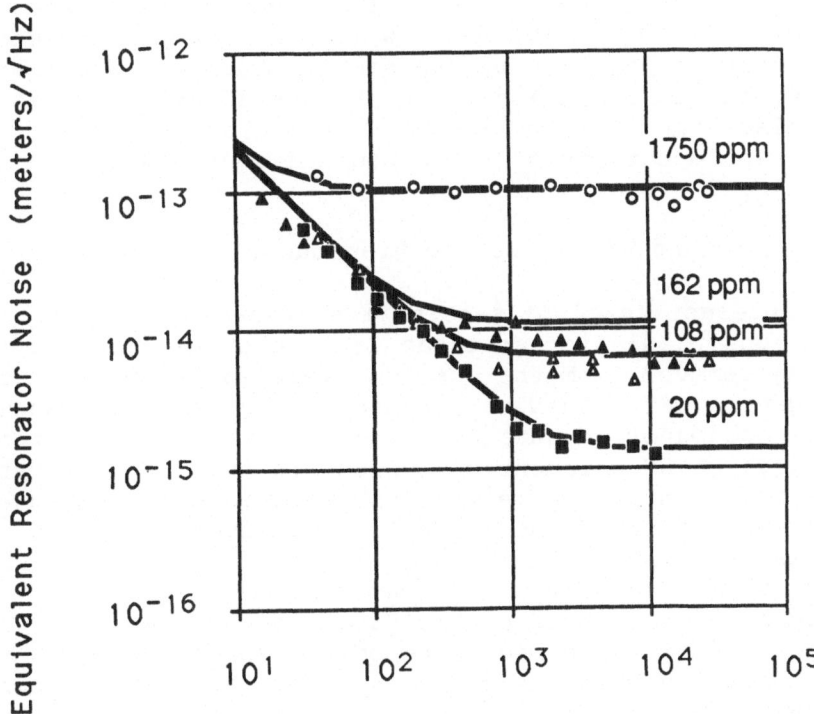

Figure 2. The equivalent mechanical resonator displacement noise is plotted versus the peak electric field amplitude in the capacitors. The four curves are a theoretical fit to the data with no free parameters; the pump noise and bridge imbalance were measured independently. The equivalent displacement noise at the best achievable bridge balance was approximately 10^{-15} meters/$\sqrt{\text{Hz}}$.

The best balance we were able to achieve was 2 parts in 10^5 which, given our moderately noisy pump, a Hewlett Packard 3325A with phase noise -120 dBc (dB below the carrier) at 2.3 kHz offset from the carrier, led to the quoted sensitivity level. If we had a pump oscillator with 20 dB less phase noise, which is readily available commercially, we could have achieved a displacement noise of 10^{-16} m/$\sqrt{\text{Hz}}$. This

sensitivity is comparable to the most sensitive transducers
developed for gravitational radiation antennae.[10]
     We have evidence that direct electrical pickup between
the pump input and the bridge read-out arm is limiting our
ability to balance the bridge. If we reduce the direct
pickup by a factor of 10, which will consequently increase
the balancing capability by the same factor, and use fixed
frequency quartz oscillators, which have 35 to 40 dB less
phase noise than the pump we used, then the pump noise
problem should not appear again until a sensitivity
approaching the quantum limit.

4.   An advantage of parametric transducers

     The present level of sensitivity achieved by
gravitational radiation antennae is some three orders of
magnitude removed from the quantum limit so it may seem that
investigating QND transducers is not yet necessary. However
there is one major advantage, aside from the QND measurement
capability, of RF parametric transducers over the DC passive
schemes which should prove to be of immediate consequence.
This is the effect of the electrical dissipation on the
mechancial resonator.
     The relaxation time of the mechanical resonator
determines the magnitude of the Brownian noise in the
following manner. The magnitude of the Langevin force
spectral density which is resposible for the Brownian motion
is given by,

$$S_f(\omega) = \frac{4kTm}{\tau_1} \tag{2}$$

where τ_1 is the relaxation time, T is the temperature and m is
the mass of the mechanical resonator; k is Boltzmann's
constant. It can be shown[11] that the relaxation time of the
mechanical resonator when it is coupled to the electrical
read-out becomes

$$\tau_1 = \left[\frac{1}{\tau_{1_0}} + f \left(\frac{\beta \omega_1^2 \tau_2}{2} \right) \right]^{-1} \tag{3}$$

where τ_{1_0} and τ_2 are the uncoupled relaxation times of the
mechanical and electrical resonators respectively. The

131

effective Brownian noise for a passive DC transducer may be calculated by substituting the coupled mechanical ralaxation time into Equation 2.[12] In contrast to this, the corresponding result for the parametric transducer may be shown to properly be

$$S_f(\omega) = 4kTm\left[\frac{1}{\tau_{1_0}} + f^2\,(\frac{\beta\,\omega_1^2\,\tau_2}{2})\,\frac{\omega_1}{\omega_2}\right] \qquad (4)$$

which contains the factor ω_1/ω_2 that effectively reduces the effect of the electrical dissipation on the mechanical Brownian noise. The effect of the electrical dissipation may thus be substantially reduced by operating the transducer at a high frequency. The passive inductive transducer scheme in wide use is presently limited by the electrical dissipation[12] and efforts are continuing to learn the source of the electrical loss. The use of a RF parametric transducer will greatly relax the requirement on the electrical dissipation.

5. Future directions

In this last section we discuss our ongoing work with RF parametric transducers. We introduce a new mechanical resonator design for a transducer, summarize our goals for the physical parameters and give the predicted sensitivity.

The new mechanical design, which is under development in Rochester, uses a torsional resonator. One draw-back of the balanced bridge transducer is that the two bridge capacitances must be modulated in opposition. It is difficult to fabricate a clamped diaphragm which has two very closely spaced and well matched gaps, one on either side of the diaphragm. The torsional oscillator has the feature that both capacitors lie in the same plane which greatly simplifies the matching of the capacitor gaps. The resonator, shown in Figure 3, has the torsion springs located at its geometrical center but material has been removed from one side of the resonator so that the center of mass is located below the location of the springs. Thus, motion normal to the plane of the transducer will be coupled to the torsional mode. This transducer may be attached to a bar detector in the same way as any other resonant transducer. We have fabricated a prototype and demonstrated that all other resonant modes of the structure are at least a factor of 3 higher in frequency than the torsional mode.

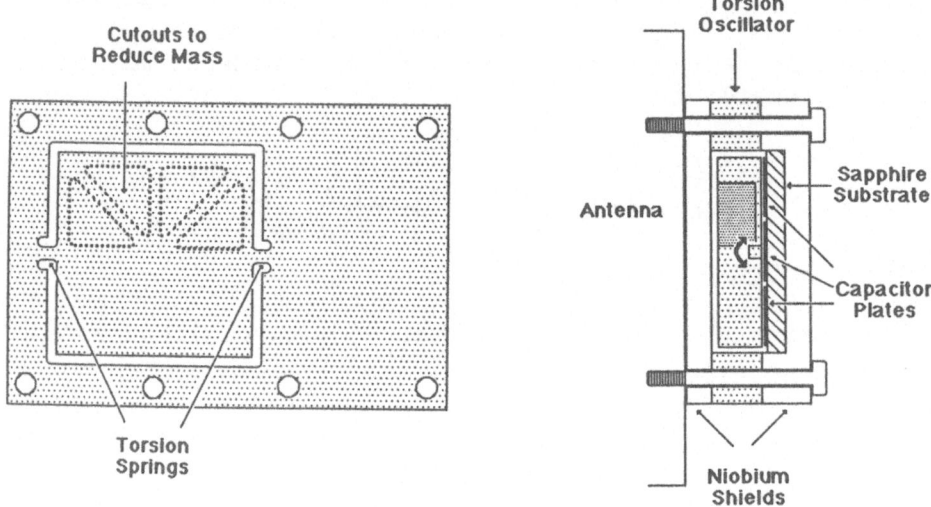

Figure 3. A torsional transducer for a bar antenna is shown above. In the view on the left one can see that part of the material is removed in a manner that will cause the motion of the antenna end face to be converted to vibration of the torsional mode of the transducer. On the right is shown a possible mounting scheme. This design has the advantage that the coupling between the transducer and the read-out is in a single plane which simplifies fabrication.

In Rochester, we are also beginning to work with an electrical read-out frequency which is a factor of 10 or more higher than that used in the past. There are two reasons for this; amplifiers made from HEMTs (High Electron Mobility Transistors) may have low noise at higher radio frequencies and a simple planar read-out circuit may be fabricated from superconducting microstrip.

HEMT amplifiers which operate at 8.4 GHz have been developed which have a noise number that is approximately a factor of 20 above the quantum limit.[13] The theory of transistor noise predicts that the noise number should be independent of frequency,[14] however a 1/f noise component always appears below some frequency. We are investigating the noise of HEMTs in the 50 to 100 MHz range to learn if the 1/f noise is a problem. It is more convenient to use a transistor amplifier than a SQUID (Superconducting QUantum Interference Device), because one can fully test the transducer at room temperature before cooling down to cryogenic temperature, but if HEMTs prove to be too noisy, then a SQUID, which has been shown to have a noise number of 100 at 100 MHz[15], could be used.

In conclusion, we summarize in Table 1, our target transducer parameters and the expected sensitivity of the transducer now under development. The calculations of the noise were made using the formulae given in Reference 8.

Table 1

Mechanical

m	effective mass	0.3	kg
$\omega_1/2\pi$	frequency	1.0	kHz
Q_m	mechanical Q	10^7	
T	temperature	1	Kelvin

Electrical

A	capacitor area	5.0×10^{-4}	meter^2
D	capacitor gap	25	micron
L	resonating inductance	63	nHenry
$\omega_2/2\pi$	read-out frequency	58	MHz
Q_e	electrical Q	10^5	
V_0	pump voltage (peak)	5.0	Volts
x_0/D	bridge imbalance	10^{-6}	
$S_\phi(\omega_1)$	pump phase noise	-155	dBc
$S_a(\omega_1)$	pump amplitude noise	-160	dBc
T_A	Amplifier noise temperature	0.061	Kelvin
N_A	noise number ($kT_A/h\omega_2 ln3$)	20	

Transducer noise

δx	r.m.s. displacement noise	9.0×10^{-18}	meters
W	bandwidth	12	Hertz
h	minimum detectable strain (on 2000 kg antenna)	1.5×10^{-19}	

Table 1 The target parameters for the parametric transducer under development at Rochester and the predicted sensitivity.

References

1. C.M. Caves, K.S. Thorne, R.W.P. Drever, V.D. Sandberg, M. Zimmermann, Rev. Mod. Phys., **52**, 341 (1980).

2. H.J. Paik, J. Appl. Phys. **47**, 1168 (1976).

3. P. Rapagnani, Nuovo Cimento, **C5**, 385 (1982).

4. V.B. Braginsky, A.B. Manukin, Measurement of Weak force in Physics Experiments, edited by D.H. Douglass, University of Chicago Press, Chicago (1977).

5. W.C. Oelfke, W.O. Hamilton, Acta Astronaut., **5**, 87 (1978).

6. D. Blair, Phys. Lett., **104A**, 197 (1982).

7. W.W. Johnson, M.F. Bocko, Phys. Rev. Lett., **47**, 1184 (1981).

8. M.F. Bocko, F. Bordoni, F. Fuligni, W.W. Johnson, in Noise in Physical Systems and 1/f Noise-1985, edited by A. D'Amico and P. Mazzetti, Elsevier Science Publishers (1986).

9. M.F. Bocko, W.W. Johnson, V. Iafolla, in preparation.

10. S. Boughn et al, Astrophys. J., **L19**, 261 (1982).

11. M.F. Bocko, PhD Thesis, University of Rochester (1984).

12. P.F. Michelson, J.C. Price, R.C. Taber, Science, **237**, 150 (1987).

13. M.W. Pospieszalski et al, IEEE Trans. Electron Dev., **ED-33**, 219 (1986).

14. H. Fukui, IEEE Trans. Electron Dev., **ED-26**, 1032 (1979).

15. C. Hilbert, J. Clarke, Journal of Low Temp. Phys., 61, 263 (1985).

SQUEEZED STATES OF LIGHT

M. H. Muendel[1], G. Wagner[1], J. Gea-Banacloche[1] and
G. Leuchs[1,2]
[1] Max-Planck-Institut für Quantenoptik, 8046 Garching,
West Germany and [2] Sektion Physik der Universität München
8046 Garching, West Germany

ABSTRACT. Gravitational wave detection with optical methods requires
interferometers with extremely high strain sensitivity. Possibilities
to increase the interferometer sensitivity using the squeezing and
recycling techniques are discussed with respect to mirror losses and
response time.

1. INTRODUCTION

A fundamental consequence of the quantization of the radiation field is
the existence of fluctuations. Minimization of these fluctuations is
constrained by Heisenbergs´s uncertainty relation. In the case of a
laser e.g. the fluctuations in two conjugate variables of the field
are equal, fulfilling the uncertainty relation symmetrically. The term
squeezed states simply refers to a different minimum uncertainty state
of the radiation field where the uncertainty relation is fulfilled
asymmetrically with unequal fluctuations in the two conjugate variables [1].

2. FIELD FLUCTUATIONS

In classical electrodynamics a monochromatic plane wave can be written
as

$$E(r,t) = \varepsilon_c \cos \omega t + \varepsilon_s \sin \omega t \qquad (1)$$

The effect of quantization can be seen immediately by remembering the
quantum mechanical harmonic oscillator [2]. There one finds two non-
commuting operators describing position and momentum, two variables
which oscillate out of phase. By analogy the two corresponding variables
in the case of the radiation field are just the amplitudes of the sine
and cosine wave (s. Eq. (1)).
    Choosing dimensionless operators $\hat{a}_c$ and $\hat{a}_s$ describing the field
mode yields

B. F. Schutz (ed.), Gravitational Wave Data Analysis, 135–143.

$$\varepsilon_{c,s} = (h\omega/V\varepsilon_o)^{1/2} \langle \hat{a}_{c,s} \rangle \tag{2}$$

with V denoting the mode volume and ω the light frequency. This leads to the commutator $[\hat{a}_c, \hat{a}_s] = \hat{1}i/2$ and the uncertainty relation

$$[\langle \Delta a_c^2 \rangle \langle \Delta a_s^2 \rangle]^{1/2} > 1/4 \tag{3}$$

With this choice the expectation values of $\hat{a}_c$ and $\hat{a}_s$ are related to the mean photon number $\langle n \rangle$ in the field mode

$$\langle n \rangle = \langle a_c^2 + a_s^2 \rangle - 1/2.$$

It is important to note, that the constraint imposed by the uncertainty relation (Eq. 3) is independent of the number of photons in the mode. This implies that minimum uncertainty states fulfilling the uncertainty relation symmetrically

$$\sqrt{\langle \Delta a_c^2 \rangle} = \sqrt{\langle \Delta a_s^2 \rangle} = 1/2,$$

have the same size of field fluctuations. Such states are called coherent states and describe both the ideal laser field in the absence of phase diffusion and the vacuum state (zero point fluctuations).

Owing to the equality of the field fluctuations the vacuum field plays a role whenever it interferes with an excited mode of the field. The photon number fluctuations are of course quite different for the coherent and the vacuum state.

It is easy to imagine field states of minimum uncertainty with unequal fluctuations in the two conjugate variables, but it took until 1985 to demonstrate experimentally the existence of these squeezed states [3]. Such field states with unequal uncertainty are refered to as non-classical light since they cannot be described in the framework of classical electrodynamics. They also exhibit quite unusual photon number distributions [4]. Other closely related types of field states are the ones that show photon-antibunching [5,6].

3. LINEAR ATTENUATION

A good way to become acquainted with non-classical states of the radiation field is to study the properties of a beam splitter [7]. Let us suppose the incoming light field is in a coherent state showing Poisson-type photon number fluctuations $n = \langle n \rangle + \Delta n$ with $\langle \Delta n^2 \rangle = \langle n \rangle$. If the beam splitter divides the incoming beam equally without introducing any noise, the mean photon number and the fluctuations in the output beams would each be divided by 2, $\langle n \rangle/2 + \Delta n'$, with $\langle \Delta n'^2 \rangle^{1/2} = 1/2 \langle n \rangle^{1/2}$. However, what one observes are Poisson-type fluctuations also in the output beams, $\langle \Delta n'^2 \rangle^{1/2} = (\langle n \rangle/2)^{1/2}$, the increase by a factor $2^{1/2}$ shows that the beam splitter introduces noise. The textbook picture is that a photon cannot be divided. Instead, the beam splitter sends the photon one way or the other with equal probability.

The discussion in terms of photons is not appropriate for squeezed states since they are defined in terms of the field amplitude. In 1980

Caves introduced the corresponding picture [8]. A field amplitude $\langle a \rangle + \Delta a$ impinges on the beam splitter. Since the amplitude is not quantized the mean amplitude $\langle a \rangle$ and the fluctuation Δa are both divided by $2^{1/2}$. (The beam splitter is 50 % for intensity). The additional noise that has to be there comes through the second usually not used input port of the beam splitter. Even if that mode of the field is not excited it will carry zero point fluctuations which interfere with the field amplitude from the main input port. This vacuum field has a large effect (it is the factor $2^{1/2}$ in this case) since the size of the zero point field fluctuations is just the same as the one of the fluctuations in a coherent state, as stated above. In the quantum treatment of a beam splitter this additional noise has always been accounted for, by writing the field operator for the output mode $\hat{b}$ as a superposition of the input field operator $\hat{a}$ and a vacuum operator $\hat{c}$, $\hat{b} = 2^{-1/2}(\hat{a} + \hat{c})$ [9]. Caves [8] deserves credit for identifying this vacuum operator with the second input port and for pointing out the consequences. The amplitude picture has the advantage that it is more general. It implies that modifying the fluctuations in the second input port using e.g. squeezed light may reduce the photon fluctuations in the output beams.

4. GENERATION OF SQUEEZED LIGHT

Minimum uncertainty states with non-equal fluctuations in the two quadratures have been produced through various effects of non-linear optics, like four wave mixing [3,10,11], second harmonic generation [12] and parametric amplification [13]. The best results so far have been obtained by Kimble and his team in Austin. In one of their experiments [14] they frequency double the light of a single frequency stabilized Nd:YAG ring laser and use it to pump an optical parametric oscillator (OPO) just below threshold. The OPO-resonator has to be resonant also at the down converted frequency. The fundamental radiation of the Nd:YAG-laser is used as a local oscillator for the phase sensitive detection of the squeezed light (60%) at the output of the OPO.

Recently Heidmann et al. [15] have observed quantum correlations between the beams of a two mode OPO operating above threshold. The measured power fluctuations were 30% below the shot noise level.

5. Nd:YAG LASER

We have set up a frequency stabilized cw Nd:YAG ring laser which is aimed not only at squeezed state generation. The flash lamp pumped laser is also being tested with respect to its potential use as an alternative master laser for a large scale laser interferometer for gravitational wave detection. The laser resonator was designed to be stable for any value of the focal length of the pump power induced thermal lens in the Nd:YAG rod (Fig. 1) [16]. All mirrors have high reflectivities. The amount of light power coupled out at the polarization dependent mirror can be optimized by adjusting the half wave plate.

In an attempt to reduce mechanical vibrations of the resonator the

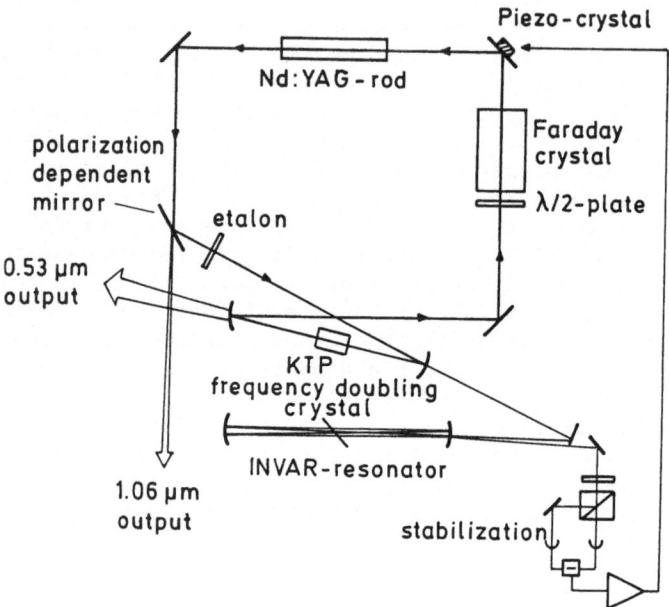

Fig. 1 Sketch of the continous wave Nd:YAG ring laser with intracavity frequency doubling and frequency stabilization.

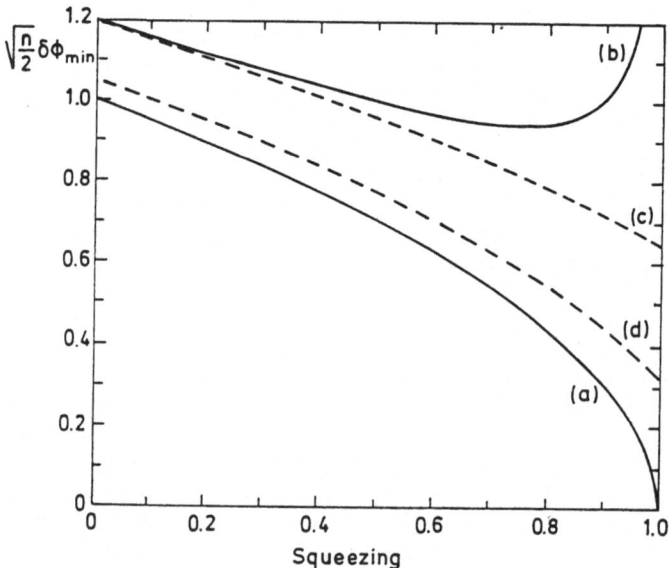

Fig. 2 Smallest observable phase change $\delta\Phi_{min}$ normalized to $(2/n)^{1/2}$ as a function of squeezing. See text for details.

Nd:YAG laser head and the rest of the optics is mounted on separate
plates. The error signal for the frequency stabilization is obtained
using the Hänsch–Couillaud scheme [17]. The linewidth of the stabilized laser was measured to be 10 kHz for an averaging time of 10^{-2} s.
Preliminary measurements of the spectral density of the frequency fluctuations gave unstabilized 10^4 Hz/Hz$^{1/2}$ near 1 kHz and stabilized
100 Hz/Hz$^{1/2}$ falling off to less than 1 Hz/Hz$^{1/2}$ above 10^5 Hz.

Operating the Nd:YAG laser head in a linear cavity the TEM$_{00}$ mul-
tilongitudinal mode output power is 10 W. With the ring laser set up
we have obtained up to 7 W of single mode radiation. A KTP (KTiOPO$_4$)
frequency doubling crystal has been placed inside the resonator at
the focus (diameter 0.1 mm). With proper temperature tuning of the
crystal a second harmonic output power of up to 2 W has been obtained,
it is routinely operated at a level of 0.5 W.

6. STRAIN SENSITIVITY OF A MICHELSON INTERFEROMETER

The precision with which a displacement $\delta\ell$ can be measured with an
interferometer depends on the precision $\delta\phi$ with which the phase of
the interference signal can be determined,

$$\delta\ell = \lambda\delta\Phi/4\pi \qquad (4)$$

If all other noise sources have been overcome the smallest observable
phase change is determined by the shot noise limit [18]

$$\delta\Phi_{min} = n_{out}^{-1/2} \qquad (5)$$

where n_{out} is the number of photons leaving the interferometer at a
maximum of interference during a sampling time interval. Eq. 5 only
holds as long as the laser power is not so high, that photon pressure
fluctuations on the mirrors are the dominant noise source [8,19]. Suppose the mirror reflectivities R are not perfect, R≠1, then the displacement sensitivity can be increased e.g. by using an optical delay
line with many (N-1) reflections in each arm of the Michelson interferometer increasing the effective arm length [18]. The formula then
changes to

$$\delta\ell = \lambda/[2\pi N(n_{in}R^{N-1})^{1/2}] \qquad (6)$$

where n_{in} is the number of photons per sampling time interval in the
incoming laser beam. The highest displacement sensitivity is obtained
by optimizing N resulting in N = 2/ln (1/R). For R close to unity one
finds

$$\delta\ell_{opt} \approx \lambda e(1-R)/(4\pi n_{in}^{1/2}) \qquad (7)$$

An interferometer operating in this way is loss limited. Only about
13.5 % (= e^{-2}) of the input laser power is put out at a maximum of
interference. With such high losses little can be gained using recycling
or squeezing [20].

However, in order to achieve this sensitivity optimized for a given mirror reflectivity the photons are stored in each arm for a time

$$t = (\ell/c) \quad (2/\ln (1/R)) \tag{8}$$

where ℓ is the armlength. For presently available mirrors (R = 0.9999) and the armlength planned for interferometric gravitational wave antennae of $\ell \approx 3$ km [21] this storage time is 0.2 s, much too long to detect gravitational wave signals around 1 KHz.

Therefore, part of this sensitivity has to be sacrificed. A shorter time response Δt of the interferometer is obtained if the number of light beams in each arm is reduced from the optimized loss limited case to

$$N = \Delta t \; c/\ell.$$

The reduced displacement sensitivity can be calculated from Eq. 6. Since the overall losses have also been reduced, it is now possible to apply both the recycling scheme and squeezed light. If the interferometer is stabilized to a minimum of interference at one output port there will be a maximum of interference at the other output port. This light, usually dumped, can be recycled giving rise to a power enhancement factor of $(1-R_{eff})^{-1}$ [22,23], where R_{eff} is given by the effective transmission of the interferometer R^{N-1}.

In addition, squeezed light may be sent through the normally not used input port of the beam splitter inside the interferometer. Ideally, for optimum squeezing the shot noise limited displacement sensitivity is lowered by the factor $(1-R_{eff})^{1/2}$ [8]. In practice, however, there are limitations to the gain in sensitivity that will be discussed in the following section. Combining the above factors for the recycling and squeezing technique, the displacement sensitivity becomes

$$\delta\ell_{rec\&squ} = \lambda(1-R^{N-1})/\left[2\pi N(n_{in}R^{N-1})^{1/2} \right] \tag{9}$$

In the limit of high mirror reflectivity R close to unity and not too large number of reflections, R^{N-1} can be approximated by $1-(N-1)(1-R)$ and the displacement sensitivity becomes independent of N,

$$\delta\ell_{rec\&squ} \approx \lambda(1-R)/(2\pi n_{in}^{1/2}) \tag{10}$$

This is essentially the same as the sensitivity obtained in the loss limited case (Eq. 7). In summary the highest displacement sensitivity of an interferometer is reached by operating in the loss limited regime. If both recycling and squeezing are employed the same sensitivity can in principle be reached even for a time response of the interferometer which is substantially higher than in the loss limited case.

7. APPLYING SQUEEZED STATES TO A REAL INTERFEROMETER

As discussed above squeezed states of light may be used to modify the shot noise limit of a Michelson interferometer when such light is coupled in through the normally not used input port of the beam splitter that divides the laser beam into the two interferometer arms. Owing to the nature of the interference at the output of the interferometer this modification may not only be on the $2^{-1/2}$ level as for the isolated beam splitter but it may reach 100 % [8,24].

As already stated in Sec. 3, any linear attenuation admixes vacuum fluctuations from other modes. Consequently, the sensitivity that can be gained using squeezed light is limited by any loss in the interferometer optics and even by a non-perfect detector quantum efficiency. If all these losses are described by an overall transmission coefficient η the shot noise limit for perfect squeezing (Eq. 5) is not changed to zero but to [8,19]

$$\delta\Phi_{min,squ} = \left[(1-\eta)/n_{out}\right]^{1/2} \tag{11}$$

In any real interferometer there are also other imperfections like wavefront distortions that lead e.g. to a fringe visibility smaller than unity i.e. a non-zero minimum of interference. As long as the goal is to reach the shot noise limit, but not to use squeezing or recycling, the fringe visibility is not too much of a concern. When using e.g. the phase modulation technique to stabilize the interferometer to a minimum of interference the shot noise can still be reached by chosing a large enough modulation index, provided the visibility is not exceedingly low [25].

For squeezing and recycling a non-unity fringe visibility is much more severe. To get a feeling for the size of this effect the appearance of a non-zero minimum of interference was modelled by assuming nonequal losses in the two arms of the interferometer [26]. When using the phase modulation technique for the stabilization of the interferometer the signal at the output arises from the interference of the remainders of the carrier with the modulation sidebands. Noise terms that appear at the same modulation frequency f_{mod} come from interference of these modulation sidebands with phaselike noise not only at the carrier frequency f_0 but also at twice the modulation frequency away from the carrier $f_0 + 2 f_{mod}$. Consequently, multimode or broad band squeezing has to be used to modify the shot noise limit [27]. Fig. 2 shows the scaled shot noise limit as a function of squeezing i.e. of the departure of the mean squared field fluctuations at the second input port in quadrature to the main laser beam from the corresponding value for the coherent state. Here the value 0.6 e.g. corresponds to 60 % squeezing in the power of the fluctuations. The ideal case is represented by curve (a).

In a recent experiment Xiao et al. [28] have nicely demonstrated a reduction of the sensitivity of a rigid Mach-Zehnder interferometer to a level 3 dB below the shot noise using squeezed light generated by optical parametric down conversion.

For the 30 m prototype laser interferometer for gravitational wave detection at Garching [22], the fringe visibility for one hundred bounces has been as good as 0.99. Curve (b) in Fig. 2 shows the modification of the shot noise limit when using broad band squeezing as predicted by the non-equal loss model for a visibility of 0.995. The increase at larger squeezing is caused by the fact that for non-equal losses the carrier does not cancel exactly at an interference minimum giving rise to photon noise terms coming from in-phase fluctuations at the first side bands interfering with the not exactly cancelled carrier.

Squeezing the in-quadrature fluctuations broad band means to squeeze them at f_0, $f_0 - 2f_{mod}$ and also at $f_0 - f_{mod}$, this goes hand in hand with a corresponding increase of the in-phase fluctuations at these frequencies. The increase of curve (b) can be avoided when either filtering out the squeezed light at $f_0 - f_{mod}$ before sending it into the second input port of the interferometer or when chosing a proper scheme for generating the squeezed light [29]. In this case one finds a modification of the shot noise given by curve (c).

At present we have extended the model to account for the finite fringe visibility not by unrealistically high non-equal losses in the two interferometer arms but by coupling part of the light in one arm to other spatial modes of the field thus modelling wavefront distortions more realistically [30]. First results indicate that there is no increase as in curve (b) one rather finds a behaviour similar to curve (c) without the necessity to modify the broad band squeezing. In any case the sensitivity is enhanced by a factor of two which equivalently could be achieved by a fourfold increase in laser power. All these numbers are based on a fringe visibility which is close to the one realized for the Garching prototype interferometer. If this visibility could be increased to 0.9998 corresponding to an interference minimum of 0.04 % the shot noise would be modified according to curve (d) of Fig. 5. The limiting value of $1-\eta$ taking into account equal losses of only 5 % is almost reached resulting in a sensitivity gain corresponding to a ten times higher laser power. This means that as long as the detector efficiency is not larger than 0.95 it does not help to improve the visibility much beyond 0.9998. The applicability of recycling is similarly affected by the fringe visibility.

In conclusion squeezed states of the radiation field will play an important role whenever an optical measurement is limited by the photon shot noise. Although there are practical limitations, the amount of sensitivity that can be gained is promising especially for the large scale laser interferometers planned for gravitational wave detection.

ACKNOWLEDGEMENT. It is a pleasure to acknowledge many helpful discussions with Marlan O. Scully. We greatly appreciate the effective text editing by Maria Schreiber.

References

1. D.F. Walls, Nature 306, 141 (1983)
2. R. Loudon, The Quantum Theory of Light, p.134, Clarendon, Oxford 1983
3. R.E. Slusher, L.W. Hollberg, B. Yurke, J.C. Mertz and J.F. Valley, Phys. Rev. Lett. 55, 2409 (1985)
4. W. Schleich and J.A. Wheeler, Nature 326, 574 (1987)
5. F. Diedrich and H. Walther, Phys. Rev. Lett. 58, 203 (1987)
6. G. Leuchs, in Frontiers of Nonequilibrium Statistical Physics, ed. by G.T. Moore and M.O. Scully, p.329, Plenum 1986

7. W. Winkler, G. Wagner and G. Leuchs in Fundamentals of Quantum Optics II, ed. by F. Ehlotzky, Springer Verlag Berlin, Heidelberg 1987

8. C.M. Caves, Phys. Rev. Lett. 45, 75 (1980); Phys. Rev. D 23, 1693 (1981)

9. p.244 in ref 2; see also S. Prasad, M.O. Scully and W. Martienssen, Opt. Comm. 62, 139 (1987)

10. M.D. Levenson, R.M. Shelby, M. Reid and D.F. Walls, Phys. Rev. Lett. 57, 2473 (1986)

11. M.G. Raizen, L.A. Orozco, M. Xiao, T.L. Boyd and H.J. Kimble, Phys. Rev. Lett. 59, 198 (1987)

12. H.J. Kimble and J.L. Hall, in Proc. of Int. Conf. on Quantum Optics, Hamilton, New Zealand, ed. by J. Harvey and D.F. Walls, Springer Verlag, Heidelberg 1986

13. L.-A. Wu, H.J. Kimble, J.L. Hall and H. Wu, Phys. Rev. Lett. 57, 2520 (1986)

14. L.-A. Wu, M. Xiao and H.J. Kimble, J. Opt. Soc. Am. B, October 1987

15. A. Heidmann, R.J. Horowicz, S. Reynaud, E. Giacobino, C. Fabre and G. Camy, Phys. Rev. Lett., to be published

16. M.H. Muendel, G. Wagner and G. Leuchs, to be published

17. T.W. Hänsch and B. Couillaud, Opt. Comm. 35, 441 (1980)

18. W. Winkler in Experimental Gravitation, Atti Dei Convegni Lincei 34, Accademia Nazionale Dei Lincei, Rome (1977)

19. R. Loudon, Phys. Rev. Lett. 47, 815 (1981); M. Ley and R. Loudon, J. Mod. Opt. 34, 227 (1987)

20. J. Gea-Banacloche, Phys. Rev. A 35, 2518 (1987)

21. See J. Hough, report on the round table discussion on interferometric gravitational wave detectors, this volume

22. K. Maischberger, A. Rüdiger, R. Schilling, L. Schnupp, D. Shoemaker and W. Winkler, Max-Planck-Institut für Quantenoptik report No. 96 (1985)

23. R.W.P. Drever and coworkers, in Quantum Optics, Experimental Gravitation and Measurement Theory, ed. by P. Meystre and M.O. Scully, Plenum Press New York 1983

24. M.I. Kolobov and I.V. Sokolov, Zh. Eksp. Teor. Fiz. 90, 1889 (1986), Sov. Phys. JETP 63, 1105 (1986)

25. D. Shoemaker, R. Schilling, L. Schnupp, W. Winkler, K. Maischberger and A. Rüdiger, "Noise behaviour of the Garching 30 m prototype gravitational wave detector", in preparation

26. J. Gea-Banacloche and G. Leuchs, J. Opt. Soc. Am. B, October 1987

27. J. Gea-Banacloche and G. Leuchs, J. Mod. Opt. 1987 special issue on squeezed light

28. M. Xiao, L.-A. Wu and H.J. Kimble, Phys. Rev. Lett. 59, 278 (1987)

29. B. Yurke, Phys. Rev. A 32, 300 (1985)

30. J. Gea-Banacloche, priv. commun.

Part 4

METHODS OF DATA ANALYSIS IN GRAVITATIONAL WAVE DETECTORS

ROUND TABLE DISCUSSION – GRAVITATIONAL WAVE DETECTORS

Main Participants
P. Bonifazi, Universita di Roma; D. Blair, University of
Western Australia; A. di Virgilio, INFN Pisa; W. Johnson,
Louisiana State University; L. Schnupp, MPI für Quanten-optik
Garching; J.C. Livas, MIT; C.N. Man, GROG/CNRS Orsay; B.J.
Meers, University of Glasgow; T.R. Stevenson, Stanford
University; J. Weber, University of Maryland; M E Zucker,
Caltech
Chairman: J. Hough, University of Glasgow

INTRODUCTION

This has been an exciting year for the people working on detector
development. There have been major advances in several areas related
to separated mass detectors using laser interferometry and the first
concrete sign of a long baseline detector being funded has appeared in
response to a proposal by the Orsay and Pisa groups to INFN in Italy.
Bar detectors have continued to develop, with the possibility of having
several bars operating in coincidence over a large part of the earth's
surface coming much closer. Analysis of the coincidence experiment
between the Stanford detector, the Rome detector (at CERN) and the
detector in Louisiana has yielded some tantalising results which,
however, will probably have to wait for future experiments to improve
the experimental statistics.

LASER INTERFEROMETER DETECTORS

Development work on and for these is spread out over laboratories at
Caltech, Glasgow, Max Planck Garching, MIT, Orsay and Pisa and the
emphasis in each laboratory tends to be complementary to that of the
others. Two different optical arrangements for measuring the phase
change of light in the arms of the interferometer, as caused by
incident gravitational radiation, are being pursued – optical delay
line systems or Fabry Perot cavities in the arms, and both are giving
most encouraging results.

Most development work on delay line systems has been carried out at the
Max Planck Institute in Garching and the operating sensitivity of this
apparatus which has arms 30m long is shown in fig. 1. It should be
noted that the performance of the apparatus is approaching that of the

B. F. Schutz (ed.), Gravitational Wave Data Analysis, 147–152.
© *1989 by Kluwer Academic Publishers.*

Stanford low temperature bar detector, a considerable technological triumph given the difficulties of developing the precision optical techniques involved. This boosts the confidence of the workers in the field that unprecedented sensitivity will be obtained when increased arm lengths are available. Recent work at Max Planck has mainly been devoted to experiments relevant to long baseline detectors. There are worries about whether the large mirrors required can be of high enough quality and Max Planck are testing whether an arrangement using a number of smaller mirrors can be used. Initial experiments seem encouraging but may suggest that tighter control of mirror pointing accuracy than previously used is required. Tests have also been carried out at Max Planck and independently at Orsay on the use of optical recycling techniques to increase the effective laser power in an interferometer. In both laboratories the systems were Michelson interferometers of short arm length. In the experiment at Max Planck where the optical components of the interferometer were individually suspended as pendulums a multiplication of incident light power by a factor of 10 was achieved and photon noise limited sensitivity was observed with 1W of stored light in the interferometer. Performance at high light level was degraded somewhat probably due to power dissipation effects in the electro optic modulators in the arms of the Michelson. Similar results using rigidly fixed mirrors were achieved at Orsay where the advantage of removing the modulators from the arms and using a sidelocking scheme (see later) as proposed by Drever at Caltech were clearly noted.

Multiple beam optical delay lines are also being developed at MIT where current emphasis is on a new prototype detector of 5m arm length. The possibility of using high power YAG lasers at 1.06μ to replace the Argon lasers is now being investigated at MIT and at Orsay. A reasonable goal on a 2 years time scale is 100W of single frequency power. A highly stabilised diode pumped low power master oscillator, to which the large laser may be injection locked has been realised in Orsay.

Suspension systems for the test mass of interferometers of novel design to allow operation of detectors down to 10Hz are being developed at Pisa. These are seven stage systems using 'gas' springs to give isolation to a 400kg mass in 3 dimensions and initial experiment on 2 stages of such a device have given encouraging results.

Work on detectors with Fabry Perot cavities in the arms is concentrated at Glasgow, Caltech and Orsay. In the Glasgow and now the Caltech detector the cavities are formed between ultra low loss mirrors optically contacted on to suspended test masses of fused silica. The use of optical contacting has greatly improved the performance of the Glasgow detector of 10m baseline where a sensitivity to displacement of $\sim 1.2 \times 10^{-18} m/\sqrt{Hz}$ has recently been achieved over a 1kHz bandwidth between 1.5kHz and 2.5kHz (fig. 2). This is the best displacement sensitivity of any interferometer detector and corresponds to a strain sensitivity close to that of the larger delay line detector at the Max

Planck Institute. Previous Glasgow work has clearly demonstrated an excess noise source due to other bonding techniques of mirrors to test masses and the need for further work on this is clearly indicated. The limitation to performance of the Glasgow detector now appears to be related to the presence of additional modes in the illuminating laser perhaps due to the intra cavity Pockels cell used, and work is underway to develop a laser frequency stabilising scheme in which pzt driven mirrors are used to control the frequency of the laser and the intra cavity Pockels cell is removed. Such an arrangement has an intrinsically narrower bandwidth of operation and the necessary stabilising loop gain can best be achieved by the implementation of a 2 loop scheme in which the laser is first stabilised to a reference cavity and then the laser/cavity combination is stabilised to one arm of the interferometer. Such a 2 loop scheme (which however retains the laser intra cavity Pockels cell) has been implemented at the Max Planck Institute. The performance of the detector at Caltech with 40m arms is also improving (fig. 3) and vigorous investigation of noise sources is underway. Work on reducing the coupling of seismic noise by means of a low power optical interferometer arrangement to lock the suspension points of the masses together at low frequency is also being carried out, as is the development of an electrostatic system to control the length of the cavities.

In both the Glasgow and Caltech detectors the light beams from the cavities in the arms are not yet being interfered together but this has to be tackled if recycling techniques are to be demonstrated for this type of detector. To make progress in this direction a system with rigidly connected masses and Fabry Perot cavities between them has been set up at Orsay. The beams reflected from the cavities are interfered together in a Michelson arrangement and the performance obtained is essentially photon noise limited as shown in fig. 4.

To summarise, exciting progress is being made on all aspects of detectors using laser interferometry and this is being greatly helped by the high degree of international cooperation towards the development of a number of long baseline instruments.

LOW TEMPERATURE BAR DETECTORS

The progress of low temperature bar detectors and the results obtained with them have been outlined in several talks given at this meeting. 3 detectors have been in serious operation, those at Stanford, Lousiana and CERN (University of Rome) and two others are close to operation (Maryland and Western Australia).

4 of the detectors use aluminium bars of mass in the range 2240kg – 4800kg at 4K and one detector uses a 1500kg Niobium bar at the same temperature (W Australia). Resonant frequencies are close to 900Hz and bandwidths are of the order of a Hz.

150

Transducer/amplifier systems are different on the various detectors, inductive transducers with SQUID amplifiers being used on the bars at Louisiana and Stanford, a capacitive transducer and SQUID being connected to the Rome detector and a superconducting parametric system being used on the Australian detector.

A coincidence run was carried out between the instruments at Stanford, Louisiana and CERN (University of Rome) over a 100 day period during April to July 1986 and an overlap time of 877 hours was achieved between the Stanford and Rome instruments. The corresponding times for Stanford Lousiana and Rome Lousiana were 145 hours and 45 hours respectively. In the first and last cases searches for pulses were made at levels of h > 2.5×10^{-18} and in the other instance at h > 5×10^{-18}. Analysis of the data using different thresholds and normalising for the directional reponse of the detectors assuming sources at the galactic centre produced two coincident events with a probability of 0.3×10^{-2} above a level of 1.25×10^{-17} between the Stanford and Rome detectors. It is difficult to interpret this result and further searches are required to improve the statistics.

Various improvements are underway on the different detectors. These include removing the refrigerator on the Stanford instrument which has been a source of excess noise, and improving its transducer, and improving the vibration isolation and helium storage capacity of the Louisiana detector.

At Maryland development work on the SQUID amplifier is underway and operation of the detector is expected soon. The Perth (Western Australia) instrument is currently in a cooled state but excess noise in its fundamental mode is currently being experienced. Development of the parametric superconducting transducer, which operates at 10 GHz, has uncovered a number of problems including the possibility of parametric instability but these have either been cured or are well in hand. New detectors of improved sensitivity and wider bandwidth are planned at Stanford and Rome. Construction of the second Stanford instrument has already begun and design work on a new 3 mode transducer, and other aspects such as the SQUID amplifier and mechanical isolation, is currently underway. This detector is predicted to have a sensitivity of 3×10^{-20} and a bandwidth of several tens of Hz.

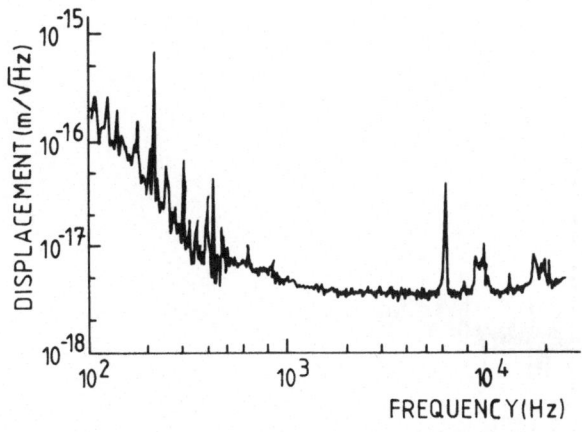

Feb. 1986
90 beams. 30m Ffo
70mA ~ 210 mWatt
I min. = 2%
SHN= 2·6 * 10 ** -18 m/rHz
h = 2·7 * 10 ** -18
    for ΔF = 1KHz

FIG. 1. DISPLACEMENT SENSITIVITY SPECTRUM—GARCHING DETECTOR

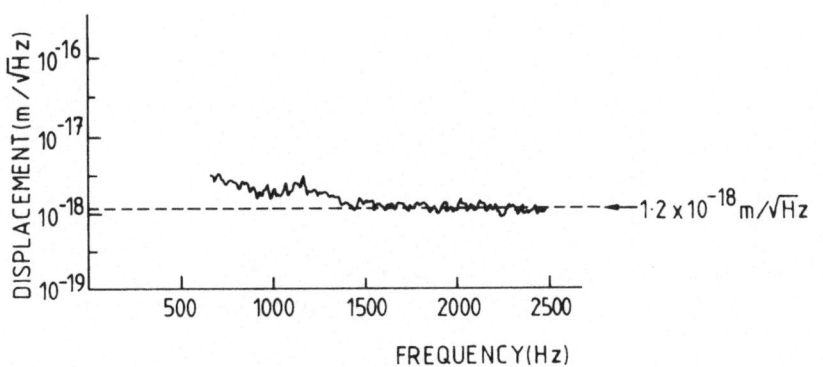

FIG 2. DISPLACEMENT SENSITIVITY SPECTRUM—GLASGOW DETECTOR.

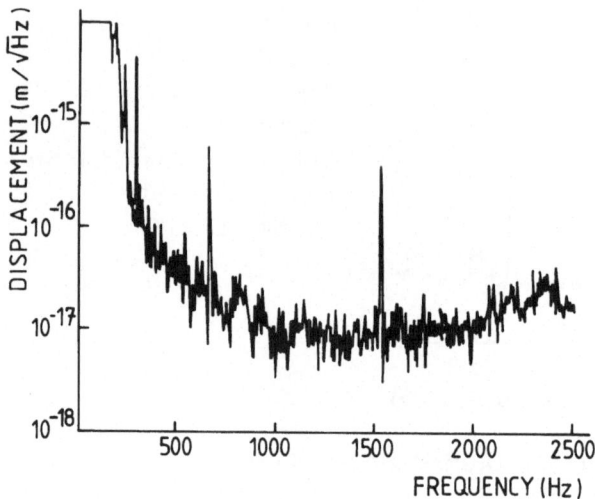

FIG. 3. DISPLACEMENT SENSITIVITY SPECTRUM-CALTECH DETECTOR.

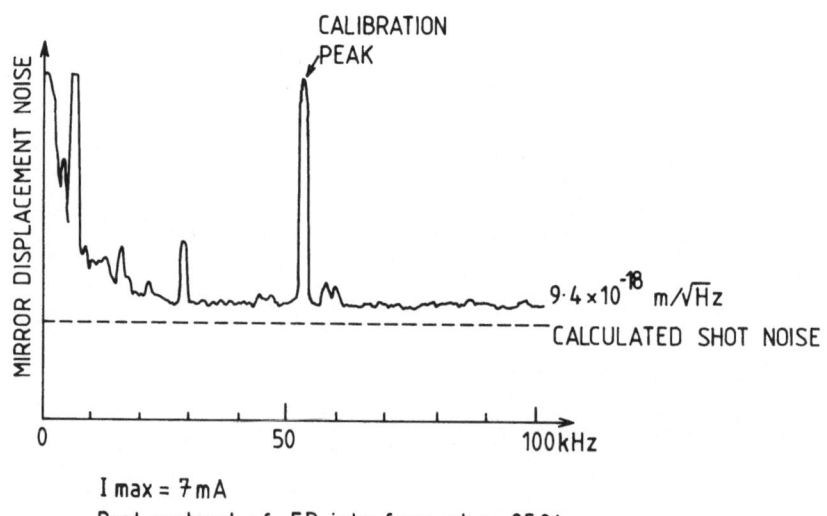

I max = 7 mA
Best contrast of FP interferometer: 95 %
Finesse 130

FIG. 4 DISPLACEMENT NOISE SPECTRUM—ORSAY
FABRY PEROT CAVITIES IN ARMS OF MICHELSON INTERFEROMETER

SPACECRAFT GRAVITATIONAL WAVE EXPERIMENTS

J. W. Armstrong
Jet Propulsion Laboratory
Mail Stop 238/737
4800 Oak Grove Dr.
Pasadena, CA 91109 USA

Doppler tracking of spacecraft can be used in broadband (frequencies $\sim 10^{-4}$-10^{-1} Hz) searches for low-frequency gravitational radiation. Although the technology involved is sophisticated, the "raw" sensitivity of current-generation experiments has been insufficient to produce a detection. To extract the maximum information from the Doppler data requires careful consideration of noise statistics, of how the noises map into the Doppler frequency observable, and of the differences between the noise and signal signatures. In this paper, the statistics of the main noise sources in spacecraft Doppler experiments are summarized along with the impulse responses that characterize their transfer to the observable. The differences in the spectral properties of the signal and noises are then used to develop signal processing approaches for three classes of signal: bursts, periodic waves, and stochastic backgrounds. Where appropriate, model probability density functions for the data under hypotheses about signal strength are developed. These procedures are then applied to existing data to demonstrate algorithm performance and limitations. Finally, the sensitivities possible in future-generation spacecraft experiments are discussed.

1. SPACECRAFT EXPERIMENTS

1.1 Response of Doppler Link to Gravitational Radiation

The Doppler frequency time series produced during tracking of a distant spacecraft can be used to search for gravitational radiation having period range ~ 10-10,000 seconds. The earth and the spacecraft act as free test masses and the Doppler tracking measures their relative dimensionless

B. F. Schutz (ed.), Gravitational Wave Data Analysis, 153–172.

velocity $\Delta v/c = \Delta f/f_0$ as a function of time, where Δf is the perturbation in the Doppler frequency and f_0 is the nominal radio frequency of the link. A gravitational wave of strain amplitude h incident on the system causes small perturbations in the tracking record. These perturbations are of order h in $\Delta f/f_0$ and are replicated three times in the Doppler data (Estabrook and Wahlquist 1975). That is, the "signal part" of the observed Doppler time series is the convolution of the gravitational waveform with the function: $[(\mu-1)/2]\ \delta(t) -\mu\ \delta[t-(1+\mu)L/c) +[(1+\mu)/2]\ \delta(t - 2L/c)$, where μ is the cosine of the angle between the earth-spacecraft vector and the gravity wavevector and L is the earth-spacecraft distance. Two of these three "pulses" are separated by the round-trip light time between the spacecraft and the earth while the third occurs at an intermediate time that depends on the wave's angle of arrival. The sum of the Doppler perturbations of the three pulses is zero. Pulses with duration longer than about the one-way light time produce overlapping responses in the tracking record and the net response then cancels to first order. The tracking system thus has a passband to gravitational excitation: the low-frequency band edge is set by pulse cancellation to be about (1/one-way light time), while thermal noise in the radio system limits the high frequency response to about (1/30 seconds). The anticipated strain amplitudes depend on the source generating the wave but are in any case expected to be small (Thorne 1987). The resulting expected perturbations in the Doppler record are thus also small so that careful attention to the noise sources and the way in which the signals are processed is required.

1.2 Noise Sources

Any signal processing approach depends on exploiting <u>differences</u> between characteristics of the signal and the noise. Substantial effort has been invested in understanding sources, strengths, and spectral properties of the principal noise sources. The main noise sources in a spacecraft gravitational wave experiment are briefly summarized in this section. (Noise sources are characterized by their power spectra--Fourier transforms of autocovariance functions--or by Allan variances. The Allan variance, $\sigma_y(\tau)$, of a random process $y(t) = \Delta f(t)/f_0$ as a function of the integration time τ, is defined as $\sigma_y^2(\tau) = 0.5 <[y'(t) - y'(t+\tau)]^2>$, where $y'(t) = \int y(a)\ da$ with the limits of integration t to $t+\tau$. Angle brackets are conceptually ensemble averages and operationally time averages.)

At frequencies higher than about 0.033 Hz, <u>thermal noise</u> dominates. This noise arises from finite signal-to-noise ratio in the Doppler tracking link. The received signal can be thought of as a phasor with additive "in-phase" and "quadrature" noises. These noises cause the observed phasor to jitter in amplitude and phase. The phase noise thus produced is typically the the rms quadrature noise divided by the magnitude of the signal part of the phasor. Since frequency is the derivative of phase, this white phase noise corresponds to a power spectrum of fractional frequency $\Delta f/f_0$ going as (Fourier frequency)^2. The strong frequency dependence causes this "blue" component to dominate at high frequencies. Its level can be reduced by increasing the signal power in the radio link or decreasing the system noise temperature.

At lower frequencies, propagation noise and instrumental instability are important. Propagation noise is caused by phase variations imposed on the radio wave as it traverses irregularities in the transmission media between the earth and spacecraft (troposphere, ionosphere, solar wind). These phase variations ("phase scintillation") map to frequency variations in the Doppler time series. Observations of the level and spectral shape of the charged particle scintillations (ionosphere and solar wind) show that they are the dominant noise source for the current generation (S-band radio link-- 2.3 GHz) experiments (Wahlquist et al. 1977; Armstrong, Woo, and Estabrook 1979; Woo and Armstrong 1979). The level of the plasma scintillation varies with sun-earth-spacecraft (elongation) angle, reaching a broad minimum in the antisolar direction. Expressed as an Allan variance of the fractional frequency deviation at an integration time of 1000 seconds, this component reaches a minimum value of about (3×10^{-15}) (8.4 GHz/radio frequency of link)^2 when the spacecraft is near solar opposition. Simultaneous observations using receivers spaced on an intercontinental baseline show correlation coefficients ~0.7 near zero time lag (Armstrong, Woo, and Estabrook 1979). This indicates significant contributions from both local ionospheric scintillation (uncorrelated over the long baseline) and solar wind plasma scintillation (approximately unity correlated over this baseline). From the observed variation of solar wind scintillations with angular distance of the spacecraft from the sun, that interplanetary plasma scintillation contamination of a gravitational wave experiment done at solar opposition clearly arises mainly from near-earth (less than ~0.5 AU) irregularities. The plasma scintillation data show a "red" spectrum varying approximately as (frequency)$^{-\alpha}$, with α typically 0.6-1.0.

Water vapor fluctuations are the dominant tropospheric noise source at microwave frequencies (Hogg et al. 1981; Resch et al. 1984), although fluctuations in the "dry component" (Shannon et al. 1979) may be important in future experiments. The index of refraction of tropospheric irregularities is independent of radio frequency (at microwave wavelengths), so that the strength of this noise source expressed as Allan variance of the fractional frequency fluctuations is also independent of radio frequency. At ~1000 second integration times at high altitude sites, the observed magnitude of the effect is parts in 10^{15}, with the available data suggesting low levels are more often observed in winter (Armstrong and Sramek, 1982). The power spectrum of tropospheric noise typically shows (frequency)$^{-\alpha}$ dependence, with $\alpha \sim 0.5$.

In addition to propagation noise, there is <u>instrumental instability</u>. Here there are a variety of contributors: clock (timekeeping) noise, instability introduced by signal distribution within the ground station, transmitter and receiver instability, mechanical stability of the antenna (e.g. dish sag changing the phase center of the antenna over time as the elevation angle changes), stability of the spacecraft's transponder, etc. In the current generation experiments (S-band radio links, early- to mid-1980's) clock-like noise is the most important instrumental noise source. ("Clock-like" here means having a spectral response in the data like timekeeping instability: approximately (frequency)$^{-1}$ intrinsic power spectrum with clock-like transfer function to the data, see below.) Considerable engineering effort within NASA's Deep Space Network (DSN) is being directed toward minimizing noise contributions in the ground system. The expected instrumental stability for the Galileo gravity wave experiment (X-band transponder, tracking with NASA's new 34-meter high-efficiency DSN antennas, 1990+) is a few parts in 10^{15} for integration times of 1000 seconds. The aggregate instrumental noise spectrum of the for the Galileo experiment is expected to be approximately (frequency)$^{-\alpha}$ at low frequencies, with $\alpha \sim 1$.

<u>Non-gravitational forces</u> (or things that mimic non-gravitational forces) also act as noise sources. Here things like spacecraft buffeting (e.g., leaking thrusters), irregularities in the spacecraft spin rate (for Doppler measurements using circularly polarized signals), microseismic disturbances, etc. can influence sensitivity. It is expected that in the Galileo era (1990+) the most important of these can be calibrated and removed with engineering telemetry to a level that is less than the propagation and instrumental noise levels.

Finally, <u>systematic errors</u> are always present at some level. These are difficult to quantify and usually influence the sensitivity to different degrees depending on the spectral characteristics of the gravitational wave being sought.

1.3 Transfer Functions of Noise Sources

Noise enters through a "transfer function" which connects the noise excitation to the Doppler observable. The transfer function depends on the type of noise. In general the Doppler system responds differently to noise than it does to gravitational radiation and these differences often suggest signal processing approaches for different classes of signals.

The impulse response for thermal noise and spacecraft buffeting is just $\delta(t)$ (i.e. the spectrum of the noise is just replicated in the Doppler spectrum through convolution with a delta function); the main virtue is that $\delta(t)$ is different from the three-pulse response to gravitational waves.

Propagation noise introduced by a thin screen of irregularities at a distance x from the tracking station perturbs the radio wave on both the uplink and the downlink. For example, a phase advance in an irregular layer advances both the uplink and the downlink radio waves passing through that layer. The result is a perturbation in the observed Doppler series that is the convolution of the noise that would have been produced by one passage through the irregular layer with the function $\delta(t) + \delta[t-2(L-x)/c]$, where L is the earth-spacecraft distance. Viewed in the frequency domain, the power spectrum of the observed perturbation is the power spectrum of the perturbing process multiplied by the square of the Fourier transform of the impulse response: $4\cos^2[\pi f 2(L-x)/c]$. Note that if x=0 (near-earth propagation noise due to troposphere or ionosphere) then $2L/c = T_2 =$ two-way light time, and the modulation of the power spectrum produces nulls at odd multiples of $f=1/(2T_2)$. (This transfer function is also appropriate for microseismic disturbance of the ground station, but the level of seismic noise in our frequency band is apparently too low to be important at our current level of sensitivity (e.g., Fix 1972)

Clock instability noise enters differently. A clock "glitch" is observed immediately as an apparent change in the velocity of the spacecraft at the current time and then again when the glitch is transponded to the receiver at time $2L/c = T_2$ later. Since the Doppler observable is the difference of the currently transmitted and currently received frequencies, the immediate and transponded effects have opposite signs: the convolving function is $\delta(t) - \delta(t-T_2)$. The corresponding modulation of the clock noise power spectrum is $4\sin^2[\pi f T_2]$, having nulls at multiples of $f = 1/T_2$.

In summary, at the level of sensitivity of current-generation experiments, the observed time series can be modeled as:

$y(t) = \Delta f/f_0 =$ gravity waves + propagation noise +
+ clock noise + thermal noise

$$= \{[(\mu-1)/2]\ \delta(t) -\mu\ \delta[t-(1+\mu)L/c) +[(1+\mu)/2]\ \delta(t - 2L/c)\}*g(t) +$$
$$+ \{\delta(t) + \delta[t-2(L-x)/c]\}*\text{propagation}(t) +$$
$$+ \{\delta(t) - \delta(t-2L/c)\}*\text{clock}(t) +$$
$$+ \text{thermal}(t)$$

where "$*$" indicates convolution, $g(t)$ is the scalar gravitational waveform (produced by the tensor wave's interaction with the earth-spacecraft system), and "propagation", "clock" and "thermal" are the noise processes. The problem is to exploit the different temporal correlation structures of the signal and the noises to maximize the probability of detection of $g(t)$ or to put the best upper limits on $g(t)$.

1.4 Pioneer 11 1983 Observations

Signal processing procedures need to be tested on real data, in part to assess "robustness" against the the vagaries of real data that are difficult or impossible to realistically include in simulations (e.g., non-stationary noise, low-level systematics, etc.) A good data set for this purpose was taken in 1983 using the Pioneer 11 spacecraft. The data were taken between days-of-year 133-136 with the spacecraft 12.9 AU from the earth at a solar elongation of about 160 degrees. A hydrogen maser frequency standard provided the time base at each of the three 64-meter antennas (California, Spain, Australia) used in this experiment. About 48.5 hours of data were obtained where the elevation angle of the radio beam of both the uplink and downlink were above 10 degrees; about 38.8 hours of data had the elevation angles above 20 degrees. The raw data--transponded electric field--were recorded at the ground stations on magnetic tape and then shipped to JPL. The phase of the received signal was later estimated in the computer using a digital phase lock loop. The time series of fractional frequency fluctuation was then computed and corrected for the known orbit of the spacecraft. Systematic variation of the frequency with elevation angle (due to the average troposphere--changing elevation causes a change in the length of the raypath that is in the neutral atmosphere and hence a change in the received phase) were estimated and removed. These observations serve as a useful test bed for the signal processing procedures discussed below.

2. SIGNAL PROCESSING APPROACHES

2.1 Sinusoidal Waveforms

Periodic waves, formed by the superposition of one or more sinusoids, are expected from condensed massive objects in binary pairs. A large fraction of the gravitational wave energy may appear in one harmonic of the orbital period (Wahlquist 1987). The prototype signals are thus sine waves. Since the response of the Earth-spacecraft system is linear, a sine wave excitation appears as a sine wave Doppler response (sum of the three phasors associated with each of the three "pulses" of the three-pulse response). The three phasors can in the best case add perfectly, producing a Doppler response that is twice the excitation; perfect destructive interference is however also possible. If the three phasors interfere destructively (for certain directions-of-arrival and frequencies) producing zero response in the observable, reobserving with another spacecraft or a slightly different two-way light time will usually remove this "blindness" to a given wave direction and frequency.

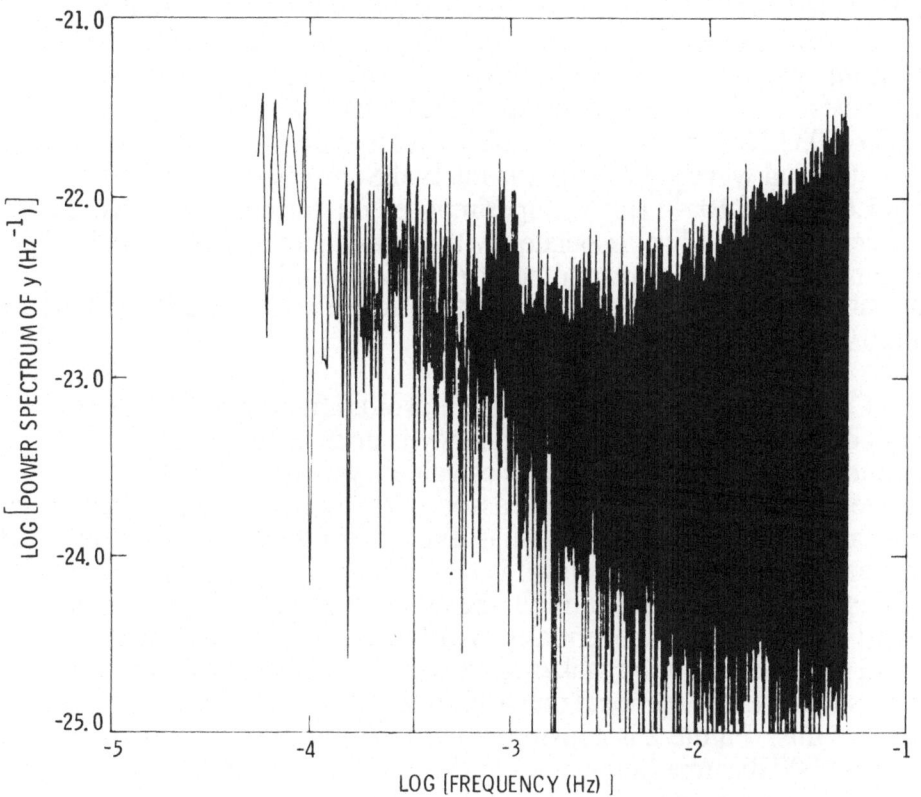

Figure 1. Power spectrum of fractional frequency fluctuations for the 1983 Pioneer 11 data

Previous searches for sinusoidal signals using spacecraft tracking data have been done by Anderson (1977) (spectra of Pioneer 9 and 10 data), Anderson et al. (1984) (looking for gravity waves at the discrete frequency of the claimed gravitational signal from Geminga) and Armstrong, Estabrook, and Wahlquist (1987) (a broadband search, 0.0005-0.033 Hz). This section discusses an approach for the search for sine waves when the frequency of the hypothetical signal is unknown.

Since by assumption we do not know the phase of the signal, it is necessary to form the power spectrum, $S_y(f)$, of the Doppler time series. Figure 1 shows this spectrum for the Pioneer 1983 data formed using standard techniques. This spectrum is two-sided; that is the variance is the integral of the spectrum over both positive and negative frequencies. The main features are "flicker" noise (spectrum roughly proportional to 1/f) at low frequencies due to propagation and clock noises, and thermal noise (spectrum approximately proportional to f^2, rounded at the higher frequencies here by an anti-aliasing digital filter). For these data the interval (0.0005-0.033 Hz) has the lowest noise level and thus the highest sensitivity to gravitational wave signals; in this interval S_y is roughly constant at 1.6 (+1.6,-0.8) X 10^{-23} Hz^{-1}. Superimposed on this continuum would be any discrete spectral line associated with a gravitational wave signal. The spectrum in Figure 1 is unsmoothed (resolution bandwidth, B, is approximately 4μHz) and statistical estimation error is evident (Jenkins and Watts 1969).

Conversion of S_y to equivalent sinusoidal fractional frequency amplitude at a given fluctuation frequency is done as follows: amplitude = $2[B\ S_y(f)]^{1/2}$. This gives values typically 1.5 X 10^{-14}. However, some spectral lines have associated amplitudes larger than this and one is faced with determining at what level any of them should be regarded as statistically significant. This level is determined by the estimation error statistics of the power spectrum. In the absence of a signal, the power spectral estimates are exponentially distributed (Jenkins and Watts 1969). The tail of this distribution makes signals difficult to detect unless the signal power is many times the local mean spectral power of the noise. Figure 2 illustrates this. Figure 2 is a histogram of the ratio of the spectral power at each of ~9000 discrete frequencies in the (0.0005-0.0333 Hz) band, Figure 1, to a local mean power (defined by a 0.5 mHz band centered on each frequency) has been formed. A normalized power of unity in Figure 2 corresponds to a typical strain of 1.5 X 10^{-14}. The distribution expected due to noise alone is exp(-P), the line drawn in Figure 2. (Note that the vertical normalization of the line is not arbitrary; the exponential model distribution and the histogram of the data each have to have unit area.)

The agreement with the "noise only" exp(-P) model is excellent, at least out to normalized spectral powers of about 8. The two features at 10 and 10.9 times the local mean correspond to single occurrences of spectral lines at 3.592 mHz and 4.431 mHz, respectively. Are these significant?

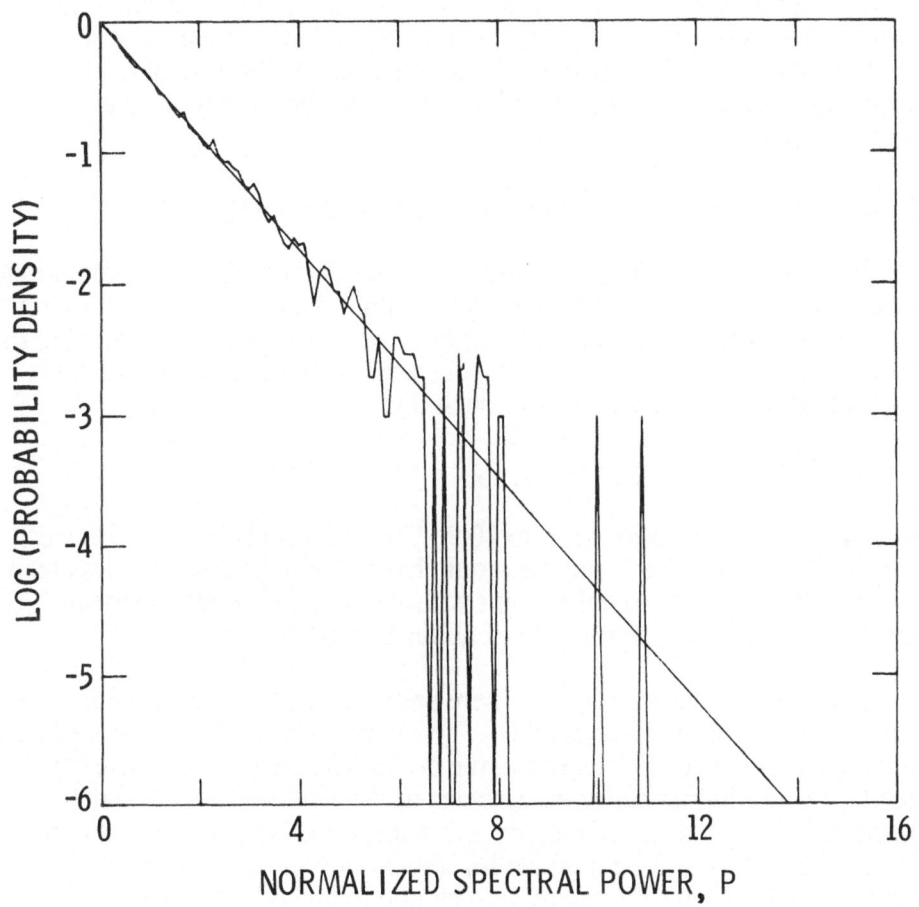

Figure 2: Histogram of normalized spectral powers from Figure 1. Line is expected dependence for estimation error only (see text).

The significance can be assessed at least two ways. First, note that 9000 spectral lines were searched for a signal. Thus the expected number of false alarms at a given threshold, A, is 9000exp(-A) for the "noise only" hypothesis. For A=10.9 (the strongest line), this is 0.17. Thus by chance alone, about 1 realization in 6 of this random process would be expected to produce a spectral line of magnitude 10.9 or greater, a not-too-improbable event. A second approach is to consider the formal probability of being able to hide a signal of amplitude s in the observed histogram, under the assumption that there is at most one signal present. If a signal is present at some given frequency, then the probability density function (pdf) for the spectral power x at that frequency is the Rice-squared distribution: $p(x) = exp[-(x+s^2)] I_0(2 s \sqrt{x})$, where s is the signal amplitude. The joint pdf of the 9000 spectral estimates in the band searched (8999 of which are independent exponential distributions) is then just the product of the individual pdf's:

$$p(x_1,...,x_{9000}) = exp[-(x_1+s^2)] I_0(2 s \sqrt{x_1}) exp(-x_2)...exp(-x_{9000}),$$

where the random variable having the signal has been called x_1 for notational convenience. We want to know the largest value of s that is consistent with the observed histogram (Figure 2): for what value of s will it be true that the probability that all 9000 spectral estimates are less than or equal to 10.9 is (say) less than 10%? We then want s such that

$$[1 - exp(-10.9)]^{8999} \int exp[-(x+s^2)] I_0(2 s \sqrt{x}) dx = 0.1$$

with the limits of integration zero to 10.9. The solution is s = 4.1. Hence there is a 10% chance that a sample would have all spectral powers less than 10.9 if there were a real signal present with level $(4.1)^2$ times the mean spectral power, or strain amplitude of about 6×10^{-14}.

How can this be improved? There are two ways: lower the noise level and reduce the resolution bandwidth. The thermal noise can be reduced by increasing the signal-to-noise ratio in the downlink, the plasma noise can be reduced by using higher radio frequencies in the Doppler link, and the instrumental noise can be reduced by using improved technology. All of these improvements will be in place for the Galileo gravitational wave experiment. The resolution bandwidth can be reduced by tracking the spacecraft for long durations; this is also planned for Galileo. (Long duration observations raise the question of amplitude and phase modulation of the Doppler observable--with subsequent line broadening--as the spacecraft-Earth "apparatus" moves with respect to the source. Calculations indicate that for the Galileo observing geometry and the planned 40 days of

:ontinuous observations this will not be a problem.) The Galileo experiment s expected to be sensitive to sinusoidal radiation at about the 10^{-16} level.

The above has ignored the important question of how upper limits to ractional frequency fluctuations map to upper limits on the gravitational vave amplitude (given that the two are related through angle of arrival, vave frequency, and polarization state--all of which are unknown). This 'equires assumptions about the statistics of wave properties (Armstrong, Estabrook, and Wahlquist 1987; Armstrong 1987)

2.2 Burst Sources

Bursts, waves that are well-defined in time, might be produced at detectable levels in the spacecraft band by, for example, collapse to form a black hole or by collisions of black holes (Thorne 1987). Previous searches for burst signals using (S-band) spacecraft tracking data have typically been done using very short arcs of data near solar opposition, where the solar plasma noise is a minimum. The only published search for bursts was conducted by Hellings et al. (1981) who looked for pulses in 1979 and 1980 using short stretches of Voyager data. Burst searches using a longer 1981 Pioneer 10 data set are in preparation by the JPL group. In this section, a signal processing approach for bursts is presented and applied to the Pioneer 11 1983 data set.

If the signal waveform is known, the signal is embedded in white, zero-mean, gaussian noise, and it is required that the detection be based on a linear operation on the data, then the maximum likelihood criterion leads to "matched filtering" (e.g. Helstrom 1968). The idea is to crosscorrelate the data with a copy of the signal waveform, the result of this correlation is itself a random process that can be thought of as having a "signal part" and a "noise part". In the absence of a signal, the output of the filter is also a zero-mean process, with a variance that depends on the filter weights and the spectrum of the noise. In the presence of a signal, the mean of the output is non-zero (this is the "signal part", proportional to the integral of the square of the waveform, for a perfect match). The observer declares that a signal is present if the absolute value of the filter output exceeds some predetermined level, G_0. Under the above assumptions about the noise G_0 can be formally computed based on the desired false-alarm/false-dismissal probabilities (Helstrom 1968).

In the practical case, the signal waveform is unknown (or perhaps only known as one of a class of candidate waveforms), the noise is not white (there are techniques for dealing with this), and low-level systematics may cause the pdf of the signal to be non-gaussian. Nonetheless, linear operations on the data can improve the signal-to-noise ratio. Let the data be $y(t) = s(t) + n(t)$, where s is the signal (nonzero in the interval $[t_s, t_s+T]$) and n is the noise. Correlation of the data set with a copy of the signal produces an output time series $z(t) = \int s(t') s(t'+t) dt' + \int s(t') n(t'+t) dt' = $ signal part + noise part. For a perfect match at $t = t_s$, the signal part is $\int s^2(t) dt$; the variance of the noise part is independent of t (assuming the statistics of $n(t)$ are stationary) and is given by $Var(z_n) = \int P_n(f) [S(f)]^2 df$, where P_n is the noise power spectrum and $S(f)$ is the Fourier transform of the signal template. A measure of the detectability of the signal is then the SNR = $\{\int s^2(t) dt\}^2 / Var(z_n)$. (If the true signal is $g(t)$ and the assumed signal is $s(t)$, then the degree of mismatch is characterized by the inner product of the two signals: $\int g(t) s(t) dt$.) It is easy to invent situations where the SNR of the filtered output is much larger than the signal-to-noise ratio in the original series $y(t)$ (e.g., pick signal and noise such that $P_n(f)$ and $S(f)$ have little spectral overlap!). However, if the spectral content of the signal and the noise are not too different, it is clear that the SNR improvement cannot be large.

To gain some practical experience with linear operations on the data, the above correlation filter was applied to real data (the 1983 Pioneer 11 data) assuming a rectangle waveform: $s(t) = $ constant if $0 < t < T$ and zero otherwise. This signal was passed through the three pulse response function (assuming angle-of-arrival) to give the signal template, which was correlated with the data. There are thus three parameters varied: arrival angle, arrival time, and burst width. Since the Fourier transform of a rectangle wave is proportional to $\sin(\pi f T)/(\pi f T)$, the asymptotic suppression of the thermal (high frequency) noise contribution is only like f^{-2}. Despite this weak high-frequency suppression, it is clear that these filters are mostly reducing the thermal noise contribution. The expected performance (one standard deviation) can be calculated by multiplying the filter transfer function by the observed power spectrum (i.e., $Var(z_n) = \int P_n(f) [S(f)]^2 df$, with $S(f)$ set by the pulse width and angle of arrival.) This filtering operation was performed on the 1983 data set varying cos(angle of arrival) between (-0.9, 0.9) in steps of of 0.1, time of arrival in steps of 100 seconds, and pulse width in the range (100, 2000 seconds) in steps of 100 sec. (Although the filter outputs will not be independent, e.g. a 2000 sec filter width stepped in 100 sec time-of-arrival increments will have much of the input data in common at adjacent time steps, the statistics of the filtered output are still unbiased). With the constraints that the elevation angle of the radio up- and downlinks be greater than 20 degrees and that there be at most 10% data missing from any filter due to data gaps, this produced slightly less than 750,000 filter outputs. (The computation time to do this was not trivial but neither was it excessive. Using early 1980's minicomputer technology

HP9000/540) and with no attempt to optimize code, these computations took bout one day for this ~three day data set.)

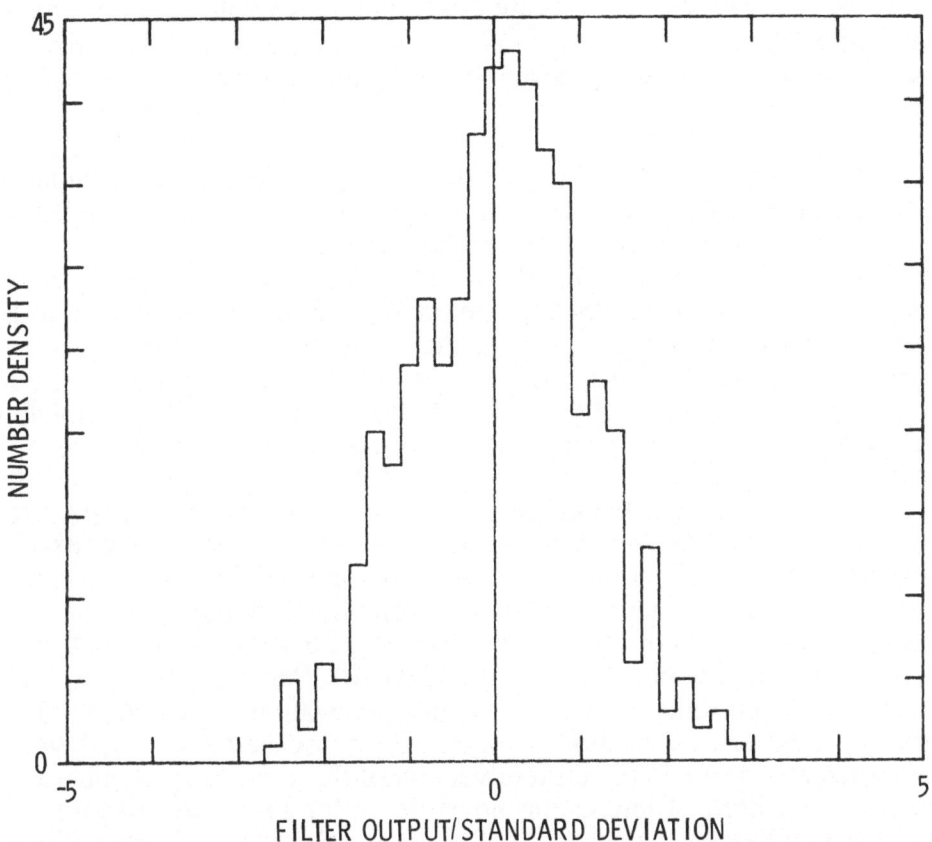

Figure 3. Histogram of filtered outputs for rectangle 'burst waveform' of width 900 seconds and cos(angle-of-arrival) = -0.9. Horizontal axis is scaled in terms of the standard deviation of the filter outputs, 7×10^{-14}.

Comparison of the one standard deviation sensitivity of this procedure calculated formally (square root of $\int P_n(f) [S(f)]^2 \, df$) with the empirical rms of the filter outputs showed excellent agreement. Since the agreement should be mathematically exact in the absence of data gaps, this indicates that data gaps were treated properly. Figure 3 shows an example of the histogram of filtered output for burst width of 900 seconds and cosine(angle of arrival) = -0.9 (close to the direction from the galactic center). Given the actual power spectrum, the calculated one standard deviation level filter output due to noise only for the parameters of this filter is 7×10^{-14}, exactly the same as the observed rms filter output. The observed one standard deviation sensitivity depended on filter parameters, ranging from a low value of 2.5×10^{-14} (burst width 2000 seconds, $\mu = -0.3$) to $\sim 2 \times 10^{-13}$ for the 100 sec burst widths (various μ's). These can be compared with the raw variance in the 0.05 Hz band of $(1.6 \times 10^{-12})^2$. These procedures have "improved SNR" by factors of 10-100 in the amplitude, but mostly by simply integrating the thermal noise.

From the viewpoint of detection the <u>extreme</u> values of filter outputs, rather than their rms outputs, are of main interest; these are the candidate signals. The statistics of the filtered outputs are expected to be approximately gaussian (the Doppler residuals that are the filter input are approximately gaussian and the integration implicit in the filter should make the output closer to gaussian). Because there is no *a priori* prejudice about the polarity of a gravity wave, a large negative filter output is just as "good" as a large positive output. Of the 743,287 filter outputs, 1790 had absolute value greater than three times the standard deviation of the appropriate associated filter, and 211 outputs were greater than 4σ in absolute value. For gaussian statistics of the filter outputs, the expected number of occurrences (two-sided) are 1933 greater than 3σ and 47 greater than 4σ. These "excess" 4σ events were studied in more detail. Analysis showed that the 4σ events came preferentially from a special (non-astronomical) population: the average elevation angle of the 4σ events was significantly lower than that of the general population of filter outputs and the 4σ events preferentially had at least one of their three pulses near the lower elevation angle cutoff (20 degrees). These excess "events" are apparently associated with very low level systematic errors in the current version of the average troposphere correction algorithm. There is thus no evidence for a reliable detection. This again emphasizes the necessity of very careful control of systematics in the data taking if the full power of analysis techniques is to be realized.

2.3 Gravitational Wave Background

A stochastic background of gravitational waves (incoherent fluctuations persisting over a time long compared with the observing time) might be produced in the spacecraft band by a variety of mechanisms (Thorne 1987). Upper limits to the gravitational wave background in this band based on spacecraft data (Hellings et al. (1981), Anderson and Mashoon (1985)) and normal mode data (Boughn and Kuhn (1984)) have been published.

As usual, the goal is to find differences in the signal and the noises that can be exploited to improve detection sensitivity. The signal spectrum results from averaging the three-pulse response function over all angles. The response to a locally white spectrum of background waves (spectral level G^2) is (Estabrook and Wahlquist 1975, equ. 27):

$$P_{bg}(f) = (G^2/2)\{ \ [(\pi f T_2)^2 - 3]/(\pi f T_2)^2 -$$
$$- [(\pi f T_2)^2 + 3] \cos(2\pi f T_2)/3 \ (\pi f T_2)^3 +$$
$$+ [2/(\pi f T_2)^3] \sin(2\pi f T_2)\}$$

For f greater than about $3/(2T_2)$, the term in braces oscillates with maxima of about 1.3 near odd multiples of $1/(2T_2)$ and minima of about 0.65 near multiples of $1/T_2$. That is, the spectral transfer function is similar to that expected from clock modulation.

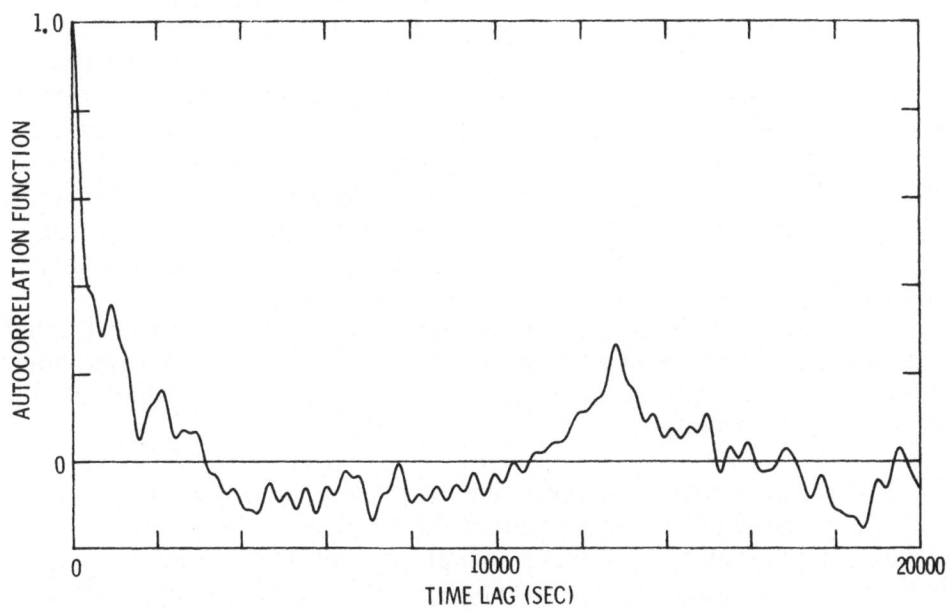

Figure 4. Autocorrelation function of the fractional frequency fluctuations for the 1983 Pioneer 11 data after filtering to remove high-frequency noise component (see text).

This modulation is fortuitous, because it is approximately orthogonal to the modulation of the leading noise source, propagation noise. Formal Wiener filters can be produced to exploit this difference (Armstrong, Estabrook, and Regier 1983), but simple examination of the second-moment descriptions of the data is perhaps more illuminating. Figure 4 shows a plot of the autocorrelation function of the Pioneer 11 1983 data, severely low-pass filtered to isolate the propagation and clock noise processes by removing the high-frequency thermal noise. The peak at time lag 12870 seconds is (within the 10 second resolution of these compressed data) exactly the two way light time (T_2 = 12867 on May 15, 1983). The positive correlation at this lag indicates that propagation noise processes dominates clock noise (clock noise would have given an <u>anti</u>correlation at the two-way light time). Since the correlation at 12870 sec lag is less than 0.5, the clock noise (or gravity wave background!) must be non-zero, however. Figure 5 shows a replot of the low frequency part of the power spectrum shown in Figure 1, this time on an expanded linear-linear scale and smoothed to a resolution bandwidth of about 15μHz. If there were only near-earth propagation noise, we would expect to see $(cosine)^2$ modulation with sharp nulls at odd multiples of $1/2T_2$. If there were only clock noise, we would expect to see $(sine)^2$ modulation with sharp nulls at multiples of $1/T_2$. In the realistic case, we have propagation noise from an <u>extended</u> medium (which smears out the peaks and nulls at the higher frequencies because the spectral modulation functions of various layers are not in phase) plus clock noise at a different spectral level. Superimposed on the spectrum in Figure 5 is a model based on extended medium propagation noise (troposphere, ionosphere, plus solar wind out to 1AU from the earth) having a Kolmogorov spectrum ($f^{-2/3}$) plus clock noise with the same spectral shape and a level 1/5 of the propagation noise. The model reproduces the main features of the observations rather well, including the local minima near $3/(2T_2)$ and $5/(2T_2)$. At these frequencies, the propagation noise is essentially zero--a propagation-limited observation has been converted to an instrumental-noise limited observation. In this case, that conversion corresponds to a factor of 5 or so reduction in spectral level of the noise.

The average spectral power where the plasma noise is a minimum (near $3/(2T_2)$ and $5/(2T_2)$), is the sum of the timekeeping instability spectrum and the gravitational wave background spectrum at these frequencies. Assuming the time-keeping spectrum is zero, an upper limit to the gravitational wave background spectral density at these frequencies is the local average of the observed spectrum. This gives $S_y(f)$ = 1.2 X 10^{-23} Hz^{-1}, or $[f\,S_y(f)]^{1/2}$ ~6 X 10^{-14}. Of course the timekeeping noise is not zero. This observed spectral level is in fact reasonable for the aggregate frequency stability (delivered to the user, averaged over all three tracking complexes) for the Deep Space Network in 1983. If the spectral level of the time-keeping instability could be measured simultaneously and independently (as will be possible in the Galileo-era tracking), the detection sensitivity for a gravitational wave background could be based on the difference in spectral level between the observed spectrum and an independently measured time-keeping spectrum. In this case the sensitivity limit for a gravitational background spectrum would depend on the accuracy with which spectral <u>differences</u> could be measured, i.e. on the estimation error statistics of the spectra. Additional factors of 5 or more improvement over the "raw" spectral level at frequencies where the plasma noise is approximately zero should be easily attainable.

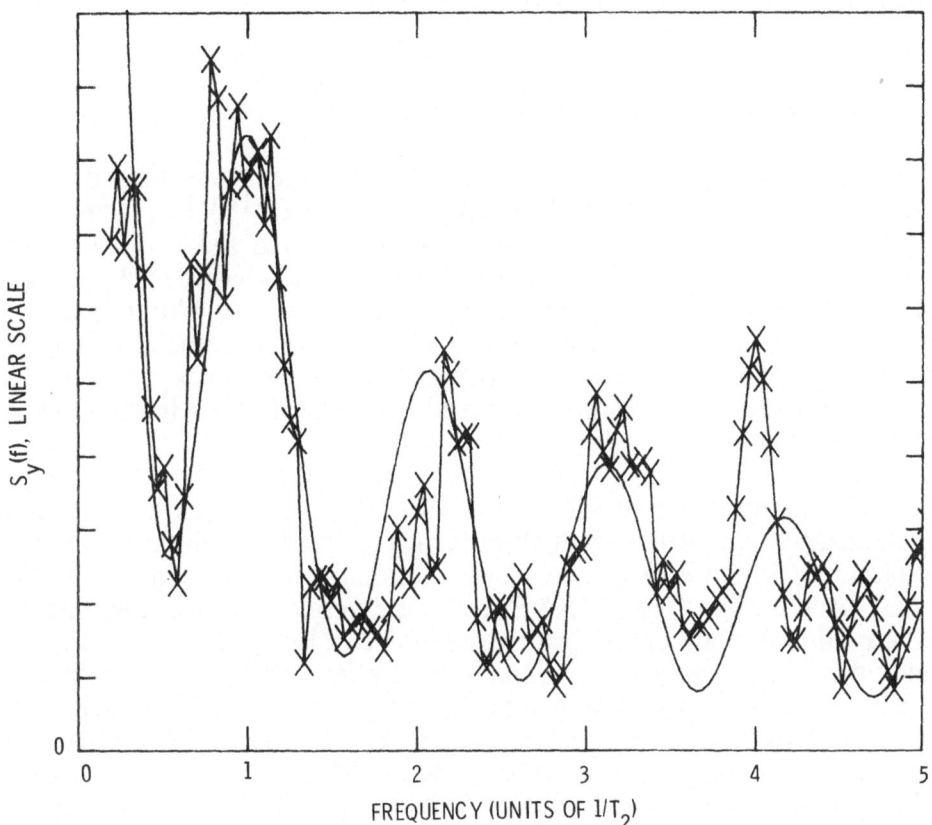

Figure 5. Low-frequency spectrum of fractional frequency fluctuations (from Figure1) on a linear-linear scale. Observed spectrum smoothed to a resolution of about 15 microhertz is shown by crosses connected by straight lines. T_2 = two-way light time between earth and spacecraft. Model calculation assuming extended medium propagation noise plus timekeeping instability noise is shown as the smooth curve. Propagation noise is essentially zero near $3/2T_2$ and $5/2T_2$; at these frequencies the spectrum is instrumental-noise limited and upper limts to a gravitational wave background spectrum can be constructed (see text).

In summary, then, the three-pulse response averaged over arrival angle produces a modulation of the spectrum that is "clock like". This is approximately orthogonal to the modulation of propagation scintillation noise. This difference between the signal and the noise can be exploited to reach a clock-noise-limited upper limit for the gravitational wave background spectrum at selected frequencies. Improvement beyond this level appears to require lower clock noise and/or independent measurements of the timekeeping noise statistics.

3. PROSPECTS FOR FUTURE EXPERIMENTS

The current generation of spacecraft experiments are limited by plasma noise for bursts (at best sensitivity of about 10^{-14}), by plasma and thermal noise plus finite duration observations for sine waves (at best sensitivity of about 10^{-14}), and by clock noise for stochastic backgrounds (at sensitivities $S_y \sim 10^{-23}$ Hz^{-1} for selected frequencies). In addition, all observations to date have been initiated after spacecraft launch, with no opportunity to influence spacecraft design to optimize the experiment. Opportunities for significantly-improved experiments exist in both the near- and intermediate terms.

Plasma noise can be minimized with higher frequency radio links or substantially eliminated with dual frequency up- and downlinks. The next generation (Galileo-era) spacecraft gravitational wave experiments will be with X-band uplink. The NASA Deep Space Network is instrumenting a high-stability X-band uplink ground stations capability; the first of the X-band uplink observations (raw sensitivity of a few times 10^{-15} for bursts, $\sim 10^{-16}$ for periodic waves, and anticipated $S_y \sim 10^{-25}$ Hz^{-1} for backgrounds) will be with the Galileo spacecraft during cruise to Jupiter. The planetary observer series (e.g. Mars Observer), and the Mariner Mark II (e.g., CRAF) series will also have X-band uplink. Although gravitational wave searches are not currently-approved experiments on these missions, these X-band uplink opportunities should be fully exploited.

Looking to the late 1990's, K-band (~32 GHz) uplink or multifrequency link gravitational wave observations can provide very high immunity to plasma noise and very sensitive gravitational wave experiments. (If flight-qualified precision timekeeping becomes practical, then the possibility of onboard Doppler extraction with multiple links offers improved ways to discriminate gravity wave and noise signatures.) To fully exploit this plasma noise immunity would require improved timekeeping on the ground and precision tropospheric monitoring; at these levels there may be significant impact on the spacecraft design (e.g., drag-free systems may be required). A tracking station in orbit, being considered by the Deep Space Network for other reasons, would eliminate many of the noise sources that might ultimately limit a ground-based experiment (e.g., uncorrectable tropospheric scintillation, motion of ground station phase center due to wind and gravitational loading, seismic noise). The utility for gravitational wave detection would depend on the tracking station's orbit. However, an appropriately instrumented spacecraft and orbiting station could push the sensitivity limit below the $\sim 10^{-17}$ level.

Acknowledgments

I thank F. B. Estabrook and H. D. Wahlquist for numerous discussions on all aspects of the spacecraft gravitational wave detection problem. I also thank J. D. Anderson, Principal Investigator of the Pioneer Celestial Mechanics Experiment, under whose auspices the Pioneer 11 data were taken, and the Pioneer Project Office at NASA/Ames Research Center and the NASA/JPL Deep Space Network for their cooperation in gathering those data. This work was supported by the NASA Astronomy/Relativity Office and was performed at the Jet Propulsion Laboratory, California Institute of Technology, under contract with NASA.

References

Anderson, A. J. Atti dei Convengni Lincei, **34**, 235 (1977).

Anderson, J. D., Armstrong, J. W., Estabrook, F. B., Hellings, R. W., Lau, E. K., and Wahlquist, H. D. Nature, **308**, 158 (1984)

Anderson, J. D. and Mashhoon, B. Ap. J., **290**, 445 (1985)

Armstrong, J. W., 'Statistical Response of Spacecraft Tracking Data to Gravitational Radiation: Detectability of Sinusoidal Waves' BAAS, **19**, 642 (1987).

Armstrong, J. W., Woo, R., and Estabrook, F. B. Ap. J., **230**, 570 (1979)

Armstrong, J. W. and Sramek, R. Radio Science, **17**, 1579 (1982)

Armstrong, J. W., Estabrook, F. B., and Regier, L. A. 'Signal Processing of Doppler Data for the Detection of a Very-Low-Frequency Gravitational Wave Background' BAAS, **15**, 673 (1983).

Armstrong, J. W., Estabrook, F. B., and Wahlquist, H. D. Ap. J.**318** ,536 (1987)

Boughn, S. P. and Kuhn, J. R. Ap. J., **286**, 387 (1984)

Estabrook, F. B. and Wahlquist, H. D. GRG **6**, 439. (1975)

Fix, J. E. Bull. Seismological Soc. Am., **62**, 1753 (1972)

Hellings, R. W., Callahan, P. S., Anderson, J. D., and Moffet, A. T.   Phys. Rev. **D23**, 844 (1981)

Hellings, R. W.,. Phys. Rev. **D23**, 832 (1981)

Hogg, D. C., Guiraud, F. O., and Sweezy, W. B. Science, **213**, 1112 (1981).

Jenkins, G. M. and Watts, D. G. Spectral Analysis and Its Applications, (San Francisco: Holden-Day), 1969.

Resch, G. M., Hogg, D. E., and Napier, P. J. Radio Science, **19**, 411 (1984)

Shannon, R. R. Smith, W. S., Metheny, W., Ceccon, C., and Philbrick, R. Science, **206**, 1267 (1979).

Thorne, K. S. 'Gravitational Radiation', in 300 Years of Gravitation, ed. S. W. Hawking (Cambridge University Press: 1987)

Wahlquist, H. D., Anderson, J. D., Estabrook, F. B., and Thorne, K. S. Atti dei Convengni Lincei, **34**, 335 (1977)

Wahlquist, H. D. GRG (1987, in press)

Woo, R. and Armstrong, J. W., JGR, **84**, 7288 (1979)

GRAVITATIONAL WAVE EXPERIMENTS WITH RESONANT ANTENNAS

G. Pizzella
Department of Physics, La Sapienza University,
P.le A.Moro 2, 00185 Rome, Italy
and
Istituto Nazionale di Fisica Nucleare, Sezione di Roma

ABSTRACT. The basic features of the resonant antennas for the search of gravitational waves are reviewed. These antennas detect the Fourier component $H(\omega)$ of the metric tensor perturbation $h(t)$ at the antenna resonance. By means of proper optimum filters, like the Wiener filter, it is possible to reach a sensitivity which is basically limited by the noise of the electronic amplifier. For one cryogenic antenna with mass M=2300 kg and temperature T=20 mK the sensitivity should be of the order of $h_{min} \simeq 5 \times 10^{-21}$. It is shown that the bandwidth of the resonant antennas is much larger than the width of the resonance curve; it can reach values up to 100 Hz if a low noise amplifier is available. Finally the Rome antenna is briefly described. This antenna has a sensitivity of the order of $h_{min} \simeq 10^{-18}$ and has operated for nine months.

1. INTRODUCTION

Gravitational waves can be detected by measuring the very small change η^m of the distance between two test bodies in free fell. The geodesic deviation law gives

$$\frac{d^2\eta^m}{dt^2} = -c^2 R^m_{ono}\xi^n_o \tag{1.1}$$

where R^m_{ono} is the Riemann tensor and ξ^n_o the original distance between the test bodies ($|\xi^m_o| \gg |\eta^m|$).

Considering a gravitational wave travelling along the x axis with polarization $h_\times$ and h_+, putting one test body in the origin of the reference system and the other body at (y,z), eq.(1.1) becomes (indicating with the dot the time derivative)

$$\ddot{\eta}_x = 0$$

$$\ddot{\eta}_y = \frac{1}{2}(y\ddot{h}_+ + z\ddot{h}_\times) \tag{1.2}$$

$$\ddot{\eta}_z = \frac{1}{2}(y\ddot{h}_\times - z\ddot{h}_+)$$

B. F. Schutz (ed.), Gravitational Wave Data Analysis, 173–194.

In order to have the test bodies in free fall one should stay in the outer space; we can think of two space probes or two celestial bodies like Earth and Moon, or two test bodies in a Spacelab environment.

In an Earth based laboratory the two bodies must be supported against their own weight. We can classify these detectors in two cathegories: a) the non resonant detectors, where the two supports for the two bodies are independent one from each other as more as possible, b) the resonant detectors, where the two bodies are linked one to each other by means of an elastic support. In the following we shall consider only the resonant detectors.

2 RESONANT ANTENNA

One way to realize an elastic support is by means of a metallic bar. Since the first pioneer experiment by J.Weber, the bar is usually an alluminium cylinder of lenght L and mass M. The bar must be thought in free fall during the interaction with the g.w., althought it is suspended by a wire across the baricentral section in order to support it against the attraction from the Earth. We take a reference system where the center of mass is at rest. If we consider two mass elements dm of the bar located along the cylinder axis symmetrically with respect to the center of mass we notice that their distance tends to change according to eq.(1.2) by effect of the g.w. h_+ and $h_\times$. Elastic and dissipative forces also act on the two mass elements due to the surrounding material. As a consequence the bar tends to dilate and contract and elastic waves propagate in it. For a thin bar these are longitudinal waves and move with velocity

$$v = \sqrt{\frac{Y}{\varrho}} \tag{2.1}$$

where Y is the Young modulus and ϱ the bar density. For alluminium at room temperature $v = 5100 m/s$.

It can be demonstrated that the bar behaves like a system of an infinite number of elementary oscillators with angular resonance frequency

$$\omega_k = (2k + 1)\frac{\pi v}{L} \qquad (k = 0, 1, 2, ...) \tag{2.2}$$

The fundamental mode oscillator has angular frequency

$$\omega_o = \frac{\pi v}{L} \tag{2.3}$$

In conclusion the study of the interaction of the bar with a g.w. is semplified if we consider one elementary oscillator at a time with resonant angular frequency ω_k. The displacement of the bar end faces at the angular frequency ω_k can be described as the displacement with respect to its equilibrium position of a point-like mass.

The mass m of this "equivalent" oscillator can be related to the mass M of the bar with the following considerations based on the energy of the process. In the bar, due to the boundary conditions of null stress on the end faces, a system of standing waves is generated. For the fundamental mode

$$\eta(t,z) = \eta_o cos\omega_o t sin\frac{\pi z}{L}$$

where z is a coordinate along the bar axis with origin in the center of mass and η (t,z) the displacement of an elementary mass located at z. The displacement is largest at the bar ends, $z = \pm L/2$. The total kinetic energy of the bar is (S is the bar section)

$$\varepsilon_{bar} = \int_M \frac{1}{2}v^2 dm = \int_{-L/2}^{L/2} \frac{1}{2}\varrho S\dot{\eta}^2 dz = \frac{1}{4}M\eta_0^2\omega_o^2 cos^2\omega_o t$$

A point-like oscillator has energy

$$\varepsilon_{os} = \frac{1}{2}mv^2 = \frac{1}{2}m\omega_0^2\eta_0^2 cos^2\omega_o t$$

Since the two energies must be the same then

$$m = \frac{M}{2} \tag{2.4}$$

Unless otherwise specied we consider only the fundamental mode oscillator at ω_0. It can be shown [1] that its unidimensional displacement $\eta(t)$ follows the equation

$$\ddot{\eta} + 2\beta_1\dot{\eta} + \omega_0^2\eta = \frac{2}{\pi^2}L\ddot{h} \tag{2.5}$$

where h indicates h_+ or $h_\times$ and the numerical factor on the right side is obtained from eq.(1.2) and by solving the problem of the continuous bar. The quantity $2\beta_1$ expresses the losses and is related to the merit factor Q by

$$2\beta_1 = \omega_0/Q \tag{2.6}$$

In the cases of interest $Q \gg 1$, typically $Q > 10^6$. The solution of (2.6) obtained by means of the Fourier transform is

$$\eta(t) = \frac{1}{2\pi}\int_{-\infty}^{\infty} \frac{2L}{\pi^2}\frac{\omega^2 H(\omega)e^{i\omega t}}{(\omega^2 - \omega_0^2) - 2i\omega\beta_1}d\omega \tag{2.7}$$

If H(ω) does not have zeros and the poles of the integrand are

$$\omega_{1,2} \cong i\beta_1 \pm \omega_0 \tag{2.8}$$

for the cases $H(\omega_1) = H(\omega_2)$ we obtain from (2.7)

$$\eta(t) = -\frac{2L}{\pi^2}e^{-\beta_1 t}H(\omega_1)\omega_1 sin\omega_0 t \tag{2.9}$$

having neglected terms of the order Q^{-1}. This formula expresses the two fundamental points of the resonant g.w. antenna: a) the antenna detects the Fourier component $H(\omega)$ of the metric tensor perturbation near its resonance frequency, b) the antenna has a "memory" for a time of the order of β_1^{-1}.

The non resonant antennas, on the other hand, detect the metric tensor perturbation $h(t)$ and do not have any memory.

Let us now consider some interesting cases.

i) G.W. burst of the δ-type, $h(t)=H_0\delta(t)$. From (2.9) we obtain

$$\eta(t) = -\frac{2L}{\pi^2}H_0\omega_0 e^{-\beta_1 t}sin\omega_o t \tag{2.10}$$

ii) G.W. short packet, $h(t)=h_0 \cos \omega_0 t$ for $|t|<\tau_g/2$, $h(t)=0$ for $|t|>\tau_g/2$. We have

$$H(\omega_1) = H(\omega_2) = \frac{h_0\tau_g}{2} \ \frac{e^{\beta_1 \tau_g/2} - e^{-\beta_1 \tau_g/2}}{\beta_1\tau_g}$$

From (2.9), using (2.6) and with τ_g small, that is $\beta_1\tau_g \ll 1$, we have for $t > \tau_g/2$

$$\eta(t) = -\frac{L}{\pi^2}h_0\tau_g\omega_0 e^{-\beta_1 t} \qquad sin\omega_0 t \tag{2.11}$$

iii) Monochromatic g.w., $h(t)=h_0 \cos \omega_0 t$. In this case it is convenient to start from (2.7) since the Fourier transform is $H(\omega) = h_0\pi[\delta(\omega + \omega_0) + \delta(\omega - \omega_0)]$ a combination of δ-functions. We obtain

$$\eta(t) = -\frac{2L}{\pi^2}Qh_0 sin\omega_0 t \tag{2.12}$$

3. THE NOISE AND THE DATA ANALYSIS

In order to detect a very small g.w. signal one has to consider all possible sources of noise.

At the energy levels produced by g.w., noises of all kind are likely to be larger or much larger than the signal: thermal, electrical, acustic, cosmic rays, seismic, electromagnetic noises, all possible phenomena occurring in the laboratory can jepardize the experiment. We classify the various possible noises in two groups. In the first group we put all those which can be estimated quantitatively employing well established mathematical models. These noises are: the brownian noise of the bar considered as an oscillator at frequency ν_0 and the current and voltage noise of the amplifier. In the second group we put all the remaining noises which are often of unknown origin. In order to get rid of these last noises, in addition to take proper precautions, one can only consider coincidences between two or more antennas.

We consider now the effects of the brownian and electronic noise only.

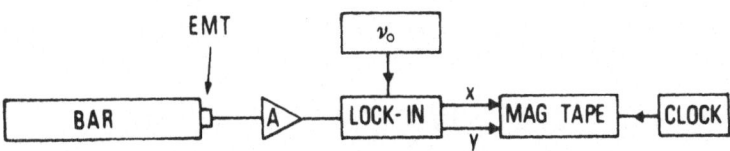

Figure 1. Schematic block diagram of the experimental apparatus.

A schematic block diagram of the antenna and associated electronics is shown in Fig.1. The bar vibrations are converted by a linear electromechanical transducer (EMT) into an electrical signal. In the following we shall consider a voltage signal V(t), but a similar treatment can be done for a current signal. The signal is amplified with a low noise amplifier (A) (for instance a dc SQUID). A lock-in (PSD), driven by a syntesizer at the resonance frequency ν_0 of the bar + EMT system, extracts the Fourier component at ν_0 which is expressed in terms of the two quantities x(t) and y(t) in quadrature. These quantities are written on a magnetic tape with sampling time Δ t for further data analysis.

The electrical amplifier noise is characterized, usually, by the power spectra of the voltage and current noise, V_n^2 and I_n^2, or the following combination of them

$$T_n = V_n I_n / k$$
$$R_n = V_n / I_n, \tag{3.1}$$

T_n is called the amplifier noise temperature (V_n^2 and I_n^2 expressed in bilateral form). R_n is the amplifier noise resistance.

It can be seen that the current noise I_n^2 exerts a force on the bar (back-action) which is stochastic in nature and, therefore, just of the type of the brownian force. Both brownian and back-action noises produce a mean square value of the voltage V(t) at the transducer output given by

$$V_{nb}^2 = \frac{2\alpha^2 k T_e}{M\omega_0^2} \tag{3.2}$$

which is called narrow band noise, expressed in terms of the transducer constant α ($V(t) = \alpha\eta(t)$) and of the equivalent temperature

$$T_e = T(1 + \frac{\beta Q T_n}{2\lambda T}) \tag{3.3}$$

In the last expression

$$\beta = \frac{2\alpha^2}{M\omega_0^3 |Z|} \tag{3.4}$$

is the ratio between the energy stored in the transducer with output impedance Z and the energy stored in the bar of mass M, and

$$\lambda = \frac{R_n}{|Z|} \tag{3.5}$$

expresses the electrical matching between transducer and amplifier.

In addition to the back-action the amplifier produces a wide band noise that, for a voltage signal, is given by

$$S_0 = V_n^2 + I_n^2 |Z|^2 = V_n^2(1 + 1/\lambda^2) = kT_n |Z| (1/\lambda + \lambda) \tag{3.6}$$

This noise adds to the narrow band noise in a way which depends on the specific algorithms used for the data analysis.

It can be shown that the autocorrelation of the lock-in output x(t) or y(t) is approximatively given by

$$R(\tau) = \frac{V_{nb}^2}{2}e^{-\beta_1|\tau|} + \frac{S_0\beta_2}{2}e^{-\beta_2|\tau|} \tag{3.7}$$

where β_2^{-1} is the integration time constant of the lock-in.

For the data analysis the key idea is to compare the actual values of x(t) and y(t) with the values that can be predicted from previous time. The difference between the two values has to be compared with the standard deviation in order to judge whether it is contained within the statistical fluctuations or is due to an innovation, say a gravitational wave.

The mathematical algorithm for this data analysis is the following. We consider the difference

$$\varrho_x(t) = x(t) - x(t - \Delta t)$$

and similarly for y(t) between two successive samplings of the voltage at the PSD output, Δt being the sampling time. We get the standard deviation of this filter, which we call ZOP (Zero Order Predictive filter), by

$$\sigma^2 \equiv \overline{\varrho_x^2(t)} + \overline{\varrho_y^2(t)} = 4R(0) - 4R(\Delta t)$$

Making use of (3.7) for $\beta_1 \ll \beta_2$ we get

$$\sigma^2 = 2V_{nb}^2\Delta t\beta_1 + 2S_0\beta_2 \tag{3.8}$$

In order to maximize the amount of information recorded on the magnetic tape it is convenient to take the sampling time equal to the integration constant of the PSD, $\Delta t = \beta_2^{-1}$. We obtain

$$\sigma^2(\Delta t) = 2V_{nb}^2\beta_1\Delta t + \frac{2S_0}{\Delta t} \tag{3.9}$$

This standard deviation of the innovation of the voltage signal can be also expressed as standard deviation of the innovation of energy of the antenna oscillations

$$\Delta\varepsilon(\Delta t) = \sigma^2(\Delta t)\frac{m\omega_0^2}{\alpha^2} \tag{3.10}$$

It is convenient to rewrite (3.10) in terms of the following dimensionless parameter

$$\Gamma = \frac{S_0}{V_{nb}^2/\beta_1} = \frac{T_n(\lambda + 1/\lambda)}{2\beta Q T_e} \tag{3.11}$$

that gives the ratio of the wide band noise in the resonance bandwidth to the narrowband noise. We have

$$\Delta\varepsilon(\Delta t) = 2V_{nb}^2\frac{m\omega_0^2}{\alpha^2}\left(\beta_1\Delta t + \frac{\Gamma}{\beta_1\Delta t}\right) \tag{3.12}$$

The greatest sensitivity is obtained when $\Delta\varepsilon$ has the smallest possible value, which occurs for

$$\Delta t_{opt} = \frac{\sqrt{\Gamma}}{\beta_1} \tag{3.13}$$

With this choice for the sampling time (and time constant of the PSD) we obtain the smallest detectable energy innovation

$$\Delta\varepsilon_{min} = 4V_{nb}^2\frac{m\omega_0^2}{\alpha^2}\sqrt{\Gamma} = 4kT_e\sqrt{\Gamma} \tag{3.14}$$

A convenient way to rewrite this equation is by making use of (3.6) and (3.3). We obtain

$$\Delta\varepsilon_{min} = 2kT_n\sqrt{(1 + \frac{1}{\lambda^2})(1 + \frac{2T\lambda}{\beta Q T_n})} \tag{3.15}$$

In obtaining this formula we have not considered the effect of the ZOP filter on the signal, but only on the noise. If the effect on the signal is taken into account the factor $2kT_n$ of eq.(3.15) becomes $2.42\, kT_n$ [2].

This expression shows the need to use an amplifier with a noise temperature T_n as small as possible. Suppose we have a very good amplifier with small T_n. In order to detect the smallest possible innovation of energy we must satisfy the two conditions

$$\lambda^2 \gg 1$$

$$\frac{\beta Q T_n}{2\lambda T} \gg 1 \tag{3.16}$$

These two are called the "matching conditions". If they are satisfied we get [3] from (3.15)

$$\Delta\varepsilon_{min} = 2kT_n \tag{3.17}$$

If the sign $\ll$ in (3.16) is replaced with $=$ we get

$$\Delta\varepsilon_{min} = 4kT_n$$

When the two matching conditions are fulfilled, from (3.13) we get

$$\Delta t_{opt} = \frac{\sqrt{\lambda^2 + 1}}{\beta\omega_o} \tag{3.18}$$

4. THE RESONANT TRANSDUCER

It is very difficult to fulfill (3.16) because it is hard to obtain large values of β. For example, for a capacitive transducer, using (3.4) with $\alpha = 5 \times 10^6 V/m$, C = 6nF, M = 2000 kg, ω = 5000 rad/s we get $\beta \sim 6x10^{-6}$.

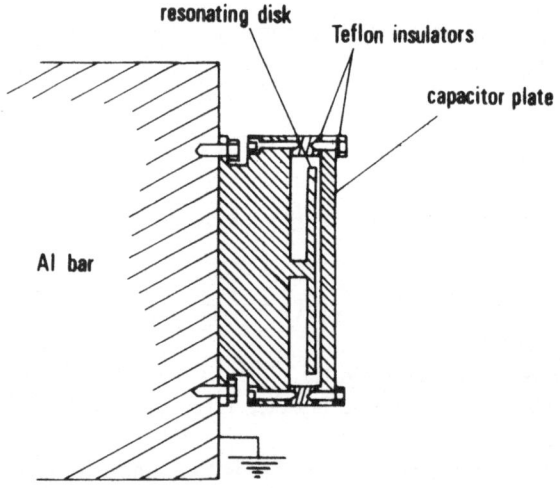

Figure 2. Scheme of the capacitive transducer of the Rome group.

For reaching much larger values of β we can use a resonant capacitive transducer [4] of the type shown in Fig.2. The disk, fixed at is center, vibrates in its flexural symmetrical mode with frequency

$$\nu_t = \frac{(2.1)^2}{4\pi} \frac{a}{R^2} \sqrt{\frac{Y}{3\varrho(1-\sigma^2)}} \tag{4.1}$$

where a and R are the thickness and radius of the disk, Y, σ and ϱ are respectively the Young and Poisson moduli and the density of the material. The thickness and radius of the disk are such that ν_t is equal to the resonance frequency ν_o of the bar.

In this way we obtain a system of two coupled oscillators, bar and transducer, which exhibits two resonances at frequencies

$$\nu_\mp = \nu_o(1 \mp \frac{\sqrt{\mu}}{2}) \tag{4.2}$$

where

$$\mu = \frac{m_t}{m} \tag{4.3}$$

is the ratio of the reduced mass of the transducer to the reduced mass of the bar given by (2.4). The displacements of the bar and of the transducer, indicated respectively with $\eta_b(t)$ and $\eta_t(t)$, neglecting the dissipations, are

$$\eta_b(t) = \frac{f_o}{2m\omega_o}(sin\omega_- t + sin\omega_+ t) = \frac{f_o}{m\omega_o}sin\omega_o t cos\Omega_b t \tag{4.4}$$

$$\eta_t(t) = \frac{f_o}{2m\omega_o\sqrt{\mu}}(sin\omega_+ t - sin\omega_- t) = \frac{-f_o}{m\omega_o\sqrt{\mu}}cos\omega_o t sin\Omega_b t \tag{4.5}$$

where f_o is the Fourier transform of an applied δ force on the bar and

$$\Omega_b = \frac{\omega_o\sqrt{\mu}}{2} \tag{4.6}$$

is the beat angular frequency. We notice that the transducer displacement is $1/\sqrt{\mu}$ larger than the bar displacement, and that the mechanical energy is tranferred back and forth between bar and transducer with the beat period $T_b = 2\pi/\Omega_b$. The transducer displacement $\eta_t(t)$ is the quantity that is measured. In this case we obtain a much larger value of β because in eq.(3.4) we must substitute M/2 with m_t. In the case of the Rome antenna $m_t = 0.348$ kg and therefore we obtain

$$\beta = 2 \times 10^{-2}$$

that is much better for fulfilling (3.16).

From (4.5) we can recognize that the system: antenna + resonant transducer, is equivalent to two sistems: antenna + non resonant transducers, resonating at frequencies ν_- and ν_+. The only difference is that the incident absorbed energy is split between the two modes. Using a non resonating transducer and a δ-force applied to the bar with Fourier transform f_o we obtain, with no dissipation,

$$\eta_b(t) = \frac{f_o}{m\omega_o}sin\omega_o t$$

The comparison of this with (4.4) shows that the displacement for the case of the resonant transducer is 1/2 of that for the case of a non resonant transducer, for each mode. Therefore, as far as the signal, we find in one mode of the transducer (ω_- or ω_+) 1/4 of the available power (the other energy is in the bar).

The brownian noise is also smaller for the case of a non resonant transducer. In fact, for the energy equipartition principle the thermal noise for one mode (say ω_+) is distributed between bar and transducer

$$\frac{1}{2}kT = \frac{1}{2}m\omega_+^2\overline{\eta_b^2} + \frac{1}{2}m_t\omega_+^2\overline{\eta_t^2}$$

From (4.4) and (4.5) we get $m\overline{\eta_b^2} = m_t\overline{\eta_t^2}$, thus

$$\overline{\eta_t^2} = \frac{kT}{2m_t\omega_+^2} \tag{4.7}$$

to be compared with (3.2), showing that the thermal noise of the resonant transducer, for one mode only, is 1/2 of that for the non resonating transducer.

As a consequence the signal to noise ratio (SNR) for one mode is 1/2 of the SNR that we have for the non resonant transducer. However, since a δ-force produces the same signal on both modes ω_- and ω_+, combining the data from the two modes we have a SNR equal to that obtained for a non resonant transducer. In conclusion, the effect of using a resonant transducer is just to obtain a larger β value while the expression of the sensitivity remains the same.

5. THE WIENER-KOLMOGOROFF FILTER

The electrical equivalent circuit of the gravitational wave detector consists, essentially, in two low-pass filters [5]. One filter, with time constant $t_o = \beta_2^{-1}$, represents the integrating part of the PSD. The other filter is due to the bar plus that part of the PSD which selects the Fourier component of the signal; the time constant of this filter is β_1^{-1}. The last assertion can be proven by inspecting formula (3.7), for instance, which shows that the autocorrelation of the brownian signal is typical of that due to a low-pass filter with time constant β_1^{-1}. In conclusion the noise coming from the bar is, first, filtered with the transfer function

$$W_b = \frac{\beta_1}{\beta_1 + j\omega} \tag{5.1}$$

and then, together with the noise arriving from the amplifier, is filtered with the transfer function

$$W_e = \frac{\beta_2}{\beta_2 + j\omega} \tag{5.2}$$

If we have a white noise with power spectrum S_{uu} at the entrance of the W_b filter we obtain the power spectrum $S_{uu}|W_b|^2$ at the output of the filter, with the autocorrelation function

$$R_{uu}^{(b)}(\tau) = S_{uu}\frac{\beta_1}{2}e^{-\beta_1|\tau|} \tag{5.3}$$

This shows that the relation between S_{uu} and V^2_{nb} is

$$V^2_{nb} = S_{uu} \frac{\beta_1}{2} \tag{5.4}$$

This reasoning shows us also that the autocorrelation (3.7) is approximate in the limit (usually well verified) $\beta_1 \ll \beta_2$.

In fact, at the end of the electronic chain, the power spectrum of the brownian noise will be $S_{uu}|W_b|^2|W_e|^2$ which provides the exact autocorrelation function

$$R^{(b+e)}_{uu}(\tau) = \frac{S_{uu}\beta_1\beta_2}{2} \frac{\beta_2 e^{-\beta_1|\tau|} - \beta_1 e^{-\beta_2|\tau|}}{\beta_2^2 - \beta_1^2} \tag{5.5}$$

Similarly if we indicate with S_{ee} the power spectrum of the noise due to the amplifier at the entrance of the PSD we have, at the output, the power spectrum $S_{ee}|W_e|^2$ with the autocorrelation

$$R^{(e)}_{ee}(\tau) = S_{ee}\frac{\beta_2}{2}e^{-\beta_2|\tau|} \tag{5.6}$$

The comparison with (3.7) gives the relation between S_{ee} and S_o

$$S_o = \frac{S_{ee}}{2} \tag{5.7}$$

We proceed now to describe the best estimation of the signal u(t) at the bar entrance, obtained by using all data measurements, past and future, weighted according to the Wiener-Kolmogoroff theory. Let us consider one component (say the real component $x(t)$) at the PSD output recorded on magnetic tape with sampling time Δt. We put

$$\hat{u}(t) = \int_{-\infty}^{\infty} x(t - \tau)w(\tau)d\tau \tag{5.8}$$

where $\hat{u}(t)$ is the estimation of $u(t)$ (concerning only one component of the signal) and $w(\tau)$ are the weights to be determined. We apply the linear mean square method. It is found that the Fourier transform $W(j\,\omega)$ of the weights $w(\tau)$ which minimizes the quantity $\overline{(\hat{u}(t) - u(t))^2}$ is

$$W(j\omega) = \frac{S_{ux}(\omega)}{S_{xx}(\omega)} \tag{5.9}$$

with the following meaning for symbols. $S_{xx}(\omega)$ is the power spectrum of x(t) and $S_{ux}(\omega)$ is the cross spectrum of u(t) and x(t). We have

$$S_{xx}(\omega) = S_{uu}|W_b|^2|W_e|^2 + S_{ee}|W_e|^2 \tag{5.10}$$

$$S_{nx}(\omega) = S_{uu}W_b * W_e* \tag{5.11}$$

where W_b* and W_e* indicate the complex conjugate of W_b and W_e. We get

$$W(j\omega) = \frac{1}{W_b W_e} \frac{1}{1 + \Gamma/|W_b|^2} \tag{5.12}$$

where $\Gamma = \frac{S_{ee}}{S_{uu}}$ has already been introduced by (3.11). In order to obtain $w(t)$ we must calculate

$$w(\tau) = \frac{1}{2\pi} \int\limits_{-\infty}^{\infty} W(j\omega) e^{i\omega\tau} d\omega$$

Performing the integration we find

$$w(\tau) = \frac{\beta_1}{\beta_2 \Gamma} \left[\delta(\tau) + \frac{(\beta_2 \mp \beta_3)(\beta_1 \mp \beta_2) e^{\mp\beta_3\tau}}{2\beta_3} \right] \tag{5.13}$$

where the sign - is for $\tau > 0$ and the sign + for $\tau < 0$.
Using (5.10) and (5.12) we have power spectrum of the estimation $\hat{u}(t)$

$$S_{\hat{u}}(\omega) = S_{xx}|W(j\omega)|^2 = \frac{S_{uu}}{1 + \Gamma/|W_b|^2} \tag{5.14}$$

We notice that the estimation is perfect if $\Gamma = 0$, corresponding to the absence of electronic noise. From (5.14) and using (5.1) we get

$$R_{\hat{u}\hat{u}}(\tau) = \frac{S_{uu}\beta_1}{2\sqrt{\Gamma(\Gamma+1)}} e^{-\beta_3|\tau|} \tag{5.15}$$

with

$$\beta_3 = \beta_1 \sqrt{\frac{\Gamma+1}{\Gamma}} \tag{5.16}$$

The variance of $\hat{u}(t)$ is obtained by taking $R_{\hat{u}\hat{u}}(0)$.

We shall now compare this variance with the signal expected from a gravitational wave signal.

Suppose we have an excitation producing a bar vibration given by (2.9). The voltage signal is

$$V_g(t) = -\alpha \frac{2L}{\pi^2} H(\omega_o)\omega_o e^{-\beta_1 t} \sin\omega_o t \tag{5.17}$$

The output of the end of the PSD turns out to be (with some calculation and considering the signal in phase with x(t))

$$v_g(t) = -\frac{2\alpha L\omega_o H(\omega_o)}{\pi^2} \frac{e^{-\beta_1 t} - e^{-\beta_2 t}}{1 - \beta_1/\beta_2} \tag{5.18}$$

Applying to this signal the Wiener-Kolmogoroff filter as derived previously

$$u_g(t) = \int\limits_{-\infty}^{\infty} v_g(t - t') w(t') dt'$$

we get, with some calculation using (5.13)

$$u_g(t) = -\alpha\frac{2L\omega_o H(\omega_o)}{\pi^2}\ \frac{\beta_1}{2\beta_2\Gamma}e^{-\beta_3 t} \tag{5.19}$$

We consider now the signal to noise ratio (SNR). This is defined as follows:

$$SNR = \frac{\overline{u_g(t)^2}}{2R_{\theta\theta}(o)} = \frac{\left(\alpha\frac{2L\omega_o H(\omega_o)}{\pi^2}\right)^2}{S_{uu}}\ \frac{e^{-2\beta_3 t}}{4\beta_1\sqrt{\Gamma(\Gamma+1)}}$$

where the factor of 2 has been introduced because both components from the PSD output contribute to the noise. Making use of (5.4) and (3.2) we get, for t=0 when SNR is largest,

$$SNR_{max} = \frac{L^2\omega_o^4 m H^2(\omega_o)}{2\pi^4 k T_e\sqrt{\Gamma(\Gamma+1)}} \tag{5.20}$$

The minimum detectable value of H(ω_o) is obtained for SNR$_{max}$=1.

$$[H(\omega_o)]_{min} = \left[\frac{2\pi^4 k T_e\sqrt{\Gamma(\Gamma+1)}}{L^2 m\omega_o^4}\right]^{1/2} \tag{5.21}$$

6. THE CROSS SECTION AND THE EFFECTIVE TEMPERATURE

In order to detect g.w. it is necessary that the detector absorbes some energy from the wave. As a first step to estimate the cross-section let us consider the energy carried by g.w. It can be shown that the energy carried per unit time across the unit area is given by

$$I(t) = \frac{c^3}{16\pi G}\left[\dot{h}_+(t)^2 + \dot{h}_\times(t)^2\right]\quad\left[\frac{joule}{s\cdot m^2}\right] \tag{6.1}$$

where h_+ (t) and $h_\times$ (t) indicate the two polarization states of the g.w. The total energy per unit area is

$$I_o = \int\limits_{-\infty}^{\infty} I(t)dt\quad\left[\frac{joule}{m^2}\right] \tag{6.2}$$

For semplicity we consider one polarization status only, $h_\times(t)$ or $h_+(t)$, and indicate it with h(t). We indicate with H(ω) the Fourier transform of h(t),

$$H(\omega) = \int\limits_{-\infty}^{\infty} h(t)e^{-i\omega t}dt \tag{6.3}$$

From (6.1) we get

$$I_o = \frac{c^3}{16\pi G} \int\limits_{-\infty}^{\infty} |\omega H(\omega)|^2 d\nu \qquad (6.4)$$

were $\nu = \omega/2\pi$ is the frequency. The quantity

$$f(\omega) = \frac{c^3}{16\pi G} |\omega H(\omega)|^2 \qquad \left[\frac{joule}{m^2 Hz}\right] \qquad (6.5)$$

is called "spectral energy density", written here in bilateral form (frequencies from $-\infty$ to $+\infty$).

As a special case, we consider a g.w. burst of duration τ_g that can be described by a sinusoidal wave with h_o amplitude and angular frequency ω_o for $|t| < \tau_g/2$ and zero value for $|t| > \tau_g/2$. Since we have $H(\omega) \cong h_o\tau_g/2$, from (6.5) we get

$$f(\omega_o) = \frac{c^3}{64\pi G} \omega_o^2 h_o^2 \tau_g^2 \qquad \left[\frac{joule}{m^2 Hz}\right] \qquad (6.6)$$

Another interesting case is a g.w. burst of the type $h(t) = h_o e^{-\beta_w|t|} \cos\omega_o t$. This wave has duration of the order of $\tau_g = 2/\beta_w$. If $\tau_g \ll 2\pi/\omega_o$ then the Fourier transform is

$$H(\omega) \cong \frac{h_o}{\beta_w} = \frac{h_o\tau_g}{2}$$

and we obtain again the result (6.6).

In order to obtain the total amount of energy per unit area we consider that the frequency bandwith for a duration τ_g is $1/\tau_g$. Multiplying (6.6) for it we get

$$I_o = f(\omega_o)/\tau_g = \frac{c^3}{64\pi G} \omega_o^2 h_o^2 \tau_g \qquad \left[\frac{joule}{m^2}\right] \qquad (6.7)$$

Finally an interesting case is also a g.w. burst of a δ-type, $h(t) = H_o\delta(t)$, $\delta(t)$ being the Dirac function. Since $H(\omega) = H_o$ we obtain from (6.5)

$$f(\omega) = \frac{c^3}{16\pi G} \omega^2 H_o^2 \qquad \left[\frac{joule}{m^2 Hz}\right] \qquad (6.8)$$

In order to compute the value of h_o on the Earth due to a g.w. burst of duration τ_g that occurs at a distance R, indicating with $M_{GW}c^2$ the total g.w. energy, we multiply (6.7) by $4\pi R^2$; we obtain

$$h_o = \sqrt{\frac{16GM_{GW}c^2}{c^3 R^2 \omega_o^2 \tau_g}} = 1.38 \cdot 10^{-17} \frac{1000Hz}{\nu} \frac{1000pc}{R} \sqrt{\frac{M_{GW}}{10^{-3}M_\odot} \frac{10^{-3}s}{\tau_g}} \qquad (6.9)$$

If we consider a sinusoidal g.w. of angular frequency ω with amplitude h_o we obtain the average power per unit area from (6.1)

$$W_o = \frac{c^3}{32\pi G}\omega^2 h_o^2 \qquad \left[\frac{watt}{m^2}\right] \tag{6.10}$$

Indicating with W the total power irradiated by the source, at distance R we obtain

$$h_o = \sqrt{\frac{8G}{c^3}\frac{W}{R^2\omega^2}} = 2.29 \times 10^{-41}\frac{1000pc}{R}\frac{1000Hz}{\nu}\sqrt{W} \tag{6.11}$$

A review about possible gravitational wave sources can be found, for instance, in ref.(6) and (7).

The cross section Σ is defined such that, multiplied by the incident spectral energy density $f(\omega_o)$, gives the energy deposited in the bar

$$\varepsilon = \Sigma f(\omega_o) \tag{6.12}$$

The energy ε is calculated from (2.9)

$$\varepsilon = \frac{1}{4}M\omega_o^2\left(\frac{2L}{\pi^2}\omega_o H(\omega_o)\right)^2 = \frac{M\omega_o^2 H(\omega_o^2)v^2}{\pi^2} \tag{6.13}$$

where $v = \omega L/\pi$ is the sound velocity in the bar. Making use of (6.5) for the spectral energy density we obtain the cross section

$$\Sigma = \frac{16}{\pi}\left(\frac{v}{c}\right)^2\frac{G}{c}M \qquad [m^2 Hz] \tag{6.14}$$

in bilateral form.

We now use the result (5.21), derived in the previous section by means of the W-K filter for the minimum detectable value of $H(\omega_o)$ Introducing (5.21) into (6.14) we get [8]

$$\varepsilon = 4kT_e\sqrt{\Gamma(\Gamma+1)} \tag{6.15}$$

which is very similar to the energy innovation $\Delta\varepsilon_{min}$ given by (3.14) and estimated with the ZOP filter, because $\Gamma \ll 1$ in all cases. The difference, essentially, is that, using W-K, there is no need to satisfy (3.13) for the sampling time; also we gain the factor 2.42/2 which was lost with the ZOP filter. However, it is found that the sampling time must satisfy

$$\Delta t < \beta_3^{-1}$$

otherwise the W-K filter breaks down. We notice that, from (3.13), we have

$$\Delta t_{opt} \cong \beta_3^{-1} \tag{6.16}$$

Finally the effective temperature is defined by

$$T_{eff} = \frac{\varepsilon}{k} = 4T_e\sqrt{\Gamma(\Gamma + 1)} \qquad (6.17)$$

We derive now another expression [9] which relates $H(\omega_o)$, the quantity measured with a resonant g.w. antenna, to T_{eff} which expresses the sensitivity of the apparatus. From (6.13) and (6.17) we obtain

$$[H(\omega_o)]_{min} = \frac{L}{v^2}\sqrt{\frac{kT_{eff}}{M}} \qquad [Hz^{-1}] \qquad (6.18)$$

In general if we detect an energy innovation $\Delta\varepsilon$ the corresponding value of $H(\omega_o)$ is

$$H(\omega_o) = \frac{L}{v^2}\sqrt{\frac{\Delta\varepsilon}{M}} \qquad (6.19)$$

It must be stressed that with a resonant antenna we measure $H(\omega_o)$ and not h(t). If we want to have a feeling about possible values for h(t) we must make assumptions on the h(t) spectrum. For instance for a flat spectrum from 0 to ν_g, which we can think due to a burst of duration $\tau_g \sim 1/\nu_g$, we can put

$$h(t) \simeq H(\omega_o)\nu_g = \frac{H(\omega_o)}{\tau_g} \qquad (6.20)$$

7.THE ANTENNA BANDWIDTH

The problem of the antenna bandwidth arises in particular when exploring the possibility to use a resonant antenna for detecting monochromatic g.w. radiation. One could, erroneously, think that the antenna is sensitive only to m.g.w. with frequency ν_g near the resonance frequency ν_o in a bandwidth of the order of ν_o/Q. We shall see now that the antenna bandwidth is, instead, equal to β_3/π and therefore much larger than ν_o/Q. It can indeed, under certain conditions, become so large as ν_o.

In order to prove this [10] we use the signal (2.12) obtained for m.g.w. The noise is of narrow band and wide band types. The narrow band noise is

$$S_{nb} = \alpha^2 \frac{2kT_e\omega_o/mQ}{(\omega_o^2 - \omega^2)^2 + \omega^2\omega_o^2/Q^2} \qquad \left[\frac{volt^2}{Hz}\right] \qquad (7.1)$$

The wide band is obtained from (3.6).

As a first step we consider the case of $\nu_g = \nu_o$. The signal to noise ratio is

$$SNR(\nu_o) = \frac{\overline{\alpha\eta(t)^2}}{[S_{nb}(\omega_o) + V_n^2(1 + \frac{1}{\lambda^2})]\,\Delta\nu} \qquad (7.2)$$

where $\Delta\nu$ is the integration band given by the total time of measurement $\Delta\nu = 1/t_m$. Introducing the parameter

$$\chi = \frac{\beta Q}{\lambda} \tag{7.3}$$

we obtain from (7.2)

$$SNR(\omega_o) = \frac{L}{\pi^4} M t_m h_o^2 \omega_o^3 \frac{Q}{2kT + kT_n\left(\chi + \frac{1}{\chi}\right)} \tag{7.4}$$

The largest SNR is obtained if the following conditions are satisfied

$$\frac{T_n}{2T} \ll \chi \ll \frac{2T}{T_n} \tag{7.5}$$

In such a case

$$SNR(\omega_o) = \frac{v^2}{2\pi^2 k} \frac{M t_m h_o^2 \omega_o Q}{T} \tag{7.6}$$

For SNR=1 we obtain the minimum detectable value of h_o

$$h_o \geq \sqrt{\frac{2\pi^2 kT}{Mv^2 t_m Q\omega_o}} = 1.87 \times 10^{-24} \times \left[\frac{T}{4.2K} \frac{2300kg}{M} \frac{10^7}{Q} \frac{900Hz}{\nu_o} \frac{1day}{t_m}\right]^{1/2} \tag{7.7}$$

We go now to compute the bandwidth by considering a m.g.w. with $\nu_g \neq \nu_o$. We evaluate SNR from (7.2) for $\omega \neq \omega_o$. With some calculation we obtain

$$SNR(\omega_g) = \left(\frac{Lh_o}{\pi^2}\right)^2 \frac{\omega_g^4 MQt_m}{2\omega_o kT_e} \frac{1}{1 + \Gamma[Q^2(1 - \frac{\omega_g^2}{\omega_o^2})^2 + \frac{\omega_g^2}{\omega_o^2}]} \tag{7.8}$$

The largest value of SNR is, obviously, for $\omega_g = \omega_o$. We calculate now for which value of ω_g the SNR reduces to one half of its maximum. Putting

$$1 + \Gamma\left[Q^2\left(1 - \frac{\omega_g^{*2}}{\omega_o^2}\right)^2 + \frac{\omega_g^{*2}}{\omega_o^2}\right] = 2$$

we find

$$\frac{\omega_g^*}{\omega_o} \cong 1 \pm \frac{1}{2Q\sqrt{\Gamma}} \tag{7.9}$$

which gives a bandwidth

$$\Delta\nu = \frac{\Delta\omega}{2\pi} = \frac{1}{2\pi} \frac{\omega_o}{Q\sqrt{\Gamma}} \cong \frac{\beta_3}{\pi} \tag{7.10}$$

We see that the parameter Γ plays a very important role. The mathematical limit for $\Delta \nu$ when Γ tends to zero is infinite. An useful expression is obtained putting (3.11) into (7.10). We get

$$\Delta \nu = \nu_o \sqrt{\frac{2\beta T_e}{QT_n(\lambda + \frac{1}{\lambda})}} \tag{7.11}$$

For obtaining a bandwidth equal to ν_o is necessary to realize an apparatus with the following characteristics:

$$Q = 10^7, T_n = 10^{-7}K, \beta = 1, T_e = 1K, \lambda = 1$$

In order to reach $\beta = 1$ it is necessary to use an active transducer (not considered here). However, already with $\beta = 10^{-2}$ and the above values for Q, T_n, T_e it should be possible to obtain $\Delta \nu \cong \nu_o/10$.

8. THE ROME GRAVITATIONAL WAVE ANTENNA

The Rome group has realized a resonant cryogenic antenna that consists of an alluminium cylinder with lenght L=3 m and mass M=2270 kg, cooled with liquid helium to a temperature T= 4.2 K [9]. The mechanical vibrations are detected by means of a resonant capacitive transducer operating with constant electrical charge; the electrical signal is amplified with a very low noise amplifier based on the use of a dcSQUID.

The antenna has been installed in Geneva, at CERN, and it has operated from November 1985 until July 1986 [11]. During these nine months of operation the antenna has produced data for about 70 % of the time with a noise temperature T_{eff} on the average between 15 and 20 mK.

In figures 3 and 4 we show the distribution of the filtered data during two periods when the antenna operation was particularly good. We note the very good exponential distribution except for a tail which could be, potentially, due to signals to be compared with signals recorded with different antennas. The value of h(t) corresponding to $T_{eff} = 12mK$ and determined with (6.19) and (6.20) is 8×10^{-19}. In figure 5 we show the hourly averages of the values of h(t). We notice that in certain periods the antenna was functioning very well, in other period there was much excess noise. For this reason the antenna was turned off after nine months of operation, in order to make some improvements.

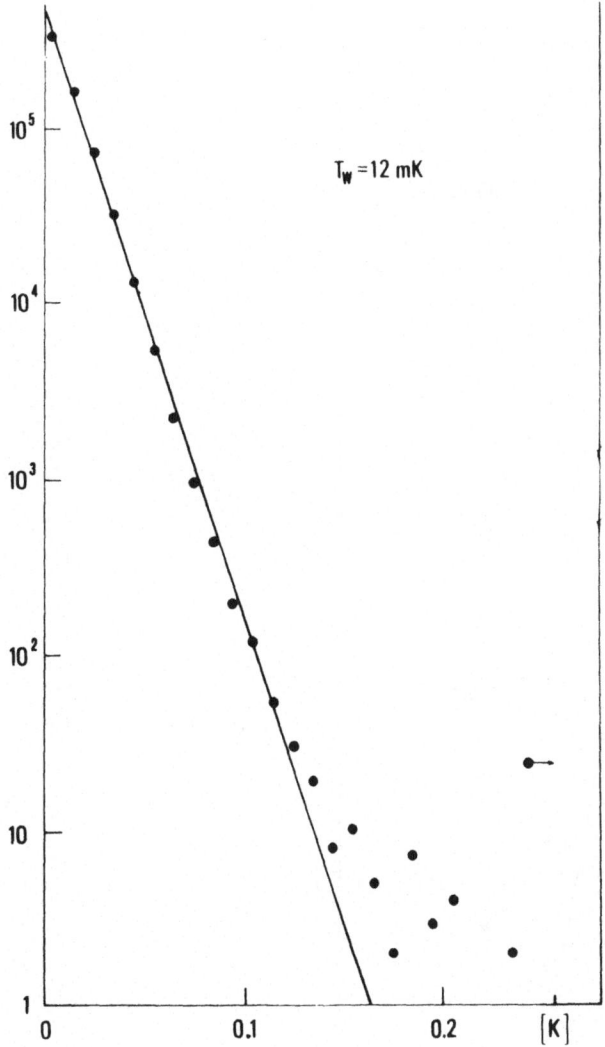

Figure 3 Differential statistical distribution of the filtered output for the period: day 121, hour 10 to day 123, hour 14, 1986 (52 continuous hours).

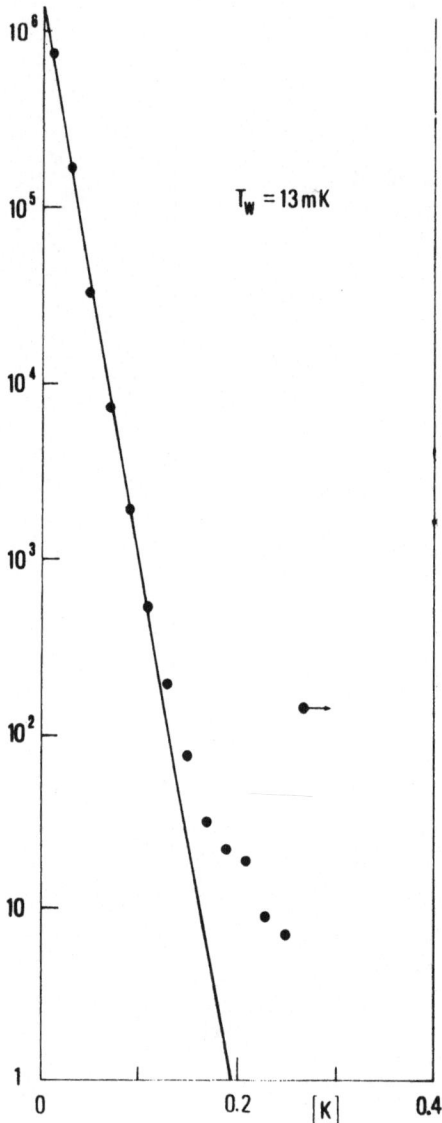

Figure 4 As in Fig.3 for day 132, hour 23 to day 136 -hour 06, 1986 (81 continuous hours).

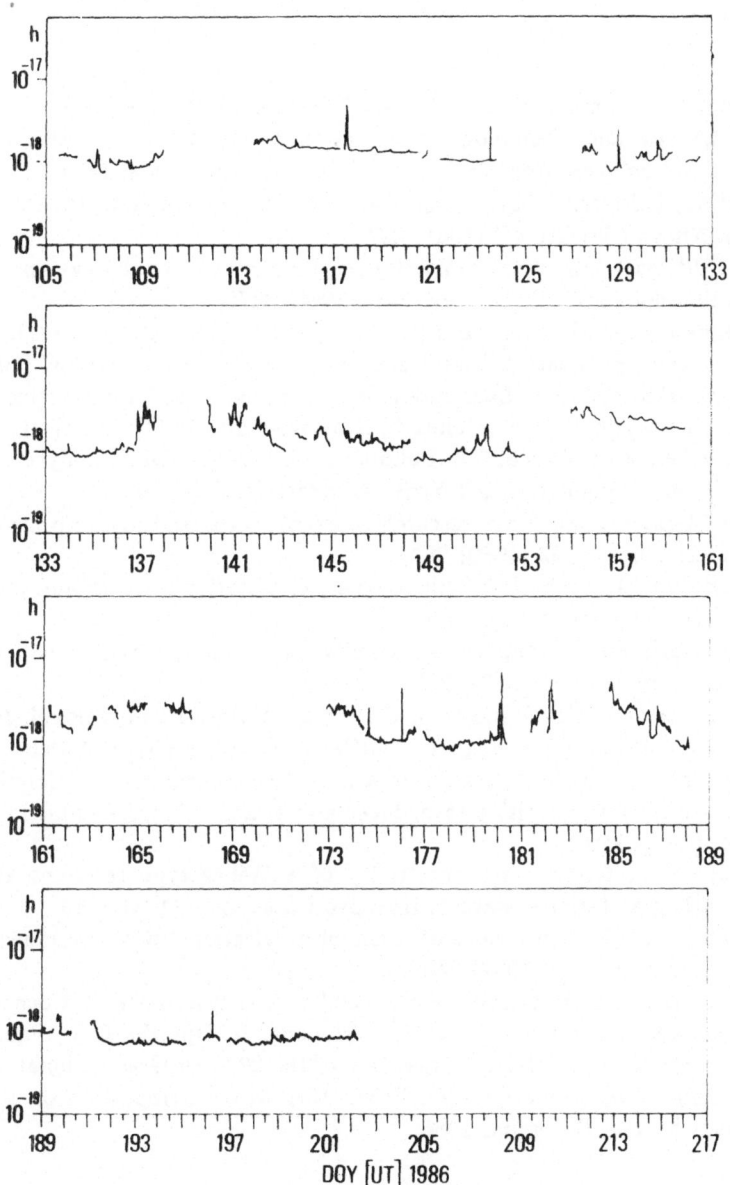

Figure 5 Hourly averages of the minimum detectable value of h(t) (SNR =1).

REFERENCES

1. See for instance: E.Amaldi and G.Pizzella "The search for gravitational waves", in Relativity, Quanta and Cosmology in the development of the scientific thought of Albert Einstein" Johnson Reprint Corporation, Academic Press, 1979
2. G.V.Pallottino, G.Pizzella, "Matching of transducers to resonant gravitational wave antenas", Il Nuovo Cimento 4C (1981) 237
3. R.Giffard, "Ultimate sensitivity limit of a resonant gravitatonal wave antenna using a linear motion detector", Phys. Rev. 14D (1976) 2478
4. (a) P.Rapagnani, "Development and test at T=4.2 K of a capacitive resonant transducer for cryogenic gravitational wave antennas", Il Nuovo Cimento 5C (1982) 385
    (b) Y.Ogawa, P.Rapagnani "Lagrangian formalism for resonant capacitive transducers for gravitational wave antennas", Il Nuovo Cimento 7C (1984) 21.
5. P.Bonifazi, V.Ferrari, S.Frasca, G.V.Pallottino, G.Pizzella, "Data analysis algorithms for gravitational wave antenna", Il Nuovo Cimento 1C (1978) 465
6. K.S.Thorne "Gravitational wave research: current status and future prospects", Review of Modern Physics, 52 (1980) 285
7. M.Rees, R.Ruffini, J.A.Wheeler: "Black holes, Gravitational waves and cosmology", Gordon and Breach, New York, 1974.
8. G.Pizzella, "Optimum filtering and sensitivity for resonant gravitational wave antenas", Il Nuovo Cimento 2C (1979) 209
9. E.Amaldi, C.Cosmelli, G.V.Pallottino, G.Pizzella, P.Rapagnani, F.Ricci, P.Bonifazi, M.G.Castellano, P.Carelli, V.Foglietti, G.Cavallari, E.Coccia, I.Modena, R.Habel, "Preliminary results on the operation of a 2270 kg cryogenic gravitational-wave antenna with a resonant capacitive transducer and a d.c. SQUID amplifier", Il Nuovo Cimento 9C (1986) 829
10. (a) G.V.Pallottino, G.Pizzella, "Sensitivity of a Weber type resonant antenna to monochromatic gravitational waves", Il Nuovo Cimento 7c (1981) 155
    (b) G.Pizzella "On the band width of resonant gravitational wave antennas", Lettere al Nuovo Cimento, 40 (1984) 240
11. E. Amaldi, P. Bonifazi, P. Carelli, M.G.Castellano, G. Cavallari, E. Coccia, C. Cosmelli, V. Foglietti, S. Frasca, R. Habel, I. Modena, R. Onofrio, G.V. Pallottino, G. Pizzella, P. Rapagnani, F. Ricci, "Operation of the 2270 kg Gravitational Wave Resonant Antenna of the Rome Group", Proc. 13th Texas Symposium on Relativistic Atrophysics, Chicago, December 1986.

GRAVITATIONAL ANTENNA BANDWIDTHS AND CROSS SECTIONS

J. Weber
University of Maryland, College Park, Maryland 20742
and
University of California, Irvine, California 92717

ABSTRACT

Elastic solid antennas are considered. It is shown that many body
coherence effects lead to cross sections for pulses,much larger than
earlier estimates. For monochromatic continuous signals, the new
approach gives the same cross sections as derived in 1960.

A new method for data recording and analysis is proposed. This
gives bandwidths several orders larger than are available by recording
the output of an optimal filter.

Means are suggested for cascading half acoustic wavelength antennas
to give increased cross sections.

Very low temperatures offer methods for selection of antenna
quantum states, which imply that the signal to noise ratio can be increased beyond any presently conceived limit.

CROSS SECTIONS

In electromagnetic theory it is proved[1] that the cross section for
scatterers in a region small compared with a wavelength, is proportional
to the square of the number of scatterers.

For a single harmonic oscillator quadrupole interacting with gravitational radiation, the cross section for monochromatic, continuous[2]
radiation is given by

$$\sigma = \frac{8\pi^3 ML^2 Q}{C^2 \lambda} \tag{1}$$

In (1) M is the reduced mass, L is the equivalent separation of the
elements of the quadrupole, C is the speed of light, λ is the wavelength,
Q is the quality factor given by

$$Q = \frac{\omega(\text{maximum stored energy})}{(\text{Dissipative Power Loss})}$$

B. F. Schutz (ed.), Gravitational Wave Data Analysis, 195–199.

For pulses the cross section is

$$\sigma_{PULSES} = \frac{8\pi^3 ML^2}{c^2\lambda} \tag{2}$$

if the quality factor Q is much greater than 1. For quality factors less than 1, the cross section is[4]

$$\sigma_{PULSES} = \frac{8\pi^3 ML^2 Q}{c^2\lambda} \tag{3}$$

The early analyses concluded that (1), (2) and (3) are valid for elastic solid antennas, described as continuous solids by the classical theory of elasticity.

An antenna is composed of atoms, coupled by chemical forces. For N quadrupoles each with reduced mass m, and quadrupole moment mr^2, in a volume with linear dimensions small compared with a wavelength, we might expect a total cross section for pulses, σ_N given by

$$\sigma_N{}_{PULSES} = \frac{8\pi^3 G}{c^2\lambda} \frac{mr^2 N^2}{} \tag{4}$$

If the quadrupoles are pairs of atoms, the total mass and the "length", are

$$M = Nm \tag{5}$$

$$L = N^{1/3}r \tag{6}$$

(4), (5), (6) give

$$\sigma_N = \frac{G \; ML^2 N^{1/3}}{c^2\lambda} \tag{7}$$

(7) exceeds (2) by about 9 orders. The model consisting of pairs of atoms is unrealistic, because a pair of atoms is not likely to have quantum states comparable with the energy levels of a large solid.

Usually the lowest longitudinal mode of a cylinder is employed. The acoustic waves are one dimensional. A slab with plane boundaries normal to the cylinder axis can be driven, at resonance, at the normal mode frequency. The slabs would exchange energy with gravitational radiation even if removed from the bar, and electromagnetic restoring forces applied. (4) then suggests

$$\sigma_{N_{\text{PULSES}}} = \frac{8\pi^3 M_{SLAB} L^2 (N_{SLAB})^2}{c^2 \lambda} \tag{8}$$

now

$$M = M_{SLAB} N_{SLAB} \tag{9}$$

$$N_{SLAB} = N^{1/3} \tag{10}$$

(8), (9), and (10) give

$$\sigma_{N_{\text{PULSES}}} = \frac{8\pi^3 GML^2 N^{1/3}}{c^2 \lambda} \tag{11}$$

(11) would be valid if the quality factor Q of a slab is large, exceeding 10. Analyses confirm our intuitive feeling that the dissipation for a single driven slab is comparable with the normal mode dissipation, but the mass is much smaller than the antenna mass. The quality factor Q_{SLAB} is given approximately by

$$Q_{SLAB} = \frac{Q_{MODE}}{N^{1/3}} \tag{12}$$

(11) and (12) then give

$$\sigma_{N_{\text{PULSES}}} = \frac{8\pi^3 GML^2 Q_{MODE}}{c^2 \lambda} \tag{13}$$

(1) implies that (13) is also valid for the monochromatic continuous case.

$$\sigma_{N\ CW} = \frac{8\pi^3 GML^2 Q_{MODE}}{c^2 \lambda} \tag{14}$$

(13) and (14) indicate that the many body analysis leads to the classical cross section for the continuous, monochromatic case, but a much larger cross section than the classical value for the pulsed case.

BANDWIDTHS

The first antennas were instrumented with optimal filters, choosing a bandwidth to give best signal to noise ratio in the presence of wideband and narrow band noise. In the absence of wideband noise,the equivalent circuit for an antenna with piezoelectric instrumentation is given by Figure 1. The upper voltage generator is proportional to the Riemann tensor, the lower one is a Brownian motion noise generator. These electromotive forces couple to the normal mode in the same way. Therefore, in absence of additional noise, the output for pulsed gravitational waves will be large at the center frequency of the normal mode,and smaller at other frequencies. The Brownian motion noise output will have the same frequency dependence. The signal to noise ratio will therefore be independent of frequency and the bandwidth infinite.

In practice the wideband noise of the antenna and electronics lead to an optimal bandwidth for pulses. It has been customary to employ an analogue filter and demodulator, recording the very narrow bandwidth on tape. Much greater bandwidth is available if the antenna output is amplified by a wideband amplifier, and the wideband output recorded on tape. A computer can then carry out the functions of demodulation and filtering, without prejudging the character of incoming signals. A sequence of different filters may be employed. Spectral regions far from the center of the lowest mode are available. The use of ferroelectric crystal center instrumentation makes available many normal modes. A spectrum analyzer was employed to study the available bandwidth of a room temperature antenna with lead zirconate titanate crystals in the center. Considerable output was observed for a bandwidth exceeding 30,000 Hertz.

LOW TEMPERATURE ANTENNA OPERATIONS

An elastic solid antenna may be cooled to very low temperatures. The thermal fluctuations decrease, the quality factors usually increase. With low noise amplifiers and back action evasion,the signal to noise ratio increases with decreasing temperature.

The increase of quality factor implies reduced interactions with the heat bath. Thermal relaxation times can be long, of the order of a day. Under these conditions an antenna can be prepared in a pure quantum state, not in thermal equilibrium. The interaction is then linear in the Riemann tensor[3]. The thermal fluctuations for non equilibrium states[4] may be much smaller than for equilibrium states. In this way the signal to noise ratio may be increased without limit.

CASCADING OF HALF ACOUSTIC WAVELENGTH ANTENNAS

The early analysis noted that cross sections decrease if the antenna length exceeds a half acoustic wavelength, because the acoustic fields change phase. Similar problems arose early in this century with electromagnetic antennas. For the electromagnetic case, it was discovered

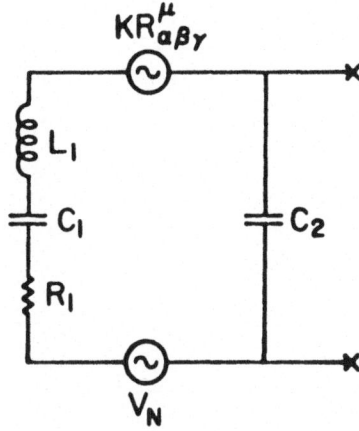

Figure 1. Antenna equivalent circuit

that half wavelength sections can be cascaded, if a network element with phase shift of π is inserted between half wavelength sections. This method of suppression of alternate phases can be employed with gravitational antennas. Half acoustic wavelength sections may be formed by a short section of a slow wave structure, to give a phase shift of π. In this way the cross section will increase as sections are added, and the signal to noise ratio may increase.

ACKNOWLEDGEMENT

We are pleased to thank the United States National Science Foundation for partial support of our research.

REFERENCES

1. J.D. Jackson, Classical Electrodynamics, formula 14.114, second edition John Wiley and Sons 1975, p. 681.
2. J. Weber, General Relativity and Gravitational Waves, chapter 8, Interscience-Wiley, New York, London, 1961.
3. J. Weber, Physics Letters 81A, 9, 542, 23 February, 1981.
4. Gravitational Radiation and Relativity, Volume 3, Proceedings of the Sir Arthur Eddington Centenary Symposium, edited by J. Weber, T.M. Karade, World Scientific, 1986.

COMPARISON OF BARS AND INTERFEROMETERS: DETECTION OF TRANSIENT GRAVITATIONAL RADIATION

Daniel Dewey
Massachusetts Institute of Technology
Room 3-253
Cambridge, MA 02139
U.S.A.

ABSTRACT. The signal-to-noise ratios for the detection of transient gravitational radiation are compared for resonant bar and laser interferometric detectors. For the detection of gravitational-wave bursts (e.g., supernovas, particle-black hole interactions), the bar antenna shows its resonant nature with a sensitivity that is broadly peaked near the bar frequency, while the interferometer shows a smooth f^{-1} frequency response (for a fixed gravitational energy emitted in the burst). Applied to the detection of the 'chirp' of radiation emitted in the decay of a compact binary system, the ratio of bar to interferometer sensitivities is independent of the source parameters. Operational bars and interferometers are compared using specific parameters of representative antennas and are seen to be operating near sensitivities allowing Galactic observations. Finally, each detector technology holds promises for increased sensitivity which will allow for collaborative detection of sources as distant as the Virgo cluster.

1.0 INTRODUCTION

The goal of this work is a quantitative comparison of the sensitivites of current and proposed bar and interferometric antennas in well-defined cases in which the antennas are located at a specified distance from and at an optimum orientation to a specified source of gravitational radiation. Using relevant waveforms and antenna models, the signal-to-noise ratios (SNRs) calculated for the antennas can then be directly compared and placed in an astrophysical context. Note that the process of averaging over a variety of parameters (source orientation, antenna orientation, source distance distribution, etc.) and considerations of event rate are left out of this comparison so that interpretation of the results is unambiguous.

201

B. F. Schutz (ed.), Gravitational Wave Data Analysis, 201–215.
© 1989 by Kluwer Academic Publishers.

As indicated above, the ingredients of the comparison
are realistic source and antenna models; these have been
presented in detail elsewhere[1] and so here I want to summarize the results and then look to the future which promises
higher sensitivities and suggests a collaboration of bar and
interferometric antennas. Models of waveforms for transient
sources (bursts and chirps) are presented in Section 2; in
Section 3 the antenna models are presented and used to calculate the response of each type of antenna to the burst and
chirp sources. To enlived the resulting equations, Section
4 presents a comparison of two operational antennas in a
Galactic context. Finally in Section 5 the potential of
gravity-wave antennas to operate at increased sensitivities
is discussed.

2.0 TRANSIENT SOURCE MODELS

2.1. Burst Sources

The expected burst sources consist of stellar collapse
events and particle-black hole interactions; examples of
predicted wavforms from these sources are given in Fig. 1.
These sources can be well modeled[8],[9] by a canonical sine
burst parameterized by the burst frequency f_g, the burst
amplitude h_o, and the number of half-cycles in the burst
N_{hc}. The amplitude of these bursts may be realistically set
by choosing a distance from source to antenna R and by
specifying the total gravitational-wave energy emitted in
the burst as ϵMc^2, where M is a mass representative of the
system and ϵ is an efficiency for gravity-wave production.
This leads to[1]

$$h_o = \frac{1}{R} [\epsilon M \frac{4G}{\pi c}]^{1/2} [N_{hc} f_g]^{-1/2} \tag{2.1}$$

$$\simeq 8\times10^{-18} \cdot [\frac{\epsilon M}{10^{-2} M_\odot}]^{1/2} [\frac{1 \text{ kHz}}{f_g}]^{1/2} N_{hc}^{-1/2} .$$

Thus the canonical burst used in this comparison is
completely specified by f_g, N_{hc}, ϵM and R.

2.2. Chirp Sources

The chirp of gravitational radiation emitted by a compact
binary system as its decays through gravitational radiation
has long been considered[10] as a potential source of detectable
gravity-waves. The waveform of emitted radiation can be
parametrized by the time-until-coalescence τ (Ref.11), in

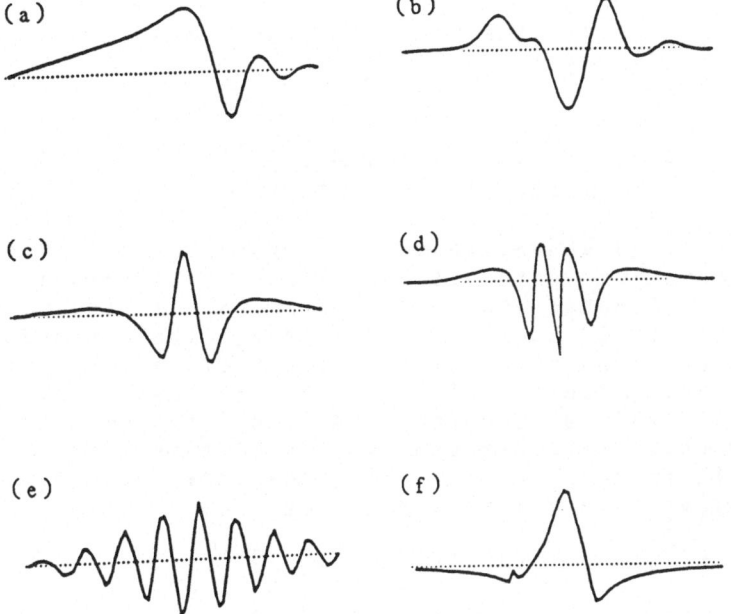

Figure 1. Waveshapes representative of the gravitational radiation expected from burst sources. The waveshapes shown here are digitized versions of published theoretical predictions for a variety of systems: (a) a test particle falling radially into a Schwarzschild black hole (Ref.2), (b) rotating stellar collapse (Ref.3), (c) a test particle scattered from a Kerr black hole (Ref.4), (d) a test particle scattered from a Schwarzschild black hole (Ref.5), (e) damped ellipsoidal stellar collapse (Ref.6), and (f) cold rotating stellar collapse (Ref.7).

terms of which the amplitude, frequency, and phase-remaining-until-coalescence are given by

$$h_{amp}(\tau) = \frac{1}{R} \left[\frac{5G^5}{c^{11}}\right]^{1/4} F^{3/4} \tau^{-1/4} \quad , \qquad (2.2)$$

$$f_g(\tau) = \frac{1}{\pi} \left[\frac{256}{5} \frac{G^{5/3}}{c^5}\right]^{-3/8} F^{-3/8} \tau^{-3/8} \quad , \qquad (2.3)$$

and

$$\phi(\tau) = \frac{16}{5} \left[\frac{256}{5} \frac{G^{5/3}}{c^5} \right]^{-3/8} F^{-3/8} \tau^{5/8} \quad , \qquad (2.4)$$

where

$$F \equiv \frac{m_1 m_2}{(m_1+m_2)^{1/3}} = \mu(m_1+m_2)^{2/3} = \mu m_T^{2/3} \quad . \qquad (2.5)$$

(Equation (2.2) above assumes a source oriented with its angular momentum vector aligned along the observer's line-of-sight.) Because the source parameters m_1 and m_2 appear in both Eq. (2.2) and Eq. (2.3) through the one combination F, an observation of a chirp, i.e., $f_g(\tau)$ and $h_{amp}(\tau)$, allows a determination of both F and $\dot{R}$; thus chirp observations will lead to independent distance determinations[12].

The chirp waveform and its second time derrivative which will be used in evaluating the interferometric and bar antenna responses are thus given by

$$h(\tau) = h_{amp}(\tau) \sin[\phi(\tau)+\phi_{arb}] \qquad (2.6)$$
$$\ddot{h}(\tau) = -h_{amp}(\tau)\omega_g^2(\tau)\sin[\phi(\tau)+\phi_{arb}] \quad ,$$

and no amplitude or efficiency assumptions are required. An example of $\ddot{h}(t)$ for a chirp waveform is seen in the lower trace of Fig. 2. The wave shape given by Eq. (2.6) is adequate for purposes of comparison, however a variety of corrections may be important in predicting real waveforms[13]. In addition to these corrections the finite size of the radiating objects will lead to deviation from these equations when the separation approaches a few object-radii. For a system of two 1.4 $M_\odot$ neutron stars this will occur when of order five cycles remain in the chirp waveform (i.e., $\phi(\tau)=5\cdot 2\pi$).

3.0 ANTENNA MODELS

The response of both bar and interferometric antennas can be presented in one framework[14]; however, present antenna types each have their own limiting noise sources and data analysis algorithms. Thus, a realistic comparison must take these into account in calculating a value for the signal-to-noise ratio (SNR) for detecting a given source waveform. The antenna models and calculations of SNRs for detecting the burst and chirps of Section 2 are given below.

3.1. Interferometric Antennas

The present limiting noise for interferometric antennas is

the laser shot noise which appears as a white noise of level
$\tilde{h}$ in the interferometer output. The gravitational-wave
signal h(t) must then be detected in the presence of this
noise. The resulting optimal (i.e., maximum) SNR for detection is given by[9,15]

$$(S/N)_I = \frac{1}{\tilde{h}} [2 \int h^2(t) \, dt]^{1/2} . \tag{3.1}$$

Note that this SNR is proportional to h(t).

3.1.1 Interferometer response to a burst. For the bursts,
substitution of the simple sine-burst waveform into Eq.
(3.1) gives the result

$$(S/N)_{I,burst} = \frac{1}{\tilde{h}} \frac{1}{R} \left(\frac{2 \epsilon MG}{\pi c}\right)^{1/2} f_g^{-1} , \tag{3.2}$$

where it is important to recall that this assumes a burst of
constant emitted gravitational-wave energy independent of
burst frequency.

3.1.2 Interferometer response to a chirp. Because an interferometric antenna typically has a noise spectrum which
rises steeply below some minimum frequency f_{min}, the chirp
waveform (which extends to the essentially-infinite past)
will only be observed when f_g is above f_{min}. Carrying out
the integration of Eq. (3.1) from τ given by $f_g(\tau)=f_{min}$ to
$\tau=0$ (coalescence) gives, to good approximation, the result

$$(S/N)_{I,chirp} = \frac{1}{\tilde{h}} \frac{1}{R} \frac{\sqrt{5}}{2\sqrt{2}} \frac{\pi^{-2/3} G^{5/6}}{c^{3/2}} F^{1/2} f_{min}^{-2/3} . \tag{3.3}$$

Note that higher SNR values can be obtained by reducing f_{min}
(increasing the total observation time) which has the additional benefit of allowing lower frequency (larger F) chirps
to be observed.

3.2. Bar Antennas

Present bar operation seeks to detect a small energy innovation in the presence of a background of thermal and transducer noises[16,17]. With the present signal processing
algorithms the detection of a gravity-wave is a two step
process: the gravity-wave deposits an amount of energy into
the bar then this energy change is detected. This picture
assumes that the burst occurs on a time scale short compared
to the sample time and the bar ring-down time. The figure

206

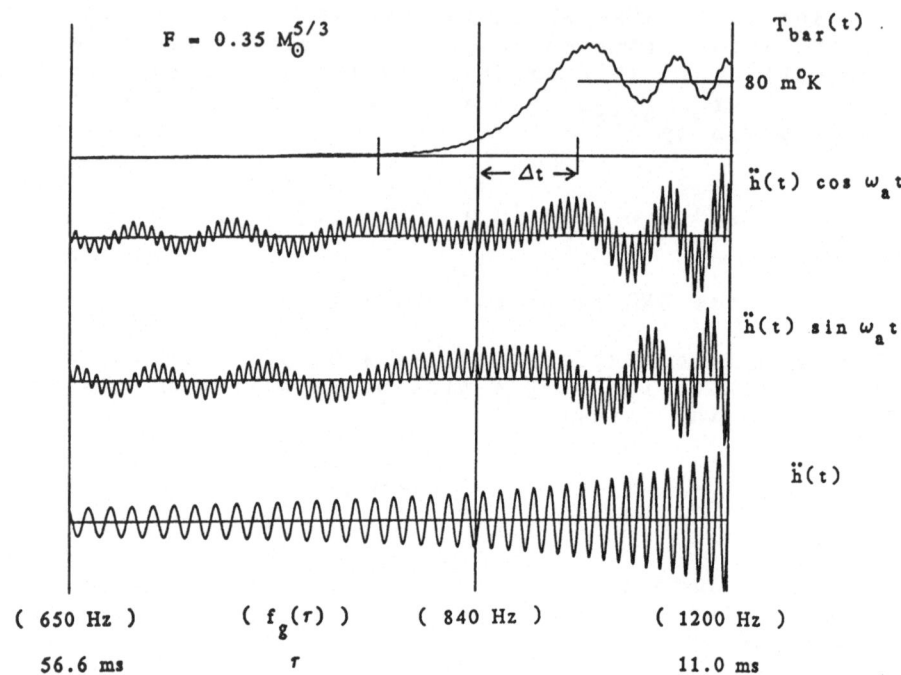

$F = 0.35 M_{\odot}^{5/3}$

$T_{bar}(t)$

80 m^{o}K

$\leftarrow \Delta t \rightarrow$ $\ddot{h}(t) \cos \omega_a t$

$\ddot{h}(t) \sin \omega_a t$

$\ddot{h}(t)$

(650 Hz)   ($f_g(\tau)$)   (840 Hz)   (1200 Hz)

56.6 ms τ 11.0 ms

Figure 2. Numerical simulation of a bar antenna's
response to a chirp waveform emitted by a compact
binary system. Plotted as a function of τ, the
time to coalescence, are the $\ddot{h}(t)$ waveform, the
integrands of Eq. (3.5), and the temperature equivalent of the energy deposited in the bar as of time
τ. The time Δt indicated is calculated from Eq.
(3.8) and the asymptotic temperature indicated has
been calculated from Eq. (3.4) and Eq. (3.9). This
chirp has $F = \mu m_T^{2/3} = 0.35$ (two $0.6 M_{\odot}$ objects); for
more massive systems the chirp moves to lower frequencies and the number of cycles of the waveform
above the (fixed) bar frequency is reduced. Other
parameters are as given in Table 1.

of merit for the bar detector, then, is the rms energy noise
in the innovation measurement, usually expressed as an
equivalent temperature $T_d = E_d/k$. If a source deposits an
energy kT_s into the bar, the amplitude signal-to-noise ratio
for detection of the source is

$$(S/N)_B = [\frac{T_s}{T_d}]^{1/2} \quad , \tag{3.4}$$

and thus the calculation of the SNR for a source requires a
calculation of T_s.

In general, T_s can be calculated from the known $\ddot{h}(t)$ waveform and the parameters of the bar using a simple harmonic oscillator model[1]; this leads to

$$(S/N)_B = \frac{1}{T_d^{1/2}} \frac{M_a^{1/2} L_a}{\pi^2 k^{1/2}} [(\int \ddot{h}(t)\sin\omega_a t \, dt)^2 \qquad (3.5)$$

$$+ (\int \ddot{h}(t)\cos\omega_a t \, dt)^2]^{1/2} ,$$

where L_a and M_a are the length and mass of the cylindrical bar, $\omega_a (=2\pi f_a)$ is the bar resonant frequency, and the bar is optimally oriented to the strain field. Note that the bracketed term in Eq. (3.5) is just the magnitude of the Fourier amplitude of $\ddot{h}(t)$ at the bar frequency.

3.2.1 Bar response to a burst. For the canonical burst of Section 2.1 evaluation of the integrals of Eq. (3.5) is straight forward (with due care of the delta functions that appear in the second derrivative of $h(t)$) and gives the result

$$\qquad (3.6)$$

$$(S/N)_{B,burst} = \frac{1}{T_d^{1/2}} \frac{1}{R} \left(\frac{4\epsilon MG}{\pi kc}\right)^{1/2} L_a (M_a N_{hc} f_g)^{1/2} C(\frac{f_a}{f_g}, N_{hc})$$

where,

$$\qquad (3.7)$$

$$C(\frac{f_a}{f_g}, N_{hc}) = \frac{\sin x}{x} + \frac{\sin y}{y} - \frac{4}{\pi N_{hc}} \cos(\frac{f_a}{f_g} \frac{\pi N_{hc}}{2})$$

with

$$x = \frac{\pi N_{hc}}{2}(1 - \frac{f_a}{f_g})$$

$$y = \frac{\pi N_{hc}}{2}(1 + \frac{f_a}{f_g})$$

and   upper of 1st and 3rd choices for N_{hc} odd

upper of 2nd choice for $N_{hc} = 1, 2, 5, 6, \ldots$

Note that the function $C()$ is equal to unity when $f_a = f_g$; thus, this tuned response is given by a simple expression (i.e., Eq. (3.6) with $C()=1$). The off-resonance response of the bar is contained in $C()$; the width of this essentially reflecting the frequency width of the source.

3.2.2. Bar response to a chirp.

Because the bar-chirp interaction is complicated and transient, it is useful to view the interaction numerically to gain insight into the interaction. In Fig. 2 the $\ddot{h}$ waveform is shown along with the integrands of Eq. (3.5) and the bar temperature, all plotted as a function of τ, the time-to-coalescence. Of importance here is the finite time-period during which energy is deposited into the bar governed by a near-constant phase relation between the bar frequency and the chirp waveform. This time scale can be calculated[11,1] and at the point in time when the chirp is radiating at f_0 it is

$$\Delta t \; \equiv \; \text{time until a } \frac{\pi}{2} \text{ phase shift} \qquad (3.8)$$

$$= \; [\; \frac{5\pi^{-8/3}}{192} \; \frac{c^5}{G^{5/3}} \;]^{1/2} \; F^{-1/2} \; f_0^{-11/6}$$

$$\simeq \; 3 \; \text{ms} \; (\frac{F_0}{F})^{1/2} \; (\frac{1 \; \text{kHz}}{f_0})^{11/6} \quad .$$

Analytic approximations to the integrands of Eq. (3.5) can be made assuming that the chirp waveform has several cycles within the period $\pm\Delta t$ centered on the bar frequency. The result of these calculations is a simple expression giving the bar SNR for detection of a chirp source[1]

$$(S/N)_{B,chirp} = \frac{1}{T_d^{1/2}} \; \frac{1}{R} \; \frac{\sqrt{10} \; G^{5/6}}{\sqrt{3k} \; \pi^{2/3} c^{3/2}} \; L_a M_a^{1/2} f_a^{5/6} \; F^{1/2} \qquad (3.9)$$

Note that, unlike the interferometer case, decreasing the bar frequency f_a in order to detect lower frequency chirps (larger F values) at face value results in a decrease in the

TABLE 1. List of source and antenna parameters used in the comparisons.

Sources	Bar (4 K)	30-m Interferometer
$f_g = 100-10$ kHz	$T_d = 20$ mK	$\tilde{h} = 2 \cdot 10^{-19}$
$N_{hc} = 4$	$f_a = 840$ Hz	$f_{min} = 500$ Hz
$\epsilon M = 0.01 \; M_\odot$	$L_a = 3.0$ m	
$R = 10$ kpc	$M_a = 4.8 \cdot 10^3$ kg	
$F = 0.35 \; M_\odot^{5/3}$		

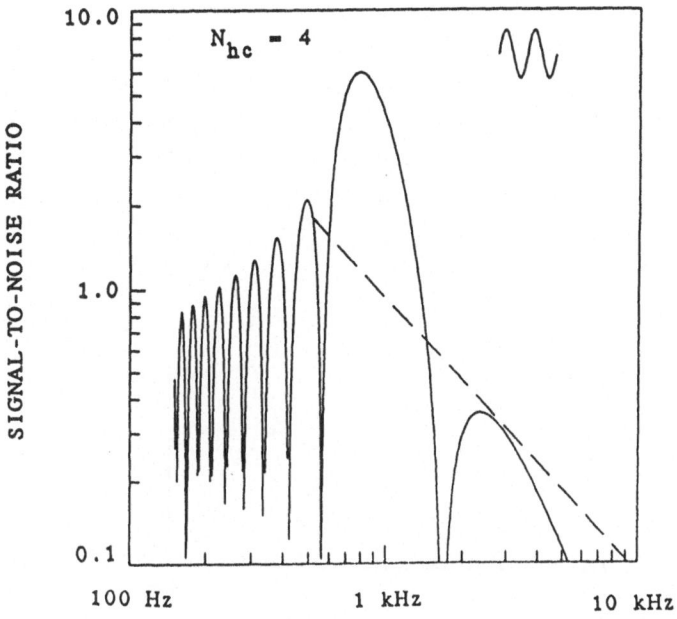

FREQUENCY OF BURST f_g

Figure 3. Signal-to-noise ratios (SNRs) for detection of a Galactic burst source as a function of
the burst frequency. The 4.2 K Stanford bar SNR is
given by the solid curve; the MPQ Garching 30-m
interferometer SNR is given by the dashed line. The
burst has N_{hc}=4 with a total emitted gravity-wave
energy held constant at $10^{-2} M_\odot c^2$. The parameters
of Table 1 have been used for these calculations.

SNR for chirp detection (with the caveat that changes in L_a
and M_a may offset the f_a term).

4.0 COMPARISON FOR TWO OPERATIONAL ANTENNAS

Numerical values of the above equations were calculated
using the operational parameters of representative antennas.
The parameters used here, Table 1, are those of the Stanford
4 K bar[18] and the MPQ Garching 30-m interferometer[19]; these
are well-documented instruments and represent the state-of-
the-art as of 1985.
        As an example of the resulting comparison for burst
source detection Fig. 3 shows the SNR for each antenna as a
function of the frequency of the burst which has N_{hc}=4, R=10
kpc ($\simeq$Galactic center), and a fairly optimistic efficiency

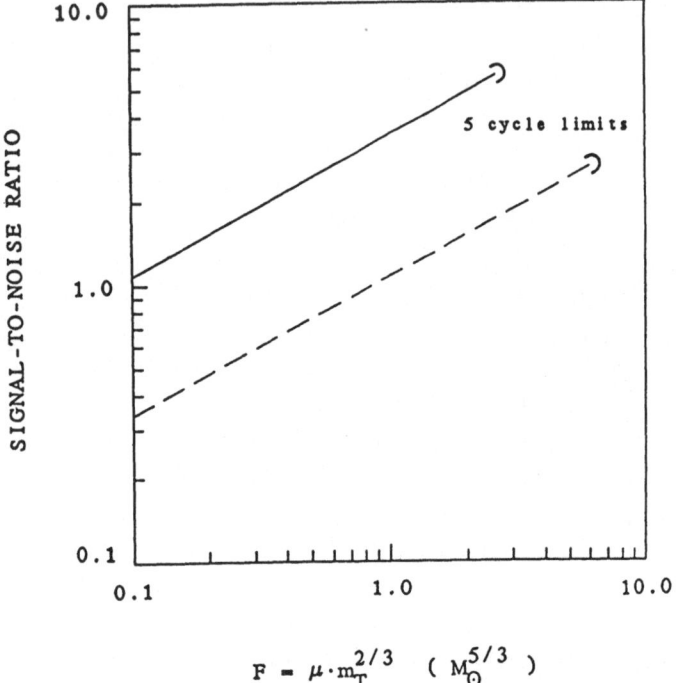

$$F = \mu \cdot m_T^{2/3} \quad (M_\odot^{5/3})$$

Figure 4. Signal-to-noise ratios (SNRs) for detection of a Galactic chirp source as a function of the source parameter $F = \mu m_T^{2/3}$. The 4.2 K Stanford bar SNR is given by the solid curve; the MPQ Garching 30-m interferometer SNR is given by the dashed line. The curves are discontinued at large F where too few cycles (<5) of the chirp waveform remain for the analytic calculations to be accurate. The parameters of Table 1 have been used for these calculations.

of $\epsilon = 0.01$. The bar response shows the frequency dependence of the source given by the C() term in Eq. (3.6). The interferometer curve has been discontinued below 500 Hz where seismic noise dominates.

A comparison of these two antennas when applied to the detection of a chirp is given in Fig. 4 where the SNRs are plotted as a function of the source parameter $F = \mu m_T^{2/3}$. As examination of equations (3.3) and (3.9) shows, both the bar and interferometer have the same dependence on the chirp source parameter F and thus the ratio of bar to interferometer sensitivity is a constant depending only on the instrumental parameters. The curves are shown discontinued at large F values where the chirp source emits too few cycles of waveform at the bar frequency f_a (or above the interferometer frequency f_{min}) for the results of Eq. (3.9) (or Eq. (3.3)) to be accurate.

5.0 FUTURE SENSITIVITIES AND FUTURE RESEARCH

There are a variety of reasons to develop more sensitive antennas. A higher SNR will allow a transition from simply detecting sources (SNR$\simeq$7) to studying source astrophysics (SNR>30). In order to increase the event rate larger volumes of space must be accessible, thus source distances will go from R=10kpc to R>10Mpc. Finally, the efficiency of ϵ=0.01 used in the calculations above is probably optimistic; to ensure detection, sources with ϵ as low as 10^{-6} should be detectable. Thus, sensitivities ten to ten thousand times those of the antennas of Table 1 are desirable. Improvements will come primarilly through reductions in $\tilde{h}$ and T_d as briefly described below[20].

For the interferometric antennas decreases in $\tilde{h}$ depend on increases in the optical storage length Nl (delay-line notation is used for clarity) and in the interferometer input power P, specifically

$$\tilde{h} = \frac{1}{Nl} \frac{\lambda}{2\pi} [\frac{2hc}{\eta P\lambda}]^{1/2} \simeq 10^{-19}/\sqrt{Hz} (\frac{1 \text{ km}}{Nl}) \frac{1}{\sqrt{P_{Watts}}} \qquad (5.1)$$

where the numerical evaluation has assumed values for λ and η of 514 nm and 0.5 respectively. Going to large baseline antennas will increase Nl from the present $\simeq$3 km to $\simeq$100 km and reduce the ratio of seismic noise to gravity-wave signal. Improvements in isolation schemes will lower the value of f_{min} improving the SNR for detecting chirp sources. Additional improvements is sensitivity will come through increases in light power P. One promising technique is the recycling[21] of light which leaves the interferometer; this may lead to power increases of order ten or more. More speculative is a reduction in shot noise through squeezing[22], which has, however, produced an increase of a factor of two in equivalent P in recent experiments[23]. Even the quantum limit (determined by the balancing of radiation-pressure noise and shot-noise) will not be accepted with out attemps to circumvent it[24].

For the bar antennas a reduction of T_d may be achieved through reducing the physical temperature of the bar T_a, increasing the bar Q, and through improvements to the transducer system expressed as an increase in the system bandwidth. There are plans to cool the bars to 10-50 mK from the current 4 K, and multi-mode bar systems[25] with increased bandwidth will allow T_d to approach the ideal value of order T_a/Q. Note that with these increased bandwidths, and thus reduced sample times, the assumption that the gravity-wave deposits energy into the bar on a timescale shorter than the sample time may no longer be valid, especially for chirp sources. Thus, new bar algorithms may be required to optimally detect these events. Finally, with a philosophy similar to squeezed states, back action evasion schemes[26] may improve present systems and allow higher sensitivities than the standard quantum limit would imply.

What kind of sensitivities can we hope to obtain with these improvements? Fig. 5 presents plots of SNR for the *hypothetical* bar and interferometer parameters listed in Table 2. In viewing this figure it is well to adopt the opinion that[27] "mere theoretical notions are ... useless, representing only sterile mental exercises." Even so, this figure does show that each detector technology holds a promise for sensitivities adequate to detect (SNR≃7) sources at Virgo cluster distances and to study (SNR>30) Galactic sources.

This comparison has focussed on transient sources and a comparison for periodic sources needs to be carried out[28]. In the context of transient sources, however, there are a variety of differences between (current) bar and interferometric antennas that have not been addressed; for example: time resolution, non-statistical event rates, antenna patterns, and frequency response. Most of these issues will become important as we analyze systems of many antennas; to this end important ground work on the problem of multiple antenna systems and coincidence rates has been carried out for bar[29] and interferometric[30] systems. In particular it will be important to consider hybrid systems of bars and

TABLE 2. Ficticious antenna systems and their parameters.

	P (W)	N	l (km)	Interferometer Systems
a)	1	2	5	low power, two pass, long arms
b)	10	30	3.3	multi-pass, medium power laser
c)	100	20	5	multi-pass, recycled (×10)

	T_d (K)	f (Hz)	M (10^3 kg)	L (m)	Bar Systems
d)	10^{-4}	800	5.5	3.1	T_a=4 K, three-mode bar
e)	10^{-6}	600	11.0	4.1	T_a=30 mK, multi-mode
f)	$3 \cdot 10^{-7}$	300	60.0	8.2	T_a=10 mK, massive bar

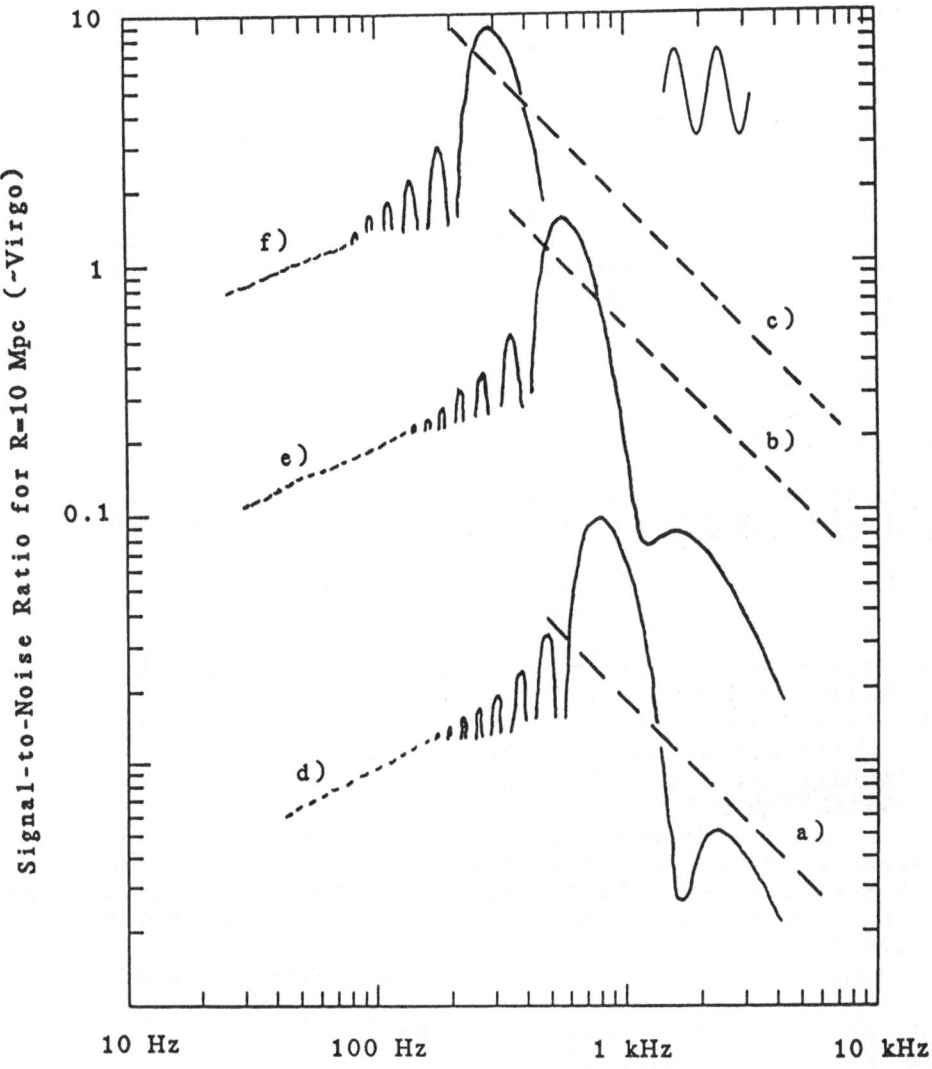

GW Burst Frequency

Figure 5. Comparison of *hypothetical* bar and
interferometer antennas to a burst source near
Virgo (10 Mpc). The bar and interferometer SNRs
are plotted as a function of the frequency of the
gravity-wave burst; other parameters of the burst
are $\epsilon M = 0.01 M_\odot$ and $N_{hc} = 4$. The parameters of the
ficticious antenna systems are given in Table 2.

interferometers leading to a grand-GRAVNET collaboration in which the antennas work together to detect and study gravitational waves and their astrophysical sources.

ACKNOWLEDGEMENTS

I gratefully acknowledge helpful discussions with W. Johnson, J.P. Richard, P.R. Saulson, and D.H. Shoemaker. This work was supported in part by the National Science Foundation under Grant 85-04836-PHY.

REFERENCES AND NOTES

[1] D. Dewey, Phys.Rev.D, **36**, 2901 (1987).

[2] S.L. Detweiler, in *Sources of Gravitational Radiation*, edited by L. Smarr (Cambridge University Press, 1979), p.211.

[3] R.F. Stark and T. Piran, Phys. Rev. Let. **55**, 891 (1985).

[4] Y. Kojima and T. Nakamura, Prog. Theor. Phys. **72**, 494 (1984).

[5] K. Oohara and T. Nakamura, Prog. Theor. Phys. **71**, 91 (1984).

[6] R. Saenz and S. Shapiro, Astrophys. J. **244**, 1033 (1981).

[7] R. Saenz and S. Shapiro, Astrophys. J. **221**, 286 (1978).

[8] G.W. Gibbons and S.W. Hawking, Phys. Rev. D **4**, 2191 (1971).

[9] D. Dewey, in *Proceedings of the Fourth Marcel Grossmann Meeting on General Relativity*, edited by R. Ruffini (Elsevier Science Publishers B.V., Amsterdam, 1986),p.581.

[10] R.L. Forward and D. Berman, Phys. Rev. Let. **18**,1071 (1967).

[11] J.P.A. Clark and D.M. Eardley, Astrophys. J. **215**, 311 (1977).

[12] B. Schutz, Nature **323**, 310 (1986).

[13] A. Krolak, in this volume.

[14] W.H. Press and K.S. Thorne, in *Annual Review of Astronomy and Astrophysics*, edited by L. Goldberg (Anual Reviews Inc., Palo Alto, 1972),p.355.

[15] D. Dewey, "Data Analysis ...", in this volume.

[16] G. Pizzella, in this volume.

[17] P.F. Michelson, J.C. Price, and R.C. Taber, Science **237**, 150 (1987).

[18] M. Bassan, W.M. Fairbank, E. Mapoles, M.S. McAshan, P.F. Michelson, B. Moskowitz, K. Ralls, and R.C. Taber, in *Proceedings of the Third Marcel Grossmann Meeting on General Relativity*, edited by H. Ning (North-Holland, Amsterdam, 1983),p.667.

[19] D. Shoemaker, W. Winkler, K. Maischberger, A. Rüdiger,

R. Schilling, and L. Schnupp, in *Proceedings of the Fourth Marcel Grossmann Meeting on General Relativity*, (Ref. 9),p.605.

[20] Many of the contributions to this volume describe planned detector systems in greater depth than is possible here.

[21] Recycling has been demonstrated by the MPQ Garching and Orsay groups as communicated by L. Schnupp and C.N. Man, respectively, at this workshop.

[22] G. Leuchs, in this volume.

[23] M. Xiao, L. Wu, and H.J. Kimble, Phys. Rev. Lett. **59**, 278 (1987).

[24] Wei-Tou Ni, Phys. Rev. D15 35, 3002 (1987).

[25] J.P. Richard, J. Appl. Phys. 60, 3807 (1986).

[26] M.F. Bocko and W. Johnson, this volume

[27] N.W. Ross, *Three Ways of Asian Wisdom* (Simon and Schuster, New York, 1966),p.94.

[28] Initial examination of the *periodic* sensitivities of bars (Ref. 16) and interferometers shows that in the frequency range where, say, the Stanford bar is operating near the thermal noise limit ($\simeq 840\pm 5$ Hz) it is two orders of magnitude more sensitive to periodic sources than the MPQ Garching interferometer!

[29] D. Blair, in this volume.

[30] B.F. Schutz and M. Tinto, Mon. Not. R. astr. Soc. **224**, 131 (1987).

BROADBAND SEARCH TECHNIQUES FOR PERIODIC SOURCES OF GRAVITATIONAL RADIATION

Jeffrey Livas
M. I. T. Physics Department
Building 20F-001
Cambridge, MA 02139
USA

ABSTRACT. Full exploitation of the potential of a laser interferometric gravitational wave detector to detect sources emitting periodic gravitational waves will require the development of specialized search techniques and place unprecedented demands upon computer memory and speed. Traditional approaches to detecting small amplitude sinusoidal signals of unknown period in the presence of broadband noise must be modified to accomodate the unusual nature of the received signal from a gravitational wave source. This paper outlines two methods that have been developed and presents the results of an application of these methods to data taken with the M.I.T. 1.5 meter prototype. Based on these results, a combined hardware/software solution to the data analysis problem is needed before full scale gravitational wave astronomy with periodic sources becomes practical.

1. INTRODUCTION

1.1 The Ideal Search

The rapid developement of gravitational wave antennas has raised the possibility of a new branch of astronomy that could provide information about strongly interacting astrophysical systems that cannot be obtained any other way. One type of detector, the laser interferometric antenna, has two special properties: it has a broadband spectral response and a non-directional spatial response. Full use of the antenna to detect sinusoidal signals is thus a two dimensional problem because both the frequency of the signal and the direction from which it comes are unknown. The ideal search technique would be a method that could recover both the frequency and the location of the source.

The search is complicated by the fact that the relative motion between the source and the detector modulates the signal in a manner that depends explicitly on the location of the source, even for a source emitting monochromatic radiation. Traditional signal processing methods for extracting a sinusoid buried in broadband noise are not applicable without modification. In addition, the sheer volume of data generated by a wide bandwidth detector presents special problems for collection, storage, and analysis.

After a brief discussion of typical sources of periodic gravitational radiation and the model of the source adopted for this discussion, two different techniques for conducting a broadband search will be presented. The first is a method for performing an ideal two dimensional search, although with a finite limit on the sensitivity. The second method has no such limit on its sensitivity, but reduces the sky coverage to a small area around a single direction.

B. F. Schutz (ed.), Gravitational Wave Data Analysis, 217–238.

1.2 Possible Astrophysical Sources

Source of gravitational waves may divided into three basic categories: periodic, impulsive, and stochastic. Several recent reviews describe each of these categories in detail.[1,2] This discussion is concerned only with periodic sources, which may be divided into two (overlapping) classes: binary star systems and rotating neutron stars.

1.2.1 *Binaries.* A binary star system is perhaps the simplest of all periodic sources because it has an intrinsically time-varying quadrupole moment that is easy to calculate. Indeed, the famous binary pulsar PSR 1913+16 is the only physical system discovered so far in which gravitational radiation provides a conclusive explanation of the observations.[3] Furthermore, it has been estimated that more than half of the stars in the galaxy have at least one stellar companion. However, the shortest known binary period is ~ 685 seconds, or $\nu = 1.5 \times 10^{-3}$ Hz, which is much lower can be achieved with ground based gravitational wave antennas. The best present prototypes have useful sensitivities down to ~ 100 Hz, and even the most optimistic predictions for future suspension designs estimate that ground motion will begin to dominate the spectrum below 10 Hz. Thus ordinary binaries are not an important source for ground based antennas.

However, a binary system may be a catalyst for an important high frequency periodic source. Wagoner[4] has proposed a model of a rotating neutron star in a close binary system in which accretion drives the rotation of the neutron star in a steady state in which the viscous damping timescale matches the timescale for the growth of gravitational wave instabilities. He estimates that monochromatic emission of gravitational radiation could occur with 200 Hz $\geq \nu \geq$ 800 Hz.

1.2.2 *Rotating Neutron Stars.* The best known example of this type of source is the pulsar. A pulsar is thought to be a rotating neutron star whose symmetry axis does not coincide with the axis of rotation. Such a system has a time changing quadrupole moment and could radiate gravitational waves. The fastest known pulsar spins at 642 Hz (1.5 msec), which is well within the bandwidth of an earth based detector.

There are approximately 450 known pulsars in the galaxy, and the total population has been estimated[5] as $70(\pm 17) \times 10^3$. The number of fast (msec) pulsars in the galaxy has been estimated[6] as 30. Thus, if pulsars produce gravitational radiation, there should be several sources.

1.2.3 *Source Model.* For the purposes of this analysis, the standard source assumed to be a simple narrowband sinusoidal emitter with a $\delta(f)$ frequency spectrum. The detection strategy is simply to integrate for as long as possible increase the signal to noise ratio (SNR). To see that a delta function source is a reasonable model, assume that the search bandwidth is 10 kHz and the integration time one year. Then if the delta function is to be a good approximation, the source must have a $Q \geq f_{max}/\Delta f_{res} \approx 3 \times 10^{11}$, where Δf_{res} is the frequency resolution bandwidth, $\sim 1/T_{int}$. Table 1 shows estimated Q's for a few periodic sources. Taken at face value, the delta function source approximation is quite reasonable.

Table 1

Typical Q's of some periodic sources.

Process	Estimated Q
Binary systems	$2\pi \left(\frac{1}{8\,\mathrm{hr}}\right)(8 \times 10^{16}\,\mathrm{sec}) = 2 \times 10^{13}$
Pulsars (in general)	$2\pi \left(\frac{1}{.6\,\mathrm{sec}}\right)(10^7\,\mathrm{yr}) = 3 \times 10^{15}$
Pulsars (msec)	$2\pi (642\,\mathrm{Hz})(3 \times 10^8\,\mathrm{yr}) = 4 \times 10^{19}$
Wagoner's model	$2\pi (500\,\mathrm{Hz})(10^7\,\mathrm{yr}) = 1 \times 10^{19}$

2. ANALYSIS TECHNIQUES

Two periodic search techniques will be discussed, both based on the same strategy: estimate the power spectrum of the data and look for large peaks.

2.1 The Naive Approach

2.1.1 *Periodogram Estimate of the Power Spectrum.* Consider a one dimensional time series of K points $y_i(t) = y(t_i)$. The naive way to find a sinusoidal signal of unknown frequency in noise would be to construct a model of the time series as a sum of sinusoids:

$$y_{model}(t) = \sum_{m=0}^{M-1} A_m e^{-i2\pi f_m t}. \tag{1}$$

where the A_m are the unknown amplitudes. The model can then be fit to the data by forming the χ^2:

$$\chi^2 = \frac{1}{K} \sum_{k=0}^{K-1} \left(y(t_k) - y_{model}(t_k)\right)^2 \tag{2}$$

and then minimizing with respect to the A_m:

$$\frac{\partial \chi^2}{\partial A_m} = 0,$$

which leads to

$$A_n = \sum_{k=0}^{K-1} y(t_k) e^{-i2\pi f_n t_k}. \tag{3}$$

This is the expression for the Discrete Fourier Transform (DFT), which can be efficiently computed with the Fast Fourier Transform (FFT) algorithm if the data has been sampled at equally spaced intervals in time.

The periodogram is just the square of the DFT

$$P_N(m\Delta f_0) = \left| \frac{1}{N} \sum_{n=0}^{N-1} y(n\Delta t_0) e^{-i2\pi nm/N} \right|^2 \tag{4}$$

and is an estimate of the power spectrum. Here Δt_0 is the sampling period and $\Delta f_0 = (N\Delta t_0)^{-1}$ is the frequency resolution of the transform. For convenience the square root of the power spectrum $S_N(m) = (P_N(m))^{1/2}$ is usually used because it represents displacement and hence strain directly. In the discussion which follows, the terms periodogram and power spectrum are used interchangeably. When it is necessary to make the distinction, the square root of the power spectrum, $S_N(m)$, will usually be used.

The periodogram is useful because the signal to noise ratio (SNR) in power of a sinusoidal signal which appears in a single frequency bin of the transfrom increases directly with the length of the transform, or by the square root of the length in amplitude.

2.1.2 *Statistical Peak Detection.* The purpose of statistical peak detection is to provide a quantitative criterion for distinguishing possible signal peaks from chance fluctuations due to noise. Each frequency resolution bin m of the periodogram contains an estimate $s = S_N(m)$ of the the square root of the amount of power at that frequency in the data. The estimate fluctuates because of the presence of noise, and the value of the estimate may be regarded as a random variable. The idea behind statistical peak detection is that the probability distribution of $S_N(m)$ when a signal is present in the data is different from the probability distribution with just noise. Figure 1 shows an example of the differential probability distribution of $S_N(m)$ with and without a signal.

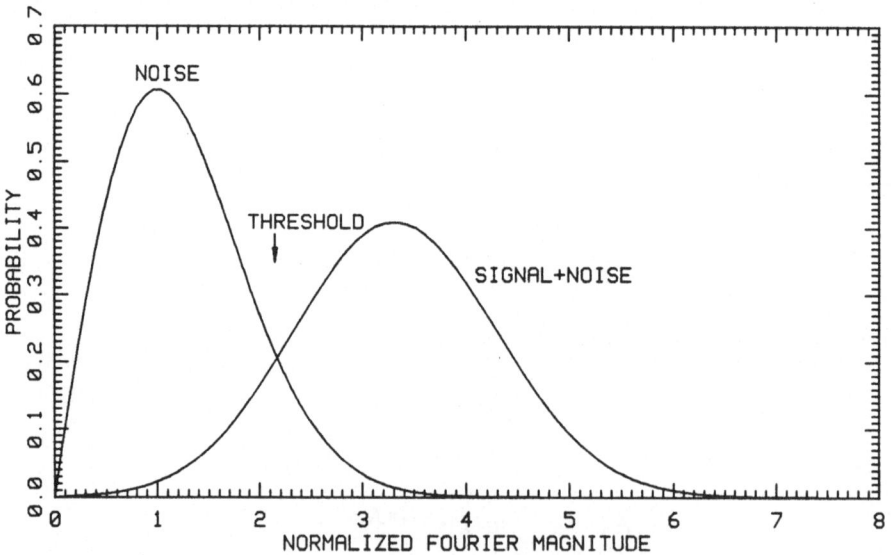

Figure 1
Illustration of the concept behind statistical peak detection. The probability distribution for the amount of power in a single bin of a periodogram changes when a signal is present. The threshold is chosen to minimize the false alarm probability.

With just noise, the $S_N(m)$ follow a χ^2 distribution with two degrees of freedom (a Rayleigh distribution). The presence of a signal means that there is a non zero expectation value for one of the DFT coefficients, so the most probable value of that coefficient (i.e. the peak of the differential probability distribution) increases. The difference between the distribution with pure noise and the distribution with a signal plus noise can be conveniently summarized with a single threshold. Any frequency bin containing more power than the threshold is judged more likely to belong to the signal plus noise distribution than the noise distribution and is flagged as a possible signal. The probability of a false alarm, that is, the probability that a chance fluctuation of the noise exceeds the threshold, is just given by the area under the noise probability curve from the threshold to infinity:

$$p_{fa} = \int_{x_{th}}^{\infty} p_n(x')dx'. \tag{5}$$

Similarly, the probability of missing a signal, which is the probability that $S_N(m)$ is less than the threshold, is given by

$$p_{ms} = \int_0^{x_{th}} p_{s+n}(x')dx'. \tag{6}$$

The probability of detecting a signal is just

$$p_{det} = \int_{x_{th}}^{\infty} p_{s+n}(x')dx'. \tag{7}$$

In principle one would choose the threshold to maximize the detection probability while simultaneously minimizing the false alarm probability. In practice, the threshold is chosen to achieve an acceptable p_{fa}.

2.2 Problems with the Naive Approach

There are three fundamental problems with the naive approach. The first is that real data is not continuous and sampled at evenly spaced intervals in time as required by the FFT algorithm. The second is that the expected signals at the antenna, even for a purely sinusoidal source, are not pure sinusoids. The third problem is that the probability density function used for statistical peak detection depends on the variance σ^2 of the data, and σ in general is an unkown function of frequency.

2.2.1 *Real Data.* Real data is not a continuous stream of samples taken at evenly spaced intervals in time. The limitations of computer memory and hardware, as well as problems with the apparatus, combine to eventually cause gaps. The seriousness of a gap depends on its length. A data collection system which is triggered from an external reference clock will keep the phase of the samples across gaps, so that the data set will in general consist of blocks of data taken at equally spaced intervals in time separated by gaps of varying widths. A very small gap could be replaced by the average value of the data without much problem. A gap that is a large fraction of the length of the data stream must be dealt with in some other way.

In the MIT prototype data collection system, there are two guaranteed types of gaps. The first type of gap is the result of having only one 1600 BPI magnetic tape drive connected to the system. After 15 minutes, a tape is full and data collection must be stopped while the tape is rewound and a new one is loaded. The gap is approximately 5 minutes. There is nothing fundamental about this type of gap. The 15 minute continuous data limit could be extended by purchasing a higher density drive, and two drives could collect data continuously.

The other type of gap is the result of ground motion induced by traffic on the street outside the laboratory. In this case the gap is approximately 16 hours, as the best time to collect data is between 22:00 and 06:00. A more remote location for the antenna and a better suspension system would increase the length of the available collection time. However, it is difficult to imagine collecting a month of continuous data without gaps of some sort.

The effect of the gaps is to limit the integration time and hence the sensitivity of the search by reducing the size of the FFT that can be performed. There is no possibility of calculating the DFT directly without using the FFT to circumvent the equally spaced samples in time requirement because the computation time required for the DFT scales as N^2. If the computation were performed on a machine capable of 10^8 operations per second, a $N = 2^{24}$ point transform (15 minutes of data on the MIT prototype) would require 2.8×10^6 seconds, or 32 *days* of cpu time. The FFT algorithm would require 4 seconds under the same conditions.

A more serious problem is the loss of information caused by the gaps. One way to think of the gaps conceptually is to consider a second time series $w(t)$ that is unity when data exists and zero in the gaps. The observed data set is then the product of a hypothetical continuous data set and $w(t)$. By the convolution theorem, a product in time space is a convolution in frequency space. The Fourier transform of the data is thus the convolution of the transform of the continuous data set and the trnsform$W(f)$ of the $w(t)$ function:

$$X_{obs}(f) = X_{cont}(f) * W(f). \tag{7}$$

$W(f)$ is called the window function. The effect of the window function is to smear out the information contained in the hypothetical continuous spectrum. For a general sequence of gaps, the smearing can be quite severe, and because the window function has zeros, deconvolution is not possible. Thus even if it were possible to perform the DFT directly, the resolution of the resulting spectrum would be degraded.

2.2.2 *Expected Signals.* The signal expected at the detector from a stationary, purely sinusoidal gravitational wave source is not sinusoidal. It is modulated by the relative motion of the source and the antenna in two distinct ways: FM and AM.

FM. The received signal is frequency modulated by the ordinary relativistic Doppler shift. The frequency observed by the antenna can be written explicitly as

$$f_{obs} = f_{emit} \cdot \gamma_0 \cdot (1 - \frac{\vec{v}}{c} \cdot \hat{r}), \tag{8}$$

where γ_0 is the usual relativistic parameter $(1 - (v/c)^2)^{-1}$, $\hat{r}$ is a unit vector from the antenna to the source, and $\vec{v}$ is the relative motion of the source and antenna. The two major contributions to the relative motion are the daily rotation of the earth, which causes a frequency modulation of $\Delta f/f \approx 10^{-6}$, and the orbital motion around the sun, for which $\Delta f/f \approx 10^{-4}$. In principle there is also a contribution from any motion of the source, but in the simplified model adopted for this discussion, source motion is neglected.

The effect of FM on the spectrum of the received signal is shown schematically in Figure 2. The emitted signal has a delta function spectrum centered at some frequency f_0. The FM spreads the power in that delta function out over a bandwidth of order $2(\Delta v/c)f_0$.

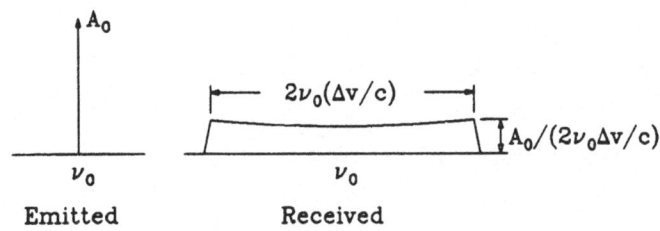

Figure 2

Schematic representation of the effect of FM on the spectrum of a sinusoidal signal.

The smearing of the signal by FM is significant only when the bandwidth over which it is spread becomes larger than the frequency resolution bandwidth of the transform. A rough idea of just when this occurs can be calculated using the following formula:

$$\underbrace{\underbrace{2f_{max}\frac{v_{rot}}{c}\frac{t_{int}}{1\,day}}_{\frac{\Delta v_{rot}}{c}} + \underbrace{2f_{max}\frac{v_{orb}}{c}\frac{t_{int}}{1\,yr}}_{\frac{\Delta v_{orb}}{c}}}_{\text{FM Bandwidth}} > \underbrace{\frac{1}{t_{int}}}_{\Delta f_{res}} \tag{9}$$

Solving for t_{int} and putting in numbers,

$$t_{int} \geq 30\,\min \tag{10}$$

The Doppler effect is not important for data records shorter than 30 minutes. However, data collected at different times will contain the signal at different center frequencies.

AM. The amplitude modulation of the received signal is also caused by the relative motion of the antenna and the source. Although the spatial sensitivity of the antenna is non-directional, it is not isotropic. The amplitude modulation is simply the result of the different sensitivity lobes of the antenna sweeping across the source position as the earth moves.

The effect of the AM on the spectrum of the received signal is illustrated by Figure 3. The emitted signal, with a delta function frequency spectrum centered at frequency f_0, develops sidebands spaced at $1/24$ hours ($=12\,\mu Hz$). The AM begins to exceed the resolution bandwidth of the transform for integration times longer than 6 hours.

2.2.3 *Unknown* $\sigma(f)$. The real and imaginary amplitudes of a single DFT follow an independent Gaussian distribution for each frequency. The distribution for

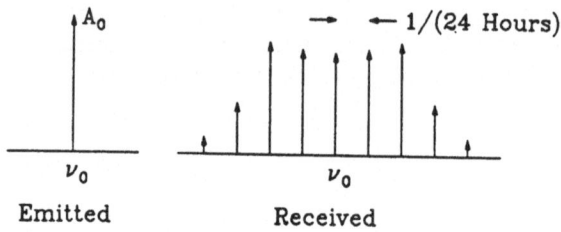

Figure 3
Schematic representation of the effect of AM on the spectrum of a sinusoidal signal.

the amplitude in a bin of the square root of the power spectrum, $s = S_N(m) = \sqrt{(real)^2 + (imag)^2}$, is given by

$$p_{noise}(s) = \left(\frac{s}{\sigma^2}\right) e^{-\frac{s^2}{2\sigma^2}}. \tag{11}$$

This is a χ^2 distribution with two degrees of freedom (Rayleigh distribution) and it depends on one only parameter, σ. To use this distribution function, σ must be known, and in general σ is a function of frequency because the spectrum is not white. The statistics of an individual frequency bin still follows a Rayleigh distribution, but the variance changes from point to point. The statistical peak detection method cannot be used to separate signals from noise unless σ can be estimated reliably.

2.3 Solutions

A solution to each of the three fundamental problems will be outlined in general in the following sections, then in discussed in more detail in Section 3 in the context of an analysis of data taken in June 1985 with the MIT prototype gravitational wave antenna.

2.3.1 Gaps.

The obvious solution to the problem of gaps in the data stream that are small compared to the length of the data stream was mentioned above – simply fill in the gap with the average value of the data.

The obvious solution to the problem of large gaps in the data is to analyze the continuous pieces separately, then combine the pieces with an r.m.s. (root mean square) average. This is easy to do, but has the disadvantage that the power SNR increases only with the square root of the integration time and not directly with the time as could be achieved with continuous data. Adding more data still improves the SNR, but not as quickly as if the data were continuous.

The gap problem for the June 1985 data taken with the MIT prototype was much more severe than it should be for a real antenna. It is worthwhile exploring methods to deal with the gaps, however, because any real data set will always have gaps of some size if it is large enough. The gap problem may become moot for another reason. The computational problems associated with large data sets may impose a separate constraint on the length of the data set that may be easily obtainable with a better site and a more sophisticated data collection system than the one used by the MIT prototype.

2.3.2 *Expected Signals.* Two different solutions to the problem of a direction dependent received signal have been developed. One solution approaches the ideal of a broadband full sky search, but with a limited sensitivity. The other solution can achieve a higher sensitivity, but only in one direction at a time. Details of the implementation of these ideas will be discussed in Section 3 of this article.

Full Sky Search. A full sky search can be achieved by making use of the FM and AM modulation as signatures to distinguish signals from local disturbances. As discussed in the Section 2.2.2, the FM modulation becomes important only for integration times longer than 30 minutes. Thus, if the data is analyzed in continuous pieces shorter than 30 minutes in length, a real signal will still appear as a delta function in frequency. However, the signal will appear at different frequencies in different pieces, and this difference in center frequency can be used to localize the source to a region of the sky. For a data set consisting of pieces spanning a week baseline, the dominant Doppler shift is that due to the orbital motion of the earth. Figure 4 shows the expected Doppler variation for a source located at the center of the galaxy in early June 1985. The vertical scale is marked off in units of the frequency resolution of a 15 minute FFT, or $\approx$ 1.5 mHz. Since the daily Doppler shift is small, the Doppler shift for the week may be characterized by a single slope, in units of Hz/Hz/sec, for each direction. Furthermore, the slope has a maximum physical value that corresponds to $\pm v_{orb}/c$. Figure 5 is a map of the sky showing areas of constant Doppler slope as a grey scale. A known value of the slope identifies a locus of points on the sky as possible source locations.

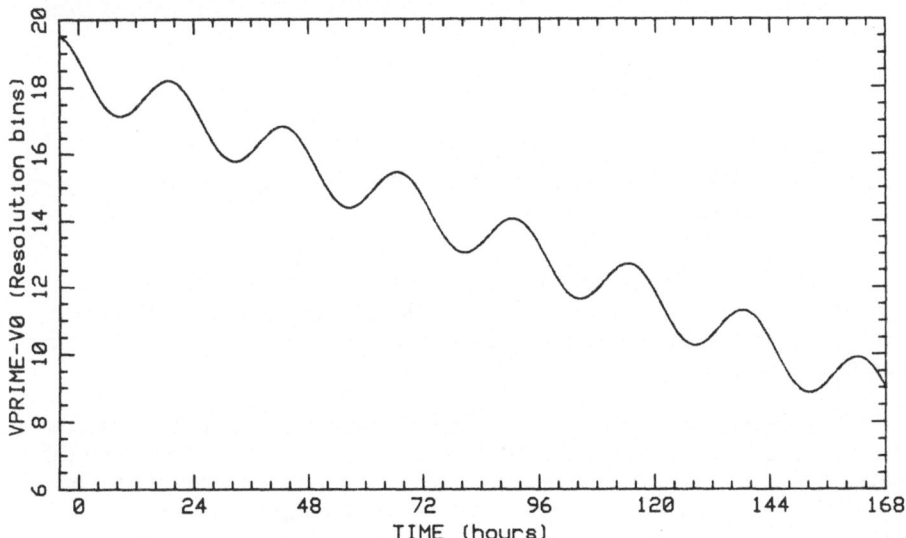

Figure 4

Doppler variation of the received frequency for a sinusoidal source at the center of the galaxy. T=0 for June 4, 1985.

The AM modulation can be used to further localize a source. As with the FM, the modulation per continuous piece is small, but the signal strength difference between pieces can be significant. The AM can be used to predict the relative signal strengths in each piece of continuous data as a function of the position of the source on the sky. A candidate signal that has been detected in more than one piece must have a physically reasonable Doppler slope and must also match the AM response pattern before it can be considered a possible astrophysical source.

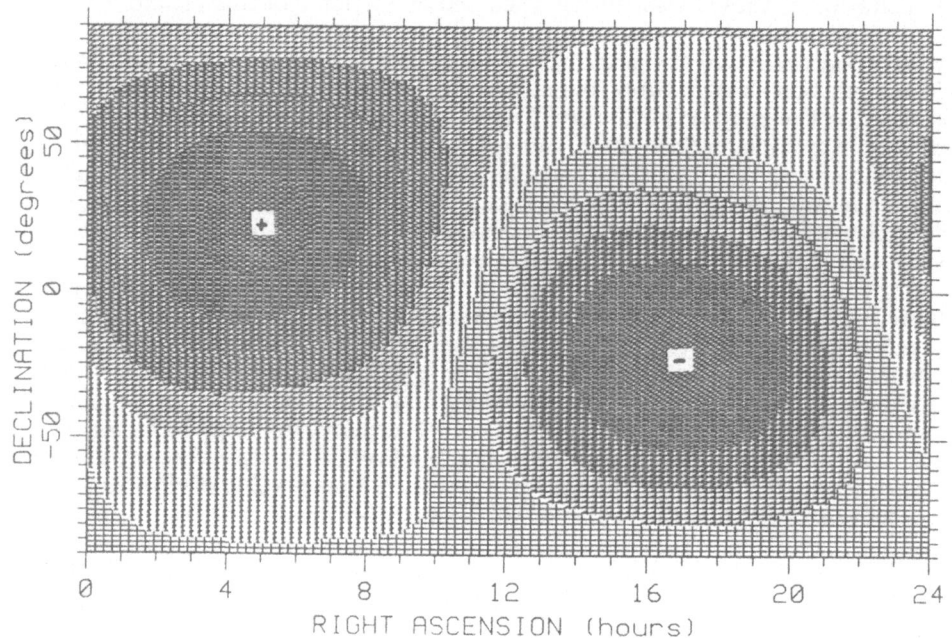

Figure 5

Doppler slope as a function of position of source on the sky, expressed as a grey scale from $+v_{orb}/c$ to $-v_{orb}/c$.

The drawback with this procedure is that the sensitivity is limited by the Doppler modulation to what can be achieved with an integration time of 30 minutes or less. Since the signal appears at different frequencies in different piece of data, averaging does not help.

Single Direction Search. Better sensitivity can be achieved by demodulating the FM, so that a candidate signal would always appear at the same frequency in different pieces. The separate pieces could then be r.m.s. averaged together to improve the SNR. The length of each piece could be much longer than 30 minutes in principle.

The idea is simple. The general form of the signal is:

$$s(t) = A \cdot \cos \varphi(t), \tag{12}$$

where φ is the phase. From the expression for the Doppler shift, the observed frequency is

$$f_{obs} = f_{emit} \cdot \gamma_0 \cdot \left(1 - \frac{\vec{v}}{c} \cdot \hat{r}\right). \tag{13}$$

The phase of the observed signal is then

$$\varphi = 2\pi \int_{t_0}^{t} f(t')dt' \tag{14a}$$

$$= 2\pi f \gamma_0 \left(t - t_0 - \frac{\hat{r}}{c} \cdot \int_{t_0}^{t} \vec{v}(t')dt'\right) \tag{14b}$$

$$= 2\pi f \gamma_0 t' \tag{14c}$$

where

$$t' = t - t_0 - \frac{\hat{r}}{c} \cdot \int_{t_0}^{t} \vec{v}(t')dt' \tag{15}$$

is called the rescaled time and depends explicitly on the position of the source. Equation (14c) is the relationship between frequency and phase if the frequency is constant in time. One rescaled time function demodulates all frequencies simultaneously, but for only one direction.

In practice, what is needed for the FFT computation is a time series that is sampled at evenly spaced intervals in rescaled time. The recorded data is a time series taken at evenly spaced intervals in laboratory time. If the recorded data is treated as though the samples were collected at unequal intervals in rescaled time and then resampled to equal intervals by interpolation, the desired result is accomplished. Figure 6 depicts the process schematically. The solid curve is the original data which was recorded at equally spaced intervals in laboratory time. It is plotted with each data point mapped into its equivalent rescaled time value, so the points are not evenly spaced. The dotted curve represents the new time series that is generated by interpolating between the unevenly spaced samples in rescaled time of the solid curve to get an evenly spaced set of samples.

Interpolation is not a clean procedure. The rescaled time and laboratory time can slip by several sampling periods over the course of 30 minutes. Since the error in interpolation increases as the point to be estimated falls further from the known point, the interpolation error puts a low frequency modulation in amplitude and phase on the original time series. The effects of the modulation may be reduced by using a more complicated interpolation algorithm or by oversampling the original time series. Neither is desirable. Oversampling increases the data acquisition and storage problems, and a complicated interpolation algorithm is computationally very slow.

2.3.3 *Unknown* $\sigma(f)$. One solution to the problem of the unknown variance as a function of frequency is to determine the variance from a local average of the data. The power spectrum is divided into pieces referred to as the averaging bandwidth. The variance of the spectrum is assumed to be constant over that bandwidth, and an estimate of σ is derived from the mean value of the spectrum in that bandwidth. Explicitly,

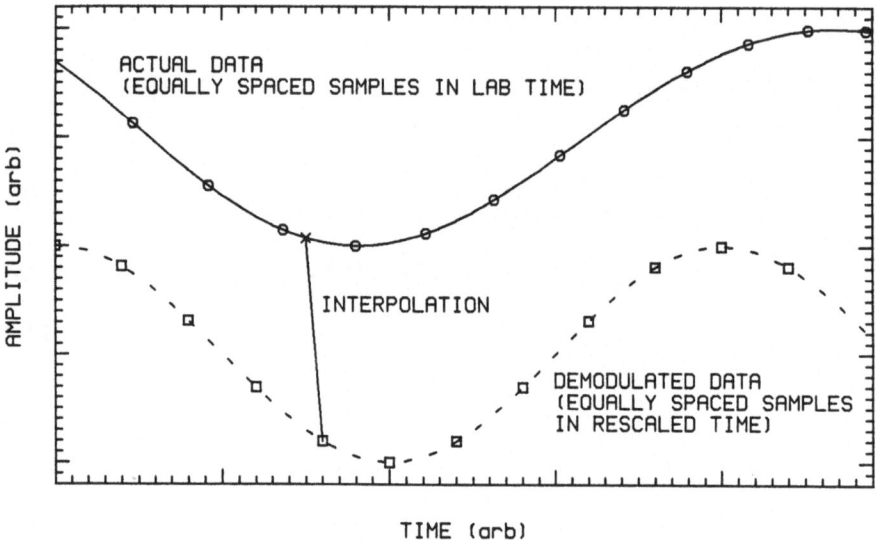

Figure 6

Schematic representation of the resampling procedure to produce a new time series sampled at equally spaced intervals in rescaled time.

$$< s^2 > = \int_0^\infty s^2 p_{noise}(s)\,ds$$

$$= 2\sigma^2 \tag{16}$$

where $s = S_N(m)$ is the amplitude of a bin in the square root of the periodogram and p_{noise} was taken from equation (11). The estimated value of σ can be used to normalize each bin in that averaging bandwidth. The normalized values of the whole spectrum may then be compared on an equal basis.

2.3.4 *Summary*. To quickly summarize the techniques discussed so far, the desired signal is assumed to be a high-Q, essentially delta function in frequency. The strategy is then to integrate for as long as possible to increase the SNR. The fact that the received signals are modulated even if emitted by monochromatic sources is handled two different ways. A full sky search can be conducted by analyzing the data in pieces and using the Doppler FM and the AM between pieces as a signature for which to search. An enhanced sensitivity search can be conducted by performing a demodulation for one specific direction at a time. The demodulation collapses the received signal back into a delta function in frequency which can then be averaged to improve the SNR. The statistical variance is estimated as a function of frequency by using local averages over a specfied bandwidth. These techniques will now be discussed in the context of a specific example – the analysis of data taken in June 1985 with the 1.5m MIT prototype antenna.

3. EXAMPLE: ANALYSIS OF JUNE 1985 DATA

3.1 Data Set

The data analysis scheme for periodic sources has been outlined conceptually. An application of these ideas to a real data set illustrates not only how well the ideas work in practice, but also points out some of the practical problems which will have a profound impact on the way data analysis for a large antenna should be conducted.

The MIT 1.5m prototype antenna and data collection system have been described in detail elsewhere.[7] The raw data consists of the demodulated output of the interference fringe lock servo, digitized at either a 20 kHz rate, or a 6 kHz rate. The digitization is triggered from a rubidium standard that is sychronized to WWV. As mentioned above, limitations in the recording hardware restrict the length of a continuous data stream to approximately 15 minutes ($N = 2^{24} \approx 16 \times 10^6$ points).

The data collected in June 1985 consists of approximately 50 magnetic tapes collected over 6 days, representing roughly 10% coverage of the total possible time. One half of the tapes were sampled at 20 kHz, the other half at 6 kHz. Eight of the 25 tapes sampled at 20 kHz were selected for the final analysis.

3.2 Full Sky Search

The first step in the analysis is to choose a value of the averaging bandwidth. For the statistical peak detection scheme to be valid, the statistics of the power in each resolution bin must agree with the theoretical Rayleigh distribution The choice of an averaging bandwidth was made by constructing an experimental probability distribution for several different values of the averaging bandwidth and choosing the value which gave the best least squares fit to the theoretical distribution.

Figure 7 shows the fit with a good choice of the averaging bandwidth. The experimental distribution is computed by normalizing the value of the each frequency bin in the square root of the periodogram to the local σ. The normalized points are then binned in a histogram which counts the number of points with a given normalized amplitude versus that normalized amplitude. The solid curve is the theoretical distribution for pure noise, $p(\xi) = \xi e^{-\xi^2}$, integrated over each bin in the histogram. A χ^2 fit of the experimental distribution to the theoretical distribution was performed, and the value of the averaging bandwidth used in the final analysis was that which gave the lowest value of the χ^2 for the fit. For this data set, that value was 0.35 Hz.

The arrow is set at a threshold of 4.97σ, which represents a false alarm probability of $p_{fa} = 5 \times 10^{-6}$. A frequency bin containing enough power to place it to the right of the threshold is a candidate signal. For this particular distribution, with $\approx 2 \times 10^6$ total resolution bins, a spectrum containing pure noise would contain on average 11 peaks above the threshold. In this case, the threshold was chosen conservatively so as not to miss any potential signals, and it was also placed at the point where the experimental distribution exhibits a non-Rayleigh tail.

Once the value of the averaging bandwidth has been chosen, the first step in the analysis is straightforward: the spectrum of each piece of data is binned separately, and candidate signals are identified with the threshold criterion. A list

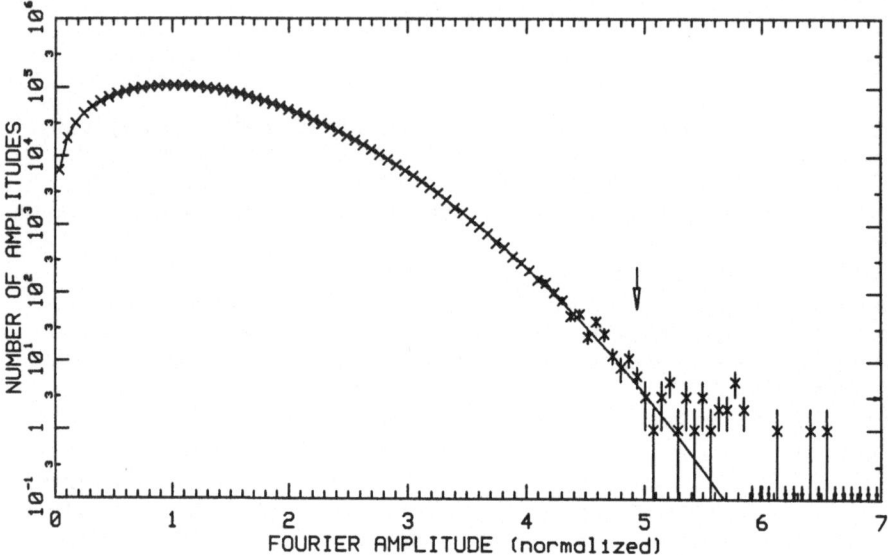

Figure 7

Experimental fit to the theoretical probability distribution for pure noise (solid curve) for a particular value of the averaging bandwidth. Normalized to $\approx 2 \times 10^6$ total points. The arrow indicates a 4.97σ threshold which gives a false alarm probability $p_{fa} = 5 \times 10^{-6}$.

of the possible signals in each piece is compiled, keeping track of the frequency of the signal, its magnitude in absolute units, and its value relative to the local σ are recorded, along with an estimate of the number of adjacent frequency bins that also qualify as possible signals (a measure of the width of the peak).

The complete spectrum of each piece extended from 0-10 kHz, but only the region from 2-5 kHz was searched. The low frequency cutoff was chosen at a point where the noise spectrum of the antenna begins to increase rapidly with decreasing frequency, and hence the sensitivity to gravitational radiation becomes small. The high frequency cutoff was chosen to avoid problems with the interpolation method used in the single direction search.

These lists of possible peaks are then examined for a series of peaks whose center frequency changes over the course of the week. The method for finding such a series is called a Doppler sieve. Figure 8 describes the operation of the sieve schematically. One piece of data is selected as the reference. If a peak in the reference list is actually an astrophysical signal, then it should appear in the lists of the other pieces, but with its center frequency shifted by a known amount. The sieve is applied by dividing the allowable range of Doppler slopes into discrete values. Each peak in the reference list is used to estimate the center frequency for each of those slopes that that peak would have if it were observed at the time the other data was collected. To get the center frequency of a peak in the jth tape:

$$f_{center}(j) = f_{center}(ref) \cdot m_D \cdot (t_j - t_{ref}) \qquad (16),$$

where m_D is the Doppler slope in units of Hz/Hz/sec. The actual list of candidate signals from the j_{th} tape is examined for anything within a given capture width Δf_{cap} to the projected center frequency. If at least one other tape has a peak which

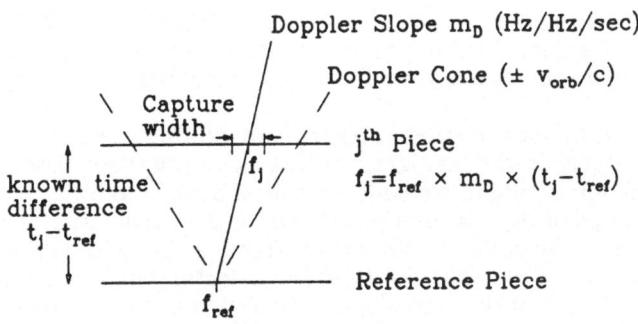

Figure 8
Schematic operation of the Doppler sieve.

satisfies this condition, the reference peak is recorded. Note that the condition for acceptance of a peak is the weakest possible. A single pair of peaks that falls along a physically plausible Doppler slope is enough to qualify. In general, there could be more than two peaks, but the received signals are also AM modulated. A weak signal is therefore most likely to be seen only in the two most favorable observation windows.

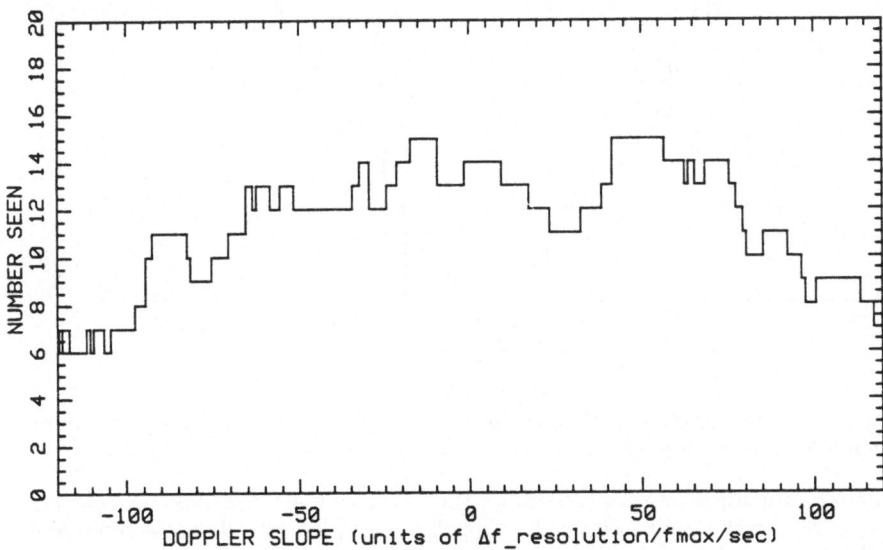

Figure 9
Typical histogram of the number of peaks in a reference list which match at least one other peak in another list versus the Doppler slope for which there is a match. Only slopes between ±50 units are physical.

For each reference piece, the peaks that fall through the Doppler sieve are binned in a histogram of the number of peaks with a given Doppler slope versus the slope. Figure 9 shows a typical histogram, with the horizontal axis in units of the quantized Doppler slope. A larger range of slopes than is physically possible is plotted to estimate the "background". The physical region in Figure 9 is between ± 50 units.

Peaks caused by local disturbances will not exhibit a Doppler shift over time and will thus appear in the histogram with a slope near zero. For this reason, any candidate peak appearing in coincidence with a peak in another list connected by a Doppler slope of $m_D = 0$ in quantized units, was eliminated from consideration. Figure 10 shows the histogram of Figure 9, but with the zero-slope bin deleted. A real source near the zenith of the detector would also exhibit no doppler shift, so the exclusion of the zero-slope bin could in principle be eliminating some real sources. If the sources are uniformly distributed on the sky, only 8% of the source would be eliminated. This is a small price to pay for the elimination of a large number of spurious coincidences. The change in the vertical scale indicates that some of the peaks eliminated appeared in the histogram with more than one Doppler slope.

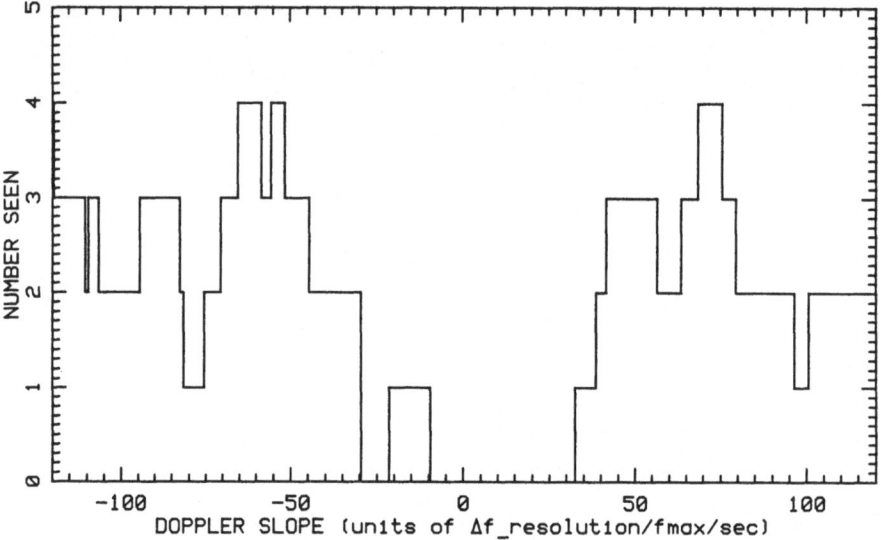

Figure 10

Histogram of the Figure 9 with peaks in the zero-slope bin deleted. Only slopes between ±50 units are physical.

The peaks that manage to fall through the Doppler sieve are then associated with a region of the sky using Figure 5. The sky is divided into discrete bins, and for each bin in the region picked out by the Doppler slope, the expected pattern of signal strengths is determined. Figure 11 is a map of the sky showing, for each bin, which pair of the particular data tapes used in the analysis should have the strongest signals based on the AM response to a source in that portion of the sky. Figure 11 is for a plus polarized wave. A similar map can be made for the cross polarization. The actual strengths can be compared against the predicted pattern.

Figure 12 summarizes the full sky search procedure. A total of three peaks out of $\approx 2 \times 10^6$ frequency bins actually passed both the Doppler sieve test and the AM signal strength pattern. None of the peaks is a good candidate signal for a number

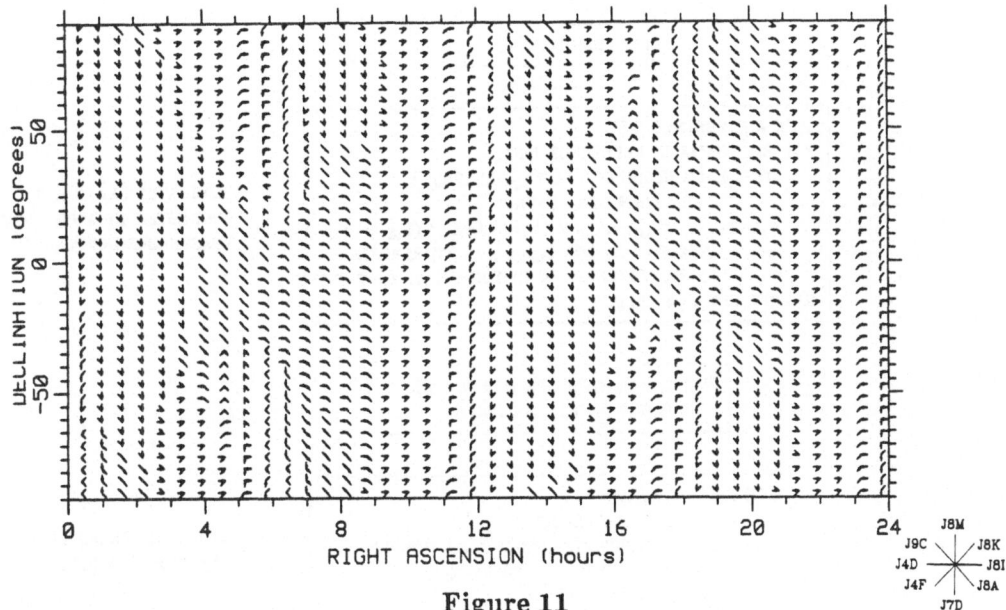

Figure 11

Map of the sky showing, for each possible source location, the pair of tapes in which the signal should be most strongly received. See key at right.

of reasons, but the primary reason is that all occured in a region of the spectrum in which there was appreciable structure in the spectrum on the same scale size as the averaging bandwidth. The local estimate of σ is therefore not a good estimate, and the statistical peak detection method fails because the experimental distribution is not a good approximation to the theoretical Rayeigh distribution.

There are several strategies that could be adopted to circumvent the failure of the statistical peak detection method. One could use a more sophisticated fit to each averaging bandwidth: instead of assuming a flat spectrum, one could allow a slope plus a constant. Another alternative would be to allow the size of the averaging bandwidth, which is chosen by a global optimization of the statistics for the entire spectrum, to vary over the spectrum and select it by some local optimization procedure. Individual peaks could also be modeled and subtracted from the spectrum. Of course, the best solution is to eliminate the noise sources in the antenna to produce a flat spectrum with which to start.

3.3 Single Direction Search

The single direction search procedure is similar to the full sky search in the initial stages of analysis. The particular direction to be searched with this data set was chosen to be the center of the galaxy. Each piece of data was demodulated according to the algorithm described in section 2.3.2, using a piecewise continuous spline interpolation procedure. An average spectrum was computed by r.m.s. averaging each averaging bandwidth, weighted by the σ for tha bandwidth, with the corresponding bandwidths in other pieces of data. The net improvement in SNR, if all of the local σ were identical, should be $\sqrt{8}$. A reference r.m.s. spectrum was also computed by following the same procedure without the Doppler demodulation.

The best value of the averaging bandwidth is chosen as before with a global

234

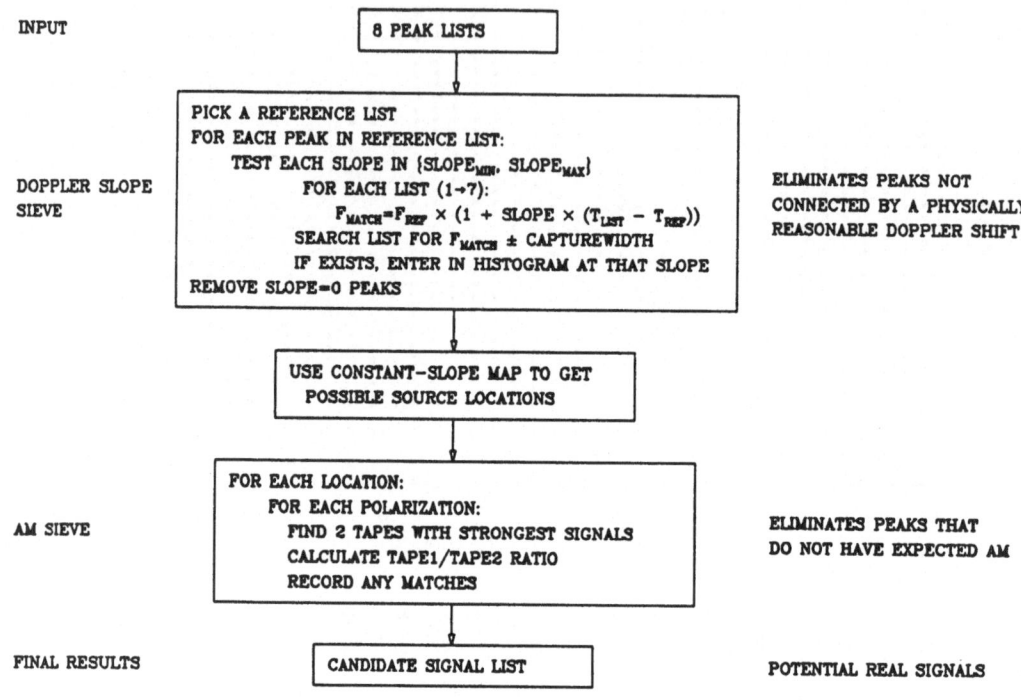

Figure 12
Summary of the full sky search procedure.

minimization of the fit of the experimental differential probability distribution to the theoretical distribution. The Doppler demodulation procedure did not change the value of the best fit averaging bandwidth. Both average spectra are then be subjected to the statistical peak detection analysis and a list of candidate peaks in each spectrum is generated. The expected probability distribution is slightly different because of the r.m.s. averaging. The expected distribution for the amplitude in a single bin of the square root of the power spectrum with only noise present in the data is a χ^2 distribution with $2q$ degrees of freedom, where q is the number of pieces combined in the average.

Real astrophysical signals should have been enhanced by the demodulation procedure, and thus any new peaks in the demodulated spectrum are potentially signals. However, local noise driven resonances which appear at the same frequency in each piece will get smeared by the demodulation procedure. To separate truly new peaks in the demodulated spectrum from local peaks that have been smeared, the two lists of peaks were compared with a procedure very similar to an inverse of the Doppler sieve. First, any peaks that appeared at exactly the same frequencies in both lists were eliminated on the grounds that such signals were not physical. Next, any peaks in the doppler corrected spectrum that had a corresponding peak in the uncorrected spectrum offset by the frequency shift expected for one of the component piece of data was eliminated. The peaks that were left over were truly

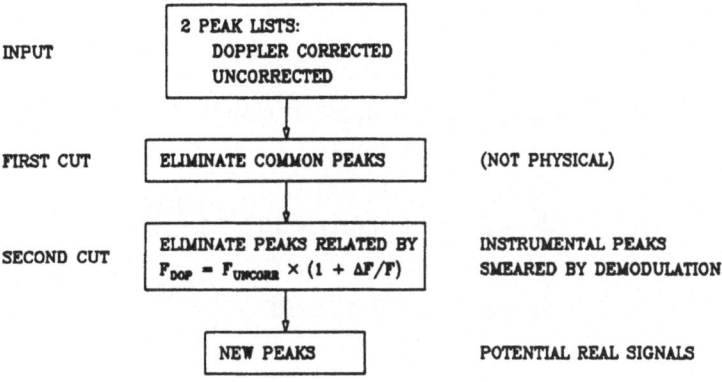

Figure 13
Summary of the single direction search procedure.

new. Figure 13 summarizes the procedure.

Only 22 peaks met the criteria for "new" peaks. For the size data set considered, 2×10^6 frequency bins, a pure noise spectrum should have generated 17^{+9}_{-5} peaks above threshold. The error bars reflect a 1% uncertainty in the value of the threshold. The results are thus consistent with a pure noise spectrum. A closer examination of the 22 peaks would require more data.

3.4 Results

The final results of the two searches for periodic signals can be stated rather simply. The full sky search did not see any periodic signals which produced a *measured* strain in the antenna above a level

$$h_{rms} = 4.6(\pm 0.9) \times 10^{-17} \left(\frac{2000}{f}\right)^2 \tag{17}$$

in a frequency band from 2-5 kHz. The indicated error is one standard deviation. The conversion of this measured strain into a source strength depends on the polarization and the direction of the source. For a source at the center of the galaxy, the limits translate into

$$h_{+,rms} = 2.1(\pm .6) \times 10^{-16} \left(\frac{2000}{f}\right)^2 \tag{18a}$$

236

$$h_{\times,rms} = 7.4(\pm 4.) \times 10^{-17} \left(\frac{2000}{f}\right)^2 \tag{18b}$$

The corresponding limits for the single direction search are

$$h_{+,rms} = 1.0(\pm 0.1) \times 10^{-16} \left(\frac{2000}{f}\right)^2 \tag{19a}$$

$$h_{\times,rms} = 5.2(\pm .8) \times 10^{-17} \left(\frac{2000}{f}\right)^2 \tag{19b}$$

The sensitivity gain for the single direction search is slightly less than the $\sqrt{8} = 2.8$ expected with eight RMS averages. Some of the discrepancy can be attributed to the unequal weighting of the individual spectra, and some is due to a choice of direction for the search that is perhaps not the best for exhibiting the best SNR increase for the method. With more data, the increase in SNR could be much larger.

3.5 Practical Implications for Future Research.

The prospects for data analysis techniques to discover periodic sources of gravitational waves may seem unduly encouraging. Two different methods have been presented to perform broadband searches. However, there are several severe practical difficulties with these methods which will make any attempt to extend these methods difficult. Both are rooted in the computational difficulties associated with large data sets.

It might be expected that single direction broadband search technique could simply be repeated for many different directions. However, the demodulation procedure correctly demodulates signals from a finite patch of sky around the desired direction. An estimate of the patch size can be generated by treating the received frequency in equation (13) as a function of the position angles α, δ, where α is the right ascension and δ is the declination. Then

$$\Delta f = \left(\frac{\partial f}{\partial \alpha}\right) \Delta \alpha + \left(\frac{\partial f}{\partial \delta}\right) \Delta \delta. \tag{20}$$

The change in either angle must be small enough to keep $\Delta f \leq \Delta f_{res}$, where Δf_{res} is the frequency resolution bandwidth of the transform. Equation (20) can be inverted to and simplified to obtain an expression for either angle of the form

$$\Delta(\alpha, \delta) = \xi(\alpha, \delta) \left(\frac{1}{f_{max}}\right) \left(\frac{1}{T}\right)$$

where $f_{max} = 1/2\Delta t_{sample}$ is the Nyquist frequency for a given sampling rate Δt_{sample}, and T is the length of the transform, and ξ is a function of α, δ. For a source at the Galactic Center, a 20 kHz sampling frequency, and a 15 minute data record, the patch size on the sky is

$$\Delta\alpha \approx 9° \left(\frac{f_{max}}{10\,\text{kHz}}\right)^{-1} \left(\frac{T}{900\,\text{sec}}\right)^{-1} \tag{21a}$$

$$\Delta\delta \approx 15° \left(\frac{f_{max}}{10\,\text{kHz}}\right)^{-1} \left(\frac{T}{900\,\text{sec}}\right)^{-1} \tag{21b}$$

If this patch size were constant over the whole sky (which it is not), this implies $\sim$ 500 patches must be searched. This does not seem like an excessive number, except that it scales as T^2. The computation time for the analysis that was performed on the MIT data is summarized in Table 2. Unpacking refers to an operation that converted the data (stored in condensed form) into Cray-2 real numbers. The volume of data was such that it could not be easily stored in unpacked form on the available disk space. Net computational speed is approximately 7.5 Mflops for the FFT.

Table 2

Computation times for the periodic source search.
Computations were performed on a Cray-2.

Operation	CPU time (secs)	% of Total	CPU time (secs)	% of Total
Unpacking	225.4	46.5	225.4	32.2
Doppler Demod.	—	—	216.0	30.8
FFT	71.3	14.7	71.3	10.2
Apodization	8.8	1.8	8.8	1.3
Peak Search	179.8	37.0	179.8	25.5
Total	485.3	100.0	701.3	100.0

Based on these execution times, a complete sky search with extended sensitivity would require $500 \times 700 = 3.5 \times 10^5$ seconds, or $\sim$ 97 hours of CPU time. This is a large amount of computation time, and any attempt to extend the analysis techniques much beyond what has already been done will require even more time. Of course, there may be physical reasons for selecting a direction in which to search, and then the number of bins required to cover the sky may not be important.

There is another computational bottleneck that is not apparent from the table. The maximum size FFT that can be computed is determined by the size of the core memory of the machine. A Cray 2 has the largest core memory of any machine currently in existence, $2^{28} \approx 268 \times 10^6$ words. Based on the measured execution time of a 2^{24} point FFT, a 2^{28} point FFT should require 1331 seconds, or 22.2 minutes of CPU time. The estimated total CPU time for a doppler demodulation analysis of this size is 3.2 hours, with 123,000 bins required to cover the sky. Any attempt to perform larger FFT's changes the problem dramatically. The computation must be done in pieces, and it becomes immediately limited by the I/O transfer rate to the disk.

The point of this discussion is that a complete solution to the data analysis problem for periodic sources is not in hand. A different approach will be needed to

238

achieve integration times of 10^6 seconds, and faster general purpose computers are not the answer. A real solution will require a combined hardware/software solution and perhaps different analysis algorithms.

ACKNOWLEDMENTS

This work was supported in part by NSF grant 85-0486-PHY, which was augmented by a separate grant of computer time at the University of Minnesota Supercomputer Center. The author would also like to thank the people of the MIT Gravity Wave group, in particular Rai Weiss, for support and encouragement.

REFERENCES

1) Thorne, Kip S., in *300 Years of Gravitation*, eds. Hawking and Israel, Cambridge, (1987).
2) Schutz, B. F., this volume.
3) Taylor, J. and Weisberg, J. *Ap. J.*, **253**, 908 (1982).
4) Wagoner, R. *Ap. J.*, **278**, 345, (1984).
5) Lyne, A. et al. *M.N.R.A.S.*, **213**, 613, (1985).
6) Henrichs, H. and van den Heuval, E. *Nature*, **303**, 213, (1983).
7) Livas, et. al. in *Proceedings of the Fourth Marcel Grossman Meeting*, ed. R. Ruffini, 591, (1986).

RESPONSE OF MICHELSON INTERFEROMETERS TO LINEARLY
POLARIZED GRAVITATIONAL WAVES OF ARBITRARY DIRECTION
OF PROPAGATION .

D. FATTACCIOLI
Groupe de Recherche sur les Ondes de Gravitation
Laboratoire de l'Horloge Atomique, Bât.104
Université Paris-Sud, 91405 ORSAY Cedex .

ABSTRACT: We derive the expression for the gravitational
Green's function of a Michelson interferometer, with
multipass delay lines or Fabry-Perot cavities, which will
be required in the analysis of gravitational wave data
from future large interferometers. We assume a linearly
polarized incoming plane gravitational wave and we analyse
the interferometer in transverse traceless (TT) coordinates.

INTRODUCTION :

     Since the gravitational signal at the output of a
Michelson interferometer is the convolution of the time
dependent waveform of the gravitational wave with the
antenna's Green function, all the informations about the
dependance of the interferometric signal on the geometrical
parameters of the gravitational wave (polarization,direction
of propagation) are contained in the Green's function. Then
it is necessary to know this Green's function in order to
be able on one hand to compute, knowing the direction of
propagation and the polarization of the wave, the response
of the interferometer to a gravitational wave of arbitrary
shape, and on the other hand to determine, knowing the
gravitational waveform and using coincidences between
several antennas,the direction of propagation and the
polarization of the wave.
     Analysing the interferometer in transverse traceless
(TT) coordinates, we first give a description of a linearly
polarized plane gravitational wave in the antenna's frame
and we find the interferometer output signal in presence of
a gravity wave, for a Michelson interferometer with a
multipass delay line or a Fabry-Perot cavity in each arm.
This leads to the characterization of the gravitational
transfer functions (i.e.,frequency response) of such
interferometers by their dependance on the direction of

239

B. F. Schutz (ed.), Gravitational Wave Data Analysis, 239–253.
© 1989 by Kluwer Academic Publishers.

propagation and the polarization of the gravity wave.
From this we finally derive the expression for the
gravitational Green's function (i.e.,impulse response) of
Michelson interferometers in terms of "box functions" whose
width depends on the position of the source of gravitational
radiation and whose height depends not only on the polar
and azimuthal angles of the source but also on the polarisation angle of the linearly polarized gravitational wave.
As examples we give some typical curves which show
the transfer functions and the Green's functions and their
dependance on the parameters of the gravity wave.

1- Description of a linearly polarized gravitational
   wave in the antenna frame :

We choose to analyse the interferometer in transverse
traceless (TT) coordinates ; consequently, the spatial
coordinates are tied to test masses (beam splitter and
mirrors of the interferometer) and the measurable is the
travel time of light between the masses.
Let (ct,X,Y,Z) and (ct',X',Y',Z') be respectively
the antenna frame and the gravitational wave frame.
In the wave frame, the polarization state of a linearly
polarized plane gravitational wave which propagates in the
Z' direction, solution of the Einstein's equations in the
weak field approximation of the General Relativity[1] can be
described by the tensor $h'_{\mu\nu}$, which characterizes one of
the two independant states of polarization, and by the
angle ϕ (angle of rotation about the wave propagation
direction Z') which specifies a general linear polarization
state ; $h'_{\mu\nu}$ takes the following form :

$$(1) \quad h'_{\mu\nu} = \begin{array}{cccc} (ct') & (X') & (Y') & (Z') \\ \left[\begin{array}{cccc} 0 & 0 & 0 & 0 \\ 0 & h(Z',ct) & 0 & 0 \\ 0 & 0 & -h(Z',ct) & 0 \\ 0 & 0 & 0 & 0 \end{array}\right] \end{array}$$

In the antenna frame, the direction of propagation of the
gravitational wave is characterized by a polar angle θ
and an azimuthal angle ψ (see Figure 1 for a geometrical
description). Then, the total metric $g_{\mu\nu}$ of the spacetime
in the antenna frame is :

$$(2) \quad g_{\mu\nu} = \eta_{\mu\nu} + h_{\mu\nu}$$

where $\eta_{\mu\nu}$ is the local inertial Minkowski metric (with the convention that the time term is negative) and where $h_{\mu\nu}$ is the gravitational wave perturbation metric ; with $h(\vec{\delta},ct)$ the time dependant amplitude of the wave, we can write :

(3) $\qquad h_{\mu\nu} = h(\vec{\delta},ct).\bar{h}_{\mu\nu}(\phi,\theta,\psi)$

where $\vec{\delta} = \sin\theta \sin\psi .\vec{X} - \sin\theta \cos\psi .\vec{Y} + \cos\theta .\vec{Z}$
is the transformed unity propagation vector, and where the non zero independant components of $\bar{h}_{\mu\nu}$ are :

(4)

$$\bar{h}_{11}=\cos2\phi(\cos^2\psi -\sin^2\psi \cos^2\theta) - \sin2\phi \sin2\psi \cos\theta$$

$$\bar{h}_{22}=\cos2\phi(\sin^2\psi -\cos^2\psi \cos^2\theta) + \sin2\phi \sin2\psi \cos\theta$$

$$\bar{h}_{33}= -\sin^2\theta \cos2\phi$$

$$\bar{h}_{21}=\bar{h}_{12}=\cos2\phi \sin2\psi \frac{(1+\cos^2\theta)}{2} + \cos2\psi \sin2\phi \cos\theta$$

$$\bar{h}_{23}=\bar{h}_{32}= \sin\theta \left[\sin2\phi \sin\psi - \cos2\phi \cos\psi \cos\theta\right]$$

$$\bar{h}_{13}=\bar{h}_{31}= \sin\theta \left[\sin2\phi \cos\psi + \cos2\phi \sin\psi \cos\theta\right]$$

In the case of a monochromatic gravitational wave of frequency $\Omega/2\pi$, the time dependant amplitude is

(5) $\qquad h(\vec{\delta},ct) = h(\Omega).e^{i(\vec{K}.\vec{r} -\Omega t)}$

where $\vec{K} = \frac{\Omega}{c}.\vec{\delta}$ is the wave vector of the plane gravitational wave.

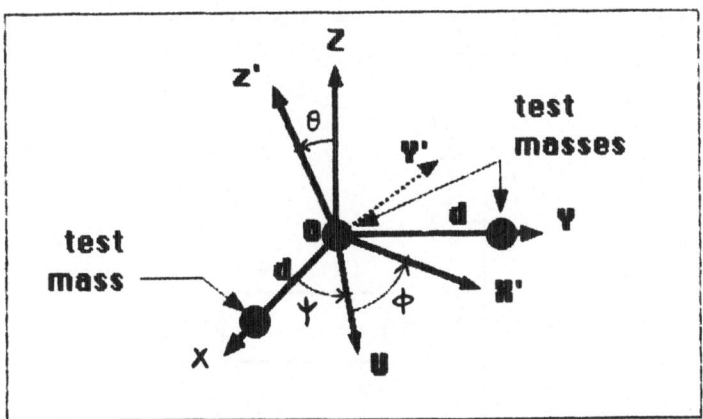

Fig.1 : Coordinates systems .(X,Y,Z) is the antenna frame and (X',Y',Z') is the wave frame .

2- Interferometer output signal in presence
of a gravity wave :

2-1 : Notations about Michelson interferometers :

We consider a Michelson interferometer (as presented
on Figure 2) which contains a multipass delay line or a
Fabry-Perot cavity in each arm. The interferometer is
assumed to be illuminated by an optical plane wave
described by the optical field $E_{in}(t)$.
We define $E_R^{(x)}(t)$ (resp. $E_R^{(y)}(t)$) as the optical field
reflected by the X-arm (resp. Y-arm) delay line (multipass
or Fabry-Perot) when the input field in this one is $E_{in}(t)$.
Assuming a dark fringe at the output of the interferometer
without a gravitational wave, we get:

$$(6) \qquad E_{out}(t) = irt(E_R^{(x)}(t) - E_R^{(y)}(t)).e^{2ikl}$$

where l is the distance between the beam splitter (whose
reflectivity and transmitivity in amplitude are respectively r and t) and the input mirror of the cavities.
Defining a normalized output signal S(t) as

$$(7) \qquad S(t) = \frac{E_{out}(t)}{irtE_{in}(t)} .e^{-2ikl}$$

we obtain

$$(8) \qquad S(t) = \frac{E_R^{(x)}(t)}{E_{in}(t)} - \frac{E_R^{(y)}(t)}{E_{in}(t)}$$

which is just the signal to noise ratio at the output of
the interferometer. In the following we will use S(t) as
the observable in the interferometric detection of gravity
waves.

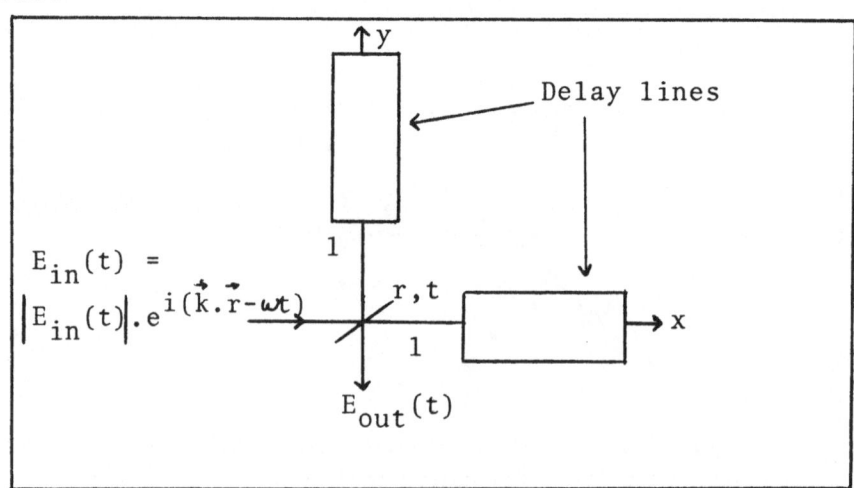

$E_{in}(t) = $

$\left| E_{in}(t) \right| . e^{i(\vec{k}.\vec{r}-\omega t)}$

2-2: Gravitational transfer functions of Michelson interferometers :

We are going to derive the expression for the gravitational transfer functions of Michelson interferometers which characterize the response of such interferometers to a monochromatic plane gravitational wave. We consider Michelson interferometers of two types:

1) Michelson interferometer with multipass delay lines (M.I.M.P.). The multipass delay lines are supposed to be constituted by two spherical mirrors (whose reflectivities in amplitude are ρ_1 and ρ_2) separated by the distance d ; we also assume that this configuration allows N round trips for the light travel in the cavities.

2) Michelson interferometer with Fabry-Perot cavities (M.I.F.P.). The Fabry-Perot cavities are supposed to be constituted by two spherical mirrors separated by the distance d ; the refectivity and transmitivity (in amplitude) of the input mirrors are respectively r_1 and t_1 while the reflectivity of the end mirrors is r_2.

For each of these two cases, the normalized signal at the output of the interferometer is :
For M.I.M.P.:

$$(9) \quad S(t) = i\,_2(-\rho_1\,\rho_2)^{N-1}.\left[e^{i\omega(t-t_{x_N})} - e^{i\omega(t-t_{y_N})} \right]$$

For M.I.F.P.:

$$(10) \quad S(t) = ir_2 t_1^2 \sum_{N=0}^{\infty}(-r_1 r_2)^N.\left[e^{i\omega(t-t_{x_{N+1}})} - e^{i\omega(t-t_{y_{N+1}})} \right]$$

where $t-t_{x_N}$ (resp. $t-t_{y_N}$) is the travel time of the light in the optical cavity for N round trips in the x-direction (resp. y-direction).
The time intervals $t-t_{x_N}$ and $t-t_{y_N}$ are found by using

$$(11) \quad \int dS^2 = 0 \quad ; \quad \int dS^2 = 0$$

$$\begin{array}{cc} \text{on light travel} & \text{on light travel} \\ \text{in the X-direction} & \text{in the Y-direction} \end{array}$$

where dS^2 is given by

$$(12) \quad dS^2 = (\eta_{\mu\nu} + h_{\mu\nu})\,dX^\mu dX^\nu \qquad (\mu,\nu = 0,1,2,3)$$

and where the components of $h_{\mu\nu}$ are given by Eq.(3),(4) and (5).
We finally find, to the first order in $h(\Omega)$ $(h(\Omega) \ll 1)$:

For M.I.M.P.:

(13) $S(t) = \omega \cdot R_{MP}(\omega) \cdot h(\Omega) \cdot F_{MIMP}(\Omega) \cdot e^{-i\Omega t}$

For M.I.F.P.:

(14) $S(t) = \omega \cdot R_{FP}(\omega) \cdot h(\Omega) \cdot F_{MIFP}(\Omega) \cdot e^{-i\Omega t}$

where

$\bullet \quad R_{MP}(\omega) = \rho_2(-\rho_1\rho_2)^{N-1} \cdot e^{2i\omega N t_{trans}}$; $t_{trans} = \dfrac{d}{c}$

R_{MP} is the complex reflectivity of each multipass delay line.

$\bullet$ $h(\Omega)$ is the gravitational wave amplitude for the pulsation Ω.

$\bullet \quad R_{FP}(\omega) = \dfrac{r_2 t_1^2 \, e^{2i\omega t_{trans}}}{1 + r_1 r_2 \, e^{2i\omega t_{trans}}}$

R_{FP} is the normalized stored amplitude which is reflected by each Fabry-Perot cavity.

$\bullet$ $F_{MIMP}(\Omega)$ is the gravitational transfer function of the M.I.M.P. antenna.

$\bullet$ $F_{MIFP}(\Omega)$ is the gravitational transfer function of the M.I.F.P. antenna.

The expressions for these transfer functions are :

(15) $F_{MIMP}(\Omega) = \dfrac{t_{trans}}{2} \cdot (\bar{h}_{22}K_y - \bar{h}_{11}K_x) \cdot$

$\dfrac{\sin N\Omega t_{trans}}{\sin \Omega t_{trans}} \cdot e^{iN\Omega t_{trans}}$

(16) $F_{MIFP}(\Omega) = \dfrac{t_{trans}}{2} \cdot (\bar{h}_{22}K_y - \bar{h}_{11}K_x) \cdot$

$\dfrac{e^{i\Omega t_{trans}}}{1 + r_1 r_2 \, e^{2i(\omega + \Omega)t_{trans}}}$

where $\bar{h}_{11}$ and $\bar{h}_{22}$ are given by Eq.(4) and where

$$(17) \quad K_x = \text{sinc}\left(\frac{\Omega t_{trans}}{2} \cdot (1-\sin\theta\sin\psi)\right) \cdot e^{\frac{i\Omega t_{trans}}{2}(1+\sin\theta\sin\psi)}$$

$$+ \text{sinc}\left(\frac{\Omega t_{trans}}{2}(1+\sin\theta\sin\psi)\right) \cdot e^{\frac{-i\Omega t_{trans}}{2}(1-\sin\theta\sin\psi)}$$

with $\text{sinc}(X) = \frac{\sin X}{X}$.

K_y is obtained by replacing $\sin\psi$ by $-\cos\psi$ in the expression of K_x.

On Fig.3,4 and 5 we give three examples for the gravitational transfer functions of M.I.M.P. and M.I.F.P. antennas for different values of the angles ϕ, θ and ψ .
The storage time of the light in the multipass delay lines and in the Fabry-Perot cavities are assumed to be the same; then, for a given set of angles, the transfer functions for the M.I.M.P. and M.I.F.P. antennas have the same value for a zero gravitational wave frequency. For these examples we have chosen

$$N = \frac{1}{1 - r_1 r_2} = 5$$

For low gravitational frequencies (i.e., smaller than the storage time of the light in the cavities) the maximum of response of the antennas occurs when the directions of polarization of the wave coincide with the arms of the interferometer (i.e., $\phi = \theta = \psi = 0$, see Fig.3).
In the Fig.4 example, the components $\bar{h}_{11}$ and $\bar{h}_{22}$ have the same value ; the only effect of the gravitational wave occurs at high frequencies and comes from a retardation effect for the gravitational wave propagating across the antenna.

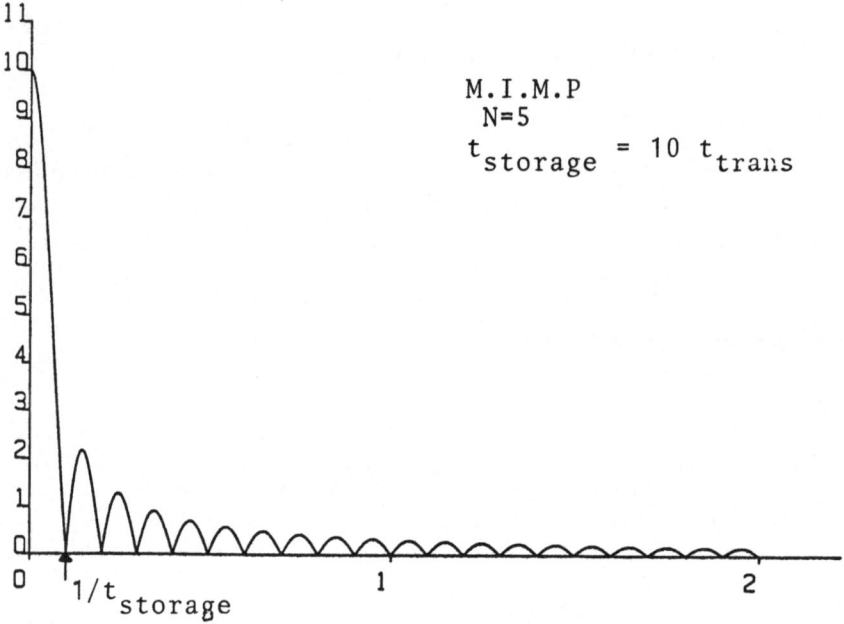

Fig.3:   TRANSFER FUNCTION (amplitude in units of t_{trans})
V.S.
GRAVITATIONAL WAVE FREQUENCY (in units of $1/t_{trans}$)
$\phi = 0$; $\theta = 0$; $\psi = 0$

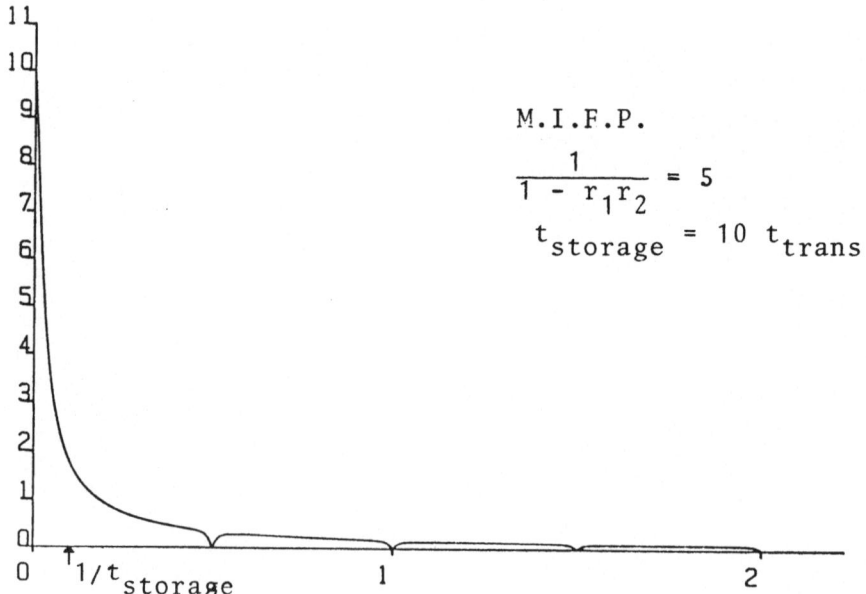

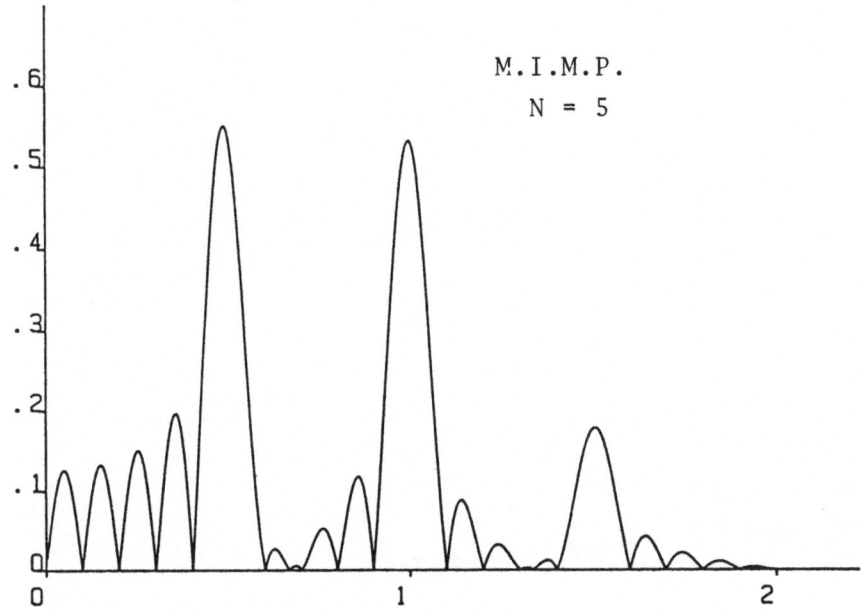

M.I.M.P.

N = 5

Fig.4:   TRANSFER FUNCTION (amplitude in units of t_{trans})

V.S.

GRAVITATIONAL FREQUENCY (in units of $1/t_{trans}$)

$\phi = 0$; $\theta = 45°$; $\psi = 225°$

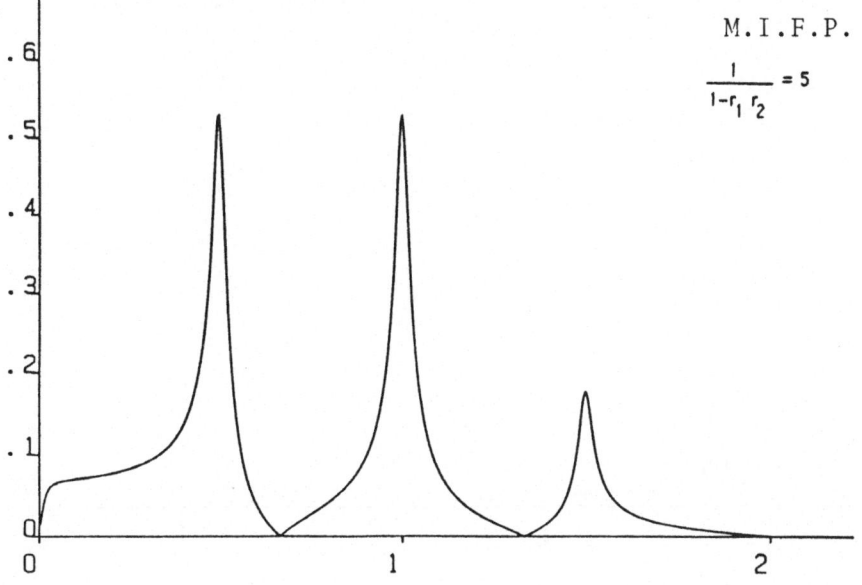

M.I.F.P.

$$\frac{1}{1-r_1 r_2} = 5$$

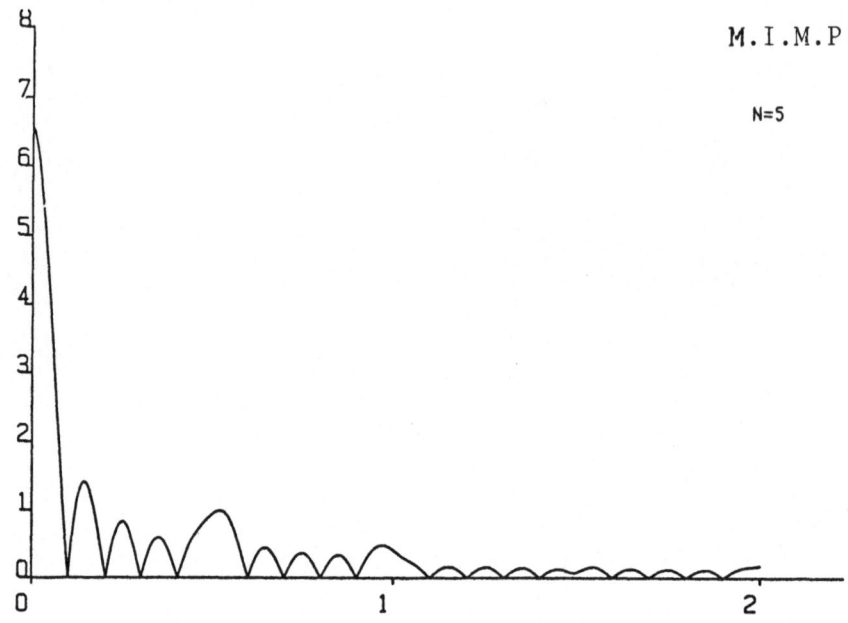

Fig.5:   TRANSFER FUNCTION (amplitude in units of t_{trans})
V.S.
GRAVITATIONAL WAVE FREQUENCY (in units of $1/t_{trans}$)
$\phi = 45°$; $\theta = 22.50°$; $\psi = 112.50°$

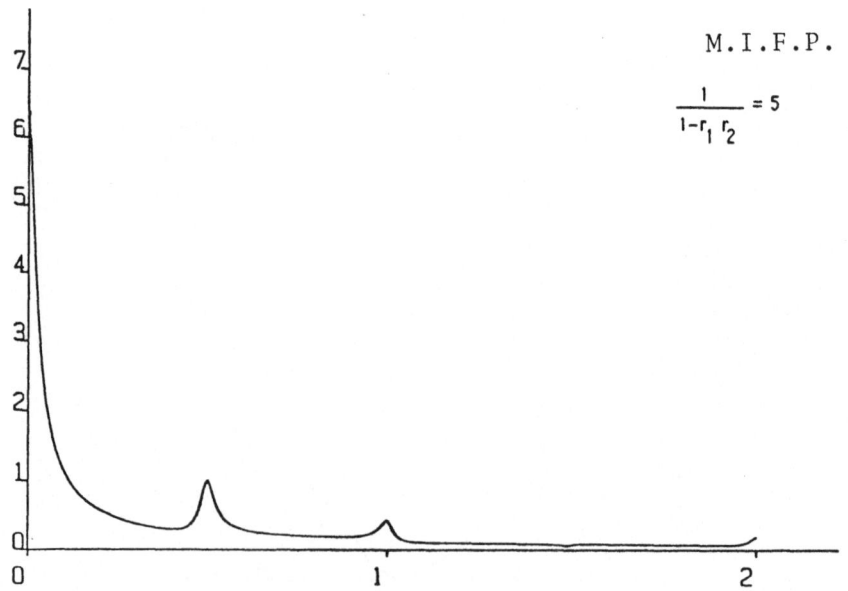

3- Gravitational Green's functions of Michelson interferometers:

The knowledge of the antenna's Green function is very useful to determine the antenna time response to any gravitational signal. From (13) and (14) we can write:

For M.I.M.P.:

$$(18) \quad S(t) = \omega \cdot R_{MP}(\omega) \cdot \int_{-\infty}^{+\infty} h(t-\tau) \cdot G_{MIMP}(\tau) \cdot d\tau$$

For M.I.F.P.:

$$(19) \quad S(t) = \omega \cdot R_{FP}(\omega) \cdot \int_{-\infty}^{+\infty} h(t-\tau) \cdot G_{MIFP}(\tau) \cdot d\tau$$

This means that G_{MIMP} and G_{MIFP} are respectively the inverse Fourier transform of the gravitational transfer functions F_{MIMP} and F_{MIFP}. Therefore, they are the Green's functions (i.e., the impulse response) of the M.I.M.P. and M.I.F.P. antennas.
Writing

$$G_{MIMP}(t) = FT^{-1}(F_{MIMP}(\Omega))$$

$$G_{MIFP}(t) = FT^{-1}(F_{MIFP}(\Omega))$$

we get:

$$(20) \quad G_{MIMP}(t) = \sum_{p=\infty}^{N-1} Z(p,t)$$

$$(21) \quad G_{MIFP}(t) = \sum_{p=0}^{} (-r_1 r_2)^p \cdot e^{2i\omega p t_{trans}} \cdot Z(p,t)$$

with, putting $a = \sin\theta \sin\psi$ and $b = \sin\theta \cos\psi$:

$$(22)$$
$$Z(p,t) = \frac{1}{2}\left\{ \frac{h_{11}}{1-a}\left[Y(t-2(p+1)t_{trans}) - Y(t-2(p+\frac{1+a}{2})t_{trans}) \right] \right.$$
$$+ \frac{h_{11}}{1+a}\left[Y(t-2(p+\frac{1+a}{2})t_{trans}) - Y(t-2p t_{trans}) \right]$$
$$- \frac{h_{22}}{1+b}\left[Y(t-2(p+1)t_{trans}) - Y(t-2(p+\frac{1-b}{2})t_{trans}) \right]$$
$$\left. - \frac{h_{22}}{1-b}\left[Y(t-2(p+\frac{1-b}{2})t_{trans}) - Y(t-2p t_{trans}) \right] \right\}$$

where $Y(t)$ is the "step function" (Heaviside function).

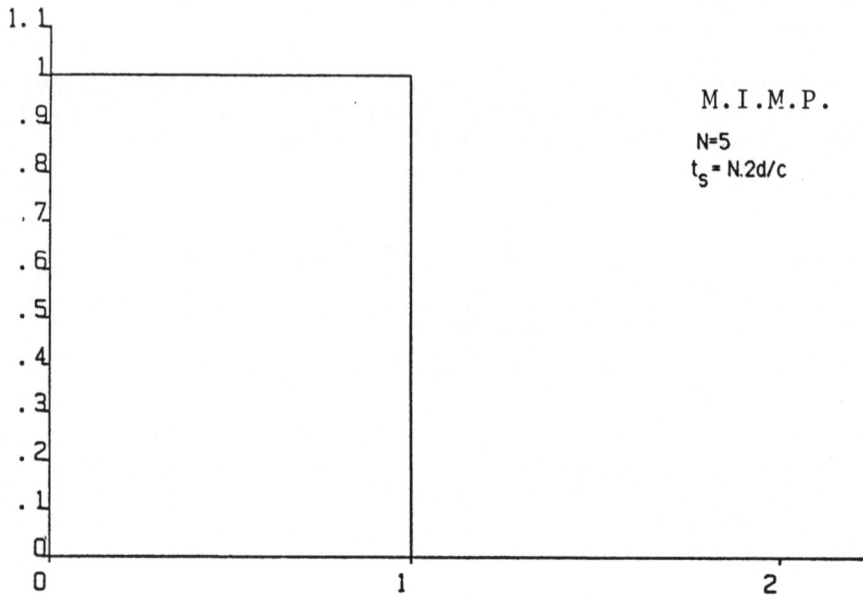

<u>Fig.6:</u>

GREEN'S FUNCTION

V.S.

TIME (in units of the storage time t_s)

$\phi = 0$; $\theta = 0$; $\psi = 0$

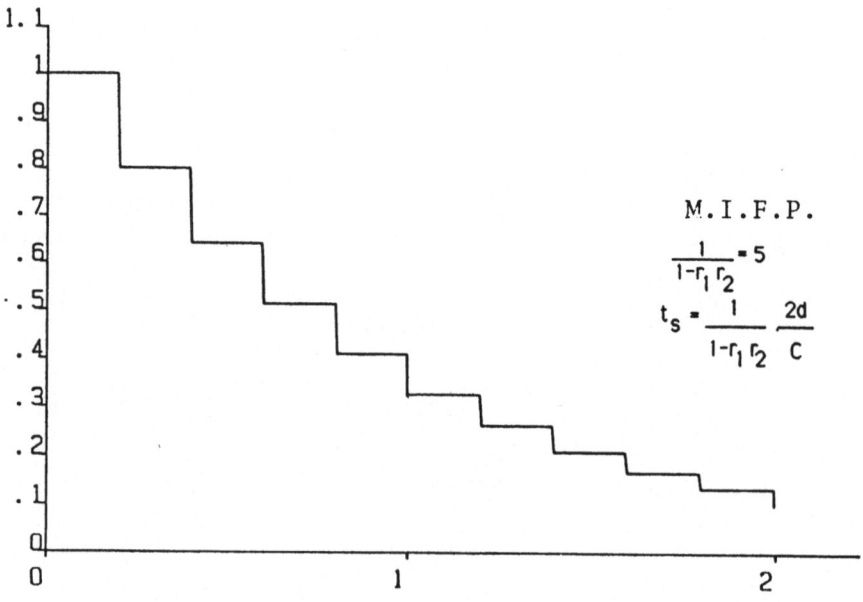

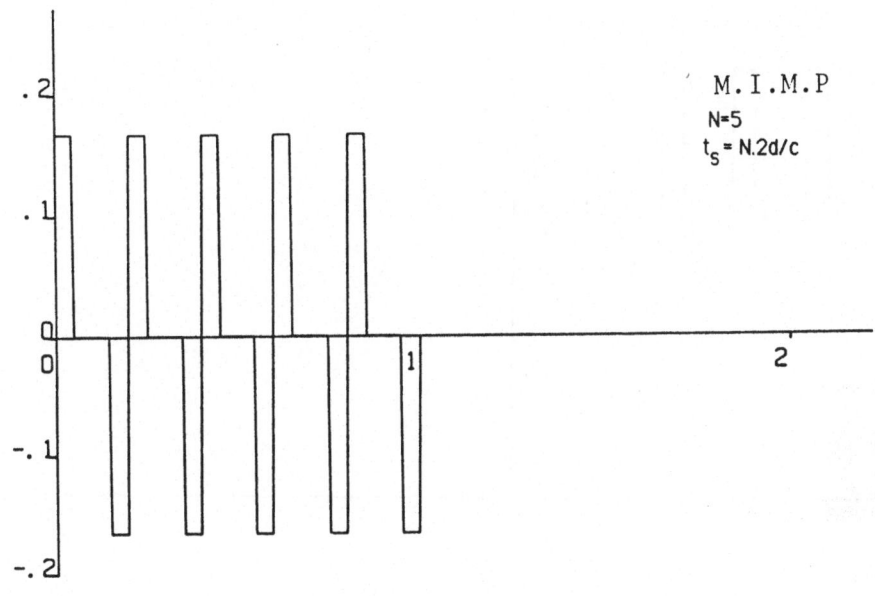

Fig.7:   GREEN'S FUNCTION
                              V.S.
                TIME (in units of the storage time t_s)

        ϕ = 0 ; θ = 45° ; ψ = 225°

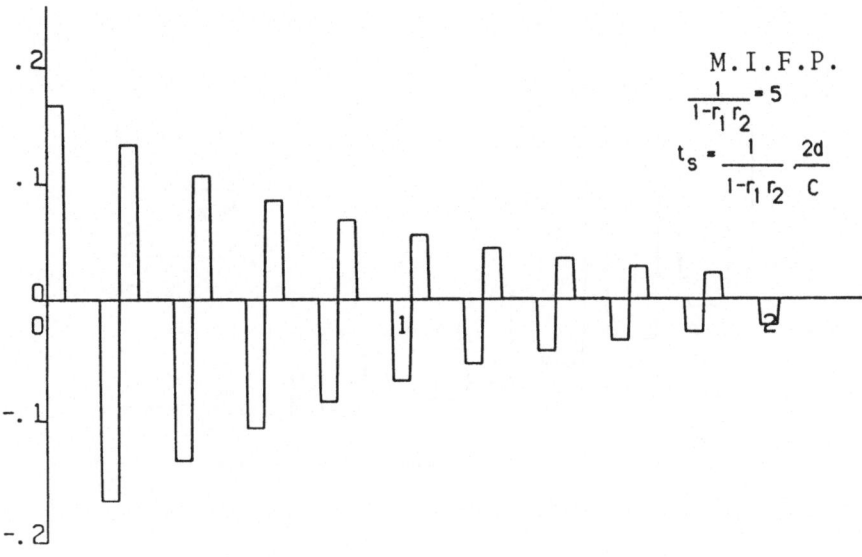

252

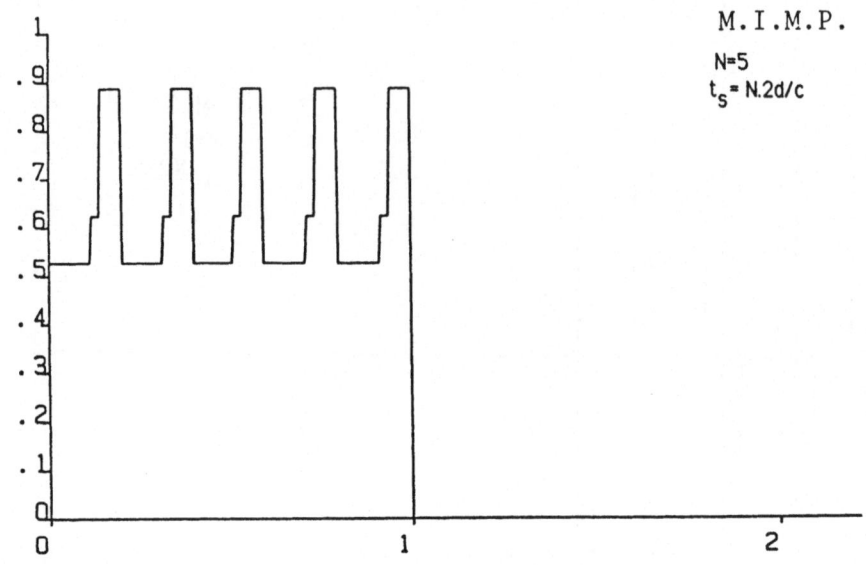

M.I.M.P.

N=5

$t_s = N.2d/c$

Fig.8: GREEN'S FUNCTION

V.S.

TIME (in units of the storage time t_s)

$\phi = 45°$; $\theta = 22.50°$; $\psi = 112.50°$

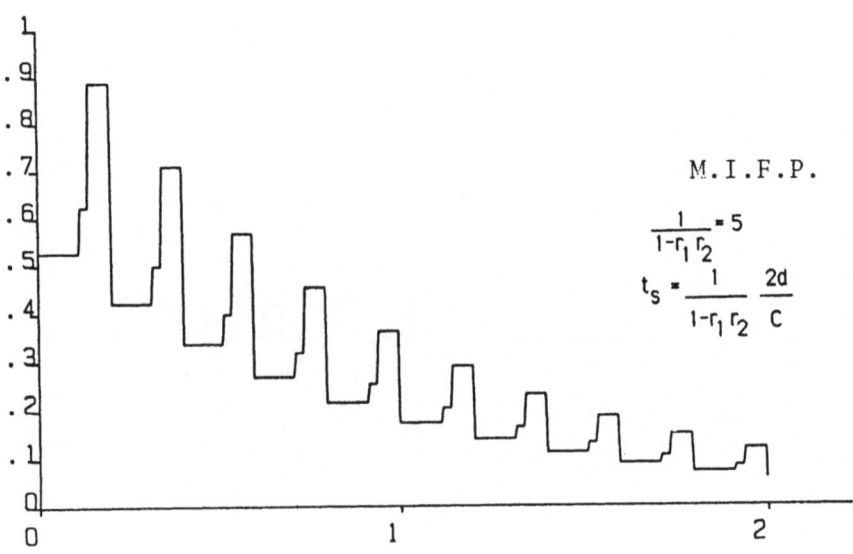

M.I.F.P.

$\frac{1}{1-r_1 r_2} = 5$

$t_s = \frac{1}{1-r_1 r_2} \frac{2d}{c}$

We see that the Green's functions of the Michelson
interferometers are a combination of "box functions" whose
width depends on the position of the source of gravitational radiation and whose height depends not only on the
polar and azimuthal angles of the source but also on the
polarization angle of the linearly polarized gravitational
wave.
With the same values for the angles ϕ, θ, and ψ than in
Fig.3,4 and 5 we give in Fig.6,7 and 8 the variations of
the Green's functions of M.I.M.P. and M.I.F.P. antennas
according to the time (expressed in units of the storage
time of the optical cavities). We can easily see the
elementary "box functions" that we have writen in Eq(22).
The Green's functions for the M.I.M.P. and the M.I.F.P.
antennas are built with the same "box functions", but in
the M.I.F.P. case a decreasing exponential factor shows
the gradual leakage of light out of the Fabry-Perot cavities.

CONCLUSION:

The work that we have presented in this paper can be
used to compare the M.I.M.P. and the M.I.F.P. antennas in
their ability to detect gravitational waves (for the
M.I.M.P. case, see also R.Weiss (1984)[2]). In this field
we have to add that the technics of recycling the light in
the interferometer do not change the dependance of the
gravitational transfer functions (and gravitational Green's
functions) of the M.I.M.P. and M.I.F.P. antennas on the
parameters of the gravitational wave (polarization,direction
of propagation) ; (full details will be published eslewhere).
The important result from this paper is that we have
obtained the analytical expressions for the instrument
functions (transfer or Green functions) which can be used
in the analysis of gravitational data from large interferometers.

REFERENCES:

1 : Misner, C.W., Thorne, K.S., Wheeler, J.A.,
    Gravitation (Freeman, San Francisco), 1973,Chapter 18.

2 : Weiss, R., and al.,in "a study of a long baseline
    gravitational wave antenna system, N.S.F. report.

DATA ANALYSIS AS A NOISE DIAGNOSTIC: LOOKING FOR TRANSIENTS IN INTERFEROMETERS

Daniel Dewey
Massachusetts Institute of Technology
Room 3-253
Cambridge, MA   02139
U.S.A.

ABSTRACT.   In addition to the ultimate goal of detecting and analyzing source waveforms, data analysis provides a framework for identifying important noise sources.   Only by developing and implementing data analysis programs designed to search for expected astrophysical sources will we uncover all relevant instrumental noise sources and have a true picture of an antenna's performance.   In particular, the characterization of an interferometric antenna solely by its noise spectrum hides the possible existence of weak and infrequent noise transients.   A search for such transients using simple matched filter templates was carried out with the MIT prototype.   An analysis of 1.7 hours of data (250 Mbytes) showed an agreement with Gaussian statistics out to an amplitude signal-to-noise ratio of 5.5.   Above this signal-to-noise level, transients identified with electrical feedthrough in our data taking system were recorded at a rate of $\simeq 70$ per hour, and transients of unidentified origin and various natures occurred at a rate of $\simeq 22$ per hour.

1. PHILOSOPHY AND INTRODUCTION

Data analysis strategies should be used at the prototype stage to uncover all relevant noise sources.

Astrophysical theory suggests two broad classes of sources: periodic sources and transient sources.   This dichotomy of source types leads naturally to (at least) two noise diagnostics: the (long-time averaged) power spectrum and the pulse-height distribution (PHD) of discrete transient events.   Measurement of an interferometer's output spectrum (typically presented as a log-log plot of equivalent strain noise ($h/\sqrt{Hz}$) vs. frequency) has been used almost exclusively by interferometer groups as the specification of instrument performance with little investigation of transient events; Forward's aural data analysis[1] is a

B. F. Schutz (ed.), Gravitational Wave Data Analysis, 255–268.
© 1989 by Kluwer Academic Publishers.

more-appropriate-than-it-first-appears exception. Conver-
sely, bar antenna groups have tended to present bar performance in terms of the distribution of transient events with
little direct measurement of the spectral properties of
their noise. To some extent these specializations reflect
the difference in instrumental bandwidths and philosophies.

This work presents results obtained by applying the
matched filter technique to search an interferometer's
output for a broad range of transient events. The basic
idea of applying matched filters has been described previously[2] and is briefly recounted in Section 2. Section 3
presents specific details of the data collection and data
analysis and the results of analyzing 1.7 hours of data are
given in Section 4. Finally, Section 5 contains some brief
comments on the technique and a discussion of future work.

2. THE MATCHED FILTER AND A SET OF TEMPLATES

2.1. The Matched Filter

The optimum signal-to-noise ratio (SNR) for the detection of
a known signal h(t) in the presence of white, Gaussian noise
characterized by a spectral density $\tilde{h}$ is a well-known result of detection theory[3,4,5] and is given by:

$$(S/N)_{opt} = [\frac{2\int h^2(t)dt}{\tilde{h}^2}]^{1/2}. \qquad (2.1)$$

How is a detection yielding this optimum (i.e. maximum) SNR
achieved? Again the result is well-known: it is obtained
through matched filter correlation in which a template of
the known signal is correlated with the input waveform to
produce an output waveform; this output is essentially the
dot product of the template with the input waveform as a
function of template position.

Because the input noise to the linear correlation
filter is Gaussian (assuming no signal is present) the resulting output time series is again Gaussian. Thus, ideally, a pulse-height distribution (PHD) of the output series
will be Gaussian and given by the probability density:

$$p(x) = \frac{1}{\sqrt{2\pi}} e^{-(x/\sigma)^2/2} \frac{dx}{\sigma} , \qquad (2.2)$$

where σ is the rms value of the filter output. Gross
deviations from this distribution represent non-Gaussian
noise, perhaps due to a non-linear component in the signal
chain; other deviations, especially 'tail events', may be

caused by the existence of transient noise sources.

It should be mentioned that these matched filter results could have been equivalently expressed in the frequency domain (i.e. using $h(\omega)$ in place of $h(t)$); however a time-domain presentation seems a more appropriate view for the understanding of discrete, transient events. Additionally, with modern digital signal processing techniques it is often more convenient to work in the time domain.

2.2 Effect of a Mismatch of Template and Signal

An initial problem in implementing the matched filter technique is the fact that the shape of the signal is not known. The effects of template-signal mismatch can be easily evaluated in the following framework. If the signal and template are viewed as made up of some number of half-cycle components N_{hc} (defined by the zero-crossings of the waveform) then a mismatch between signal and template can be viewed in terms of a mismatch of: 1) the shape of each component pulse, 2) the amplitude of each component pulse, 3) the spacing of the component pulses, and 4) the total number of component pulses. In general, if a template $h_1(t)$ is used to detect a signal of shape $h_2(t)$ the reduction in SNR from an ideal matched filter is[6]

$$\frac{(S/N)}{(S/N)_{opt}} = \frac{\int h_1(t)h_2(t+\tau)dt}{[\ \int h_1^2(t)dt \cdot \int h_2^2(t)dt\ ]^{1/2}}, \qquad (2.3)$$

where τ is chosen to maximize the numerator. Note that because of the symmetry of Eq. (2.3) the roles of template and signal may be interchanged. Through Eq.s (2.1) and (2.3) the effect of the four types of mismatch can be determined.

2.2.1 <u>Shape mismatch</u>. In general a shape mismatch of component pulses will have only a slight effect on the SNR for detection. For example, detecting a triangular waveform using a square wave gives $(S/N)/(S/N)_{opt} \simeq 87\%$; for a three-level waveform (Fig. 1b) and a sine wave, it's 96%.

2.2.2 <u>Amplitude mismatch</u>. It can be seen from Eq. (2.1) that a single component pulse of amplitude h_{cp} makes a contribution to the integrand that is proportional to $h_{cp}^2 \Delta t$, where Δt is the length of the component pulse. Thus, only the largest amplitude component pulses are important.

2.2.3 <u>Number mismatch</u>. A mismatch in the number of component pulses leads to a reduction in SNR given by:

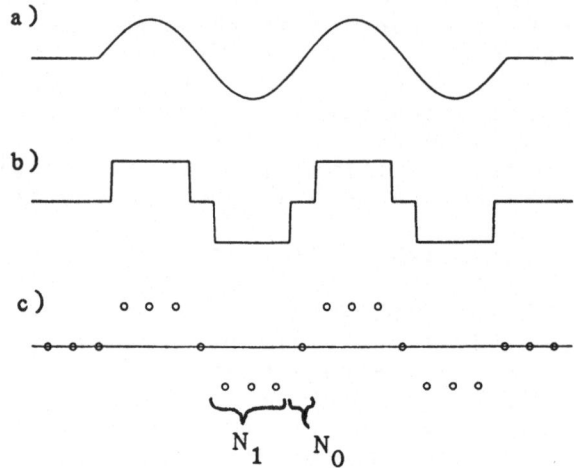

Figure 1. From a canonical burst to a digital template. The four half-cycle burst in a) is re-placed by a three level approximation in b) and finally a discrete-time template in c); this tem-plate is specified by the three integers $N_1=3$, $N_0=1$, and $N_{hc}=4$, more compactly written as 314.

$$\frac{(S/N)}{(S/N)_{opt}} = [\ \frac{\min\{N_{hc,1},N_{hc,2}\}}{\max\{N_{hc,1},N_{hc,2}\}}\ ]^{1/2} . \qquad (2.4)$$

For example, detecting a 5 half-cycle waveform with a 6 half-cycle template results in $(S/N)/(S/N)_{opt} \simeq 91\%$.

2.2.4 Spacing (or frequency) mismatch. The sensitivity of the SNR to a mismatch in the spacing or frequency of compo-nent pulses is a function of the number of half-cycles in the template; the bandwidth decreases with increasing N_{hc}. For templates with $N_{hc}=2$ a frequency range of $\pm 30\%$ is covered with a minimum $(S/N)/(S/N)_{opt}$ of 80%, while the $N_{hc}=6$ templates cover only a $\pm 12\%$ frequency range when main-taining a comparable sensitivity.

From the mismatch discussions above it becomes clear that to ensure a high SNR for detection the parameters it is most important to match between template and waveform are the frequency and number of half-cycles.

2.3 The Set of Templates

Many of the expected waveforms from gravitational wave burst sources can be well approximated by several (1 to 8) half-cycles of a sine wave at a constant frequency[2]; one such canonical burst, made up of four half-cycles, is shown in Fig. 1a. The frequency scale for burst sources (e.g.

stellar collapse and particle-black hole encounters) is predicted to be in the range of 1 kHz or less to as high as 10
kHz. Here, the goal in choosing a set of templates is to
reduce the chance of a "false dismissal" by choosing a small
set of templates that covers the space of frequency and
number-of-half-cycles without excessive SNR reduction.

In order to avoid (computationally expensive) multi-
plications in the correlation operation, a three level approximation to the sine wave[7], as shown in Fig. 1b, is used.
This introduces little reduction in sensitivity: the SNR for
detecting the original burst by the clipped version is 96%
of optimum. This three-level waveform is then converted to
the digitized form as shown in Fig. 1c where N_1 specifies
the number of digital points per half-cycle during which the
template takes the value +1 (or -1) with an equivalent definition of N_0; N_0 is chosen to be of order $N_1/3$ to obtain
the best SNR. The center frequency of the template is a
function of the digital sample rate and is roughly given by:

$$f_{template} \simeq \frac{f_{sample}}{2(N_1+N_0)} - \frac{f_{Nyquist}}{N_1+N_0} \quad . \qquad (2.5)$$

Thus, the range in frequency spanned by a given template or
set of templates depends on the digital sample rate. Note
that larger values of N_1 and N_0 allow for finer frequency
resolution in the location of templates; however, there is a
tradeoff as smaller values reduce the number of points per
template, reduce the amount of required computation, and
make full use of the available digital bandwidth.

The set of digital templates used here are listed in
Table 1., categorized by their frequency and number of half-

TABLE 1. Set of templates used in the analysis,
organized by frequency and number of half-cycles
(based on $f_{Nyquist}=10$ kHz).

Center frequency	Template Code ($N_1N_0N_{hc}$)			
	$N_{hc}=1$	$N_{hc}=2$	$N_{hc}=4$	$N_{hc}=6$
0.9 kHz		832	834	
1.0 kHz				736
1.3 kHz			624	626
1.4 kHz		522		
1.7 kHz	601		424	426
2.0 kHz				416
2.5 kHz		312	314	316
3.3 kHz	301		214	216
5.0 kHz		202	204	206
$\simeq$10 kHz	101	102	104	

cycles. This set was chosen to cover a frequency range of 900 Hz to 5 kHz (when used with data sampled at 20 kHz) and N_{hc}=1-8 with an SNR response better than 80% of maximum for most bursts in these ranges. The increased frequency density of larger N_{hc} filters needed to ensure adequate overlap of response is visible in the table. The N_{hc}=1 templates have response at DC and were included primarily for their computational use as mentioned in Section 3.2.3. Note that template 101 is the identity template in that filtering by it reproduces the original waveform.

3. DATA COLLECTION AND ANALYSIS

3.1. The Data Taking System

The data were taken spaced throughout the night of June 9, 1985 and consist of ten tapes each recording 15 minutes of interferometer output signal at a rate of 20 ksamples/s giving a Nyquist frequency of 10 kHz. The data were taken with a system comprising an A/D with an analogue anti-aliasing filter (four poles with -6 dB at 6.5 kHz), a dual-port memory, and tape drive all under control of an 11/23 computer[8].

One detail that will become important in the results is the operation of the data taking system as it records data to tape. Following standard dual-port operation two buffers are set aside in memory, and the A/D data are written into one buffer while the previously filled buffer is transferred to tape. One complete cycle of the two buffers is here called a 'chunk' of data and consists of 32,608 A/D samples and 160 words of housekeeping information and represents $\simeq$1.6 seconds of data. Thus, for each chunk of data the tape drive performs two data transfers, one at the beginning of the chunk and one at the middle of the chunk.

3.2. Processing of the Data

The data processing steps for the data analysis are briefly outlined below. Typically these steps were performed on one fourth of a tape at a time, producing four PHDs (Section 3.2.4) and four lists of high-SNR events (Section 3.2.5) from each tape.

3.2.1. <u>Check for proper antenna operation</u>. Using either the interferometer output itself (i.e., checking for A/D saturation) or through housekeeping information, times during which the interferometer is unlocked are determined and not analyzed.

3.2.2. Highpass filter the data. Because the interferometer noise is non-white and rises steeply with decreasing frequency below a noise corner frequency of about 1 kHz[8], the data are digitally highpass filtered at 1 kHz using a 61 point FIR filter apodized with a Kaiser window[9]; this is essentially a pre-whitening step.

3.2.3. Template filter the data. The highpass filtered data are then filtered by all 22 templates of Table 1, resulting in 22 output time series. Because the templates are three level the correlation can be performed with only additions and subtractions. Additionally, by saving past values of the templates it is often possible to generate a new template output simply by adding two previously determined values. For example, to generate the ith value of templates 301, 312, 314 and 316 from the new data point $h(i)$ the following four additions are performed:

$$
\begin{aligned}
301(i) &= 201(i-1) + h(i) \\
312(i) &= 301(i) - 301(i-4) \\
314(i) &= 312(i) + 312(i-8) \\
316(i) &= 312(i) + 314(i-8) \quad .
\end{aligned}
\tag{3.1}
$$

3.2.4. Form pulse-height distributions. The ouput values of the 22 templates are binned in 22 pulse-height distributions for later examination. These PHDs contain all points output from the template, however once a point is binned its identity is lost.

3.2.5. Save location of high-SNR events. In order to be able to go back and examine the time series of events with high values of SNR, a list for each template is formed of the location and SNR of the 128 events with the highest SNR values detected by that template. This scheme for saving events, thus, does not set a predetermined SNR threshold but rather floats the threshold. If the data were completely Gaussian in character (with no non-statistical tail events), the 128 events in $\simeq 4$ minutes of data would set an SNR threshold of approximately 4.2; in practice the threshold varied from 4.5 to 5 becauses of tail events.

3.2.6. Exclude brief high event rate regions. By plotting the location in time of the high-SNR events, striking departures from a uniform distribution of event arrival times may show up. In particular, occasional clumping of events on a time scale of tens of seconds was seen in the data analyzed here and suggests that the clump of events was due to momentary excitation of a component of the antenna, e.g., excitation of the string resonances of the mass suspension wires. These 'regions of excitation' were noted and high-SNR events in them were vetoed from further processing.

3.2.7. <u>Remove multiple detection of high-SNR events</u>. If a single high-SNR burst appears in the data it is likely that it will be detected several times by a single template and, furthermore, be detected by more than one template. For example, a burst of five half-cycles at 1 kHz having an SNR of 10 may be detected by template 736 (see Table 1) with SNR≈9.1 (according to Section 2.2.3) at each of the two times one half-cycle apart when the template 'covers' the burst. When the template-burst overlap is four half-cycles a detection at an SNR of ≃8.2 will be made, etc. Additionally, templates 834, 624, and 626 will also detect the burst at significant SNR levels. Thus these multiple detections of a single burst must be removed from the data to present a PHD in which *tail events correspond to single independent bursts*. This is accomplished by 1) searching all templates for the event with the largest SNR value and saving it, then 2) removing all other events in all other templates which fall within ±30 points (≃3 ms) of this event and 3) continuing this process until the largest SNR values of remaining events are 5 to 6; thus ensuring that only the tail events are subject to this processing.

3.2.8. <u>Combine all saved events into one distribution</u>. Finally, with multiple detections removed, all high-SNR events from all templates can be combined into one PHD which gives information on the agreement of the data with Gaussian statistics. Because the multiple detection removal has essentially classified events according to template, the PHD of events from each template alone can be viewed and together these provide a rough classification of the types of bursts that were detected.

Using non-optimized code the steps above take ≃30 hours of VAX/730 CPU time (typically as four ≃8 hour overnight runs) to process 15 minutes of data, or 6 ms per data point. Much of this time is spent performing the digital highpass filter correlation (essentially 61 real multiply-accumulates per point) and could be avoided through the use of an analogue (or dedicated digital) highpass filter before data recording.

4. RESULTS

The pulse-height distribution (PHD) obtained by applying the data analysis steps outlined above to the 1.7 hours of data is presented in Fig. 2a, plotted as log number-of-events vs. SNR^2 so that the Gaussian distribution of Eq. 2.2 is a straight line. Here many non-Gaussian tail events are visible above SNR≈5.6. Examining the individual PHDs showed that most of these tail events were detected (at the highest

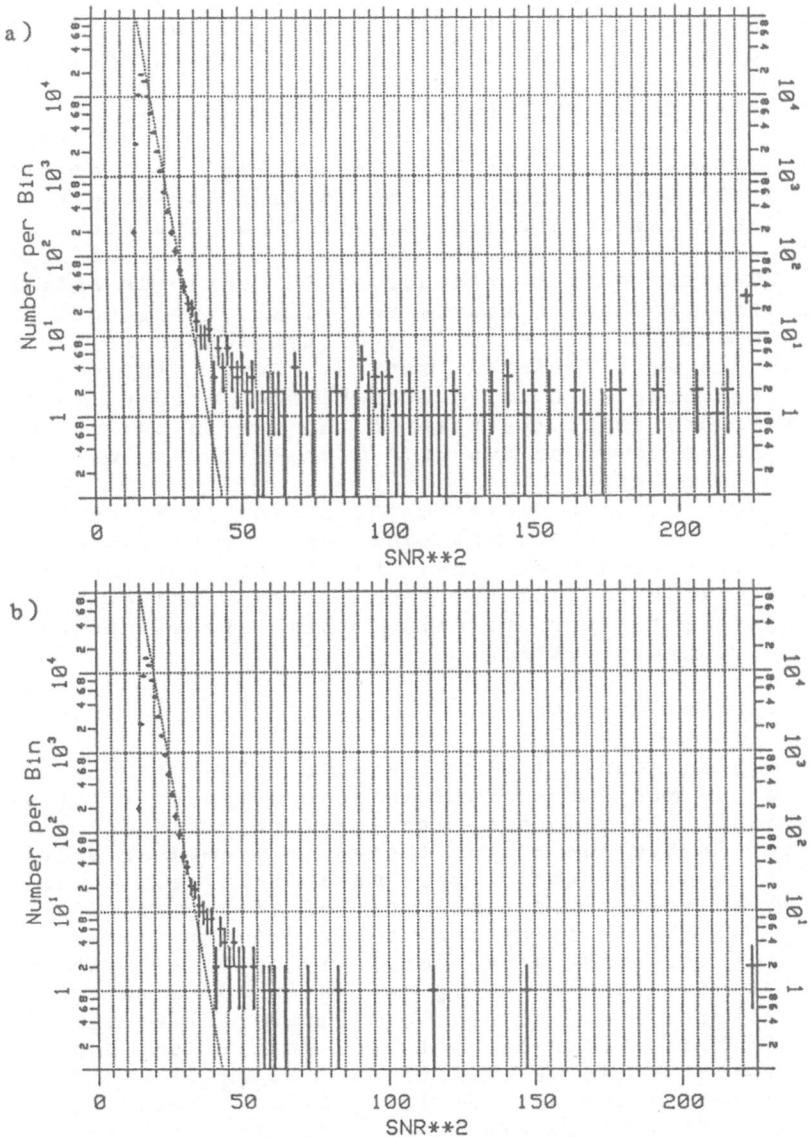

Figure 2.   Pulse-height distributions of events.
In a) all high-SNR events from all templates are
binned.   In b) the events from template 102 and the
N_{hc}-1 templates (which showed no tail events) have
been removed leaving a pulse-height distribution
indicative of the instrument operation.   In both
cases, multiple detection of high-SNR events have
been removed (Section 3.2.7).

SNR level) by template 102; if the events detected by template 102 (and those of templates with $N_{hc} = 1$) are excluded, the PHD of Fig. 2b results. The template-102 events and the Fig. 2b events are described in the following sections.

4.1. The Template-102 Events

As the difference between Fig. 2a and Fig. 2b shows, the template-102 events dominate the total pulse-height distribution. Looking at the time series of a '102 event, Fig. 3a, reveals that it consists of a single displaced data point. Because of the presence of the analogue anti-aliassing filter this noise source most likely originates at or after the A/D input. A variety of hypotheses regarding the noise source were explored, many of these (single bit errors in the recording of data, A/D conversion errors) made specific predictons about the size of the pulses as recorded on tape (i.e., a power of two in A/D units); this was not supported by the data.

However, when a distribution of the location in time of the template-102 events was made modulo one chunk of data, Fig. 3b, it became clear that the events were most likely associated with feedthrough from the tape drive as it started each write-to-tape operation; these operations occurring twice per chunk of data.

4.2. The Other Events

The remaining non-statistical events, those in Fig. 2b, were examined and many were found to occur at nearly the same time (within ± 25 ms) and, like the more extensive clumps discussed in Section 3.2.6, these are viewed as being caused by a single disturbace. There were, however, several events which were isolated in time; most of these were detected at the highest SNR level by the $\simeq 1$ kHz 736 template. The largest of these isolated events was detected with an SNR of 23.5 (!) and is shown in Fig. 4a as it appeared to the template filters, i.e., after the highpass filtering. The cause of these events has not been determined, however a gravity-wave origin is unlikely as the amplitude of the tail events detected corresponds to $h_o > 5 \cdot 10^{-14}$ which would require an efficient supernova at a distance closer than that of Proxima Centauri.

To summarize, all non-statistical, non-template-102 events observed in the 1.7 hours of data are categorized in Table 2. An example of a glitch within a region excluded from pulse-height binning is shown in Fig. 4b; this may represent a temporary or near unlocking of the interferometer; however, its origin is also unknown.

a)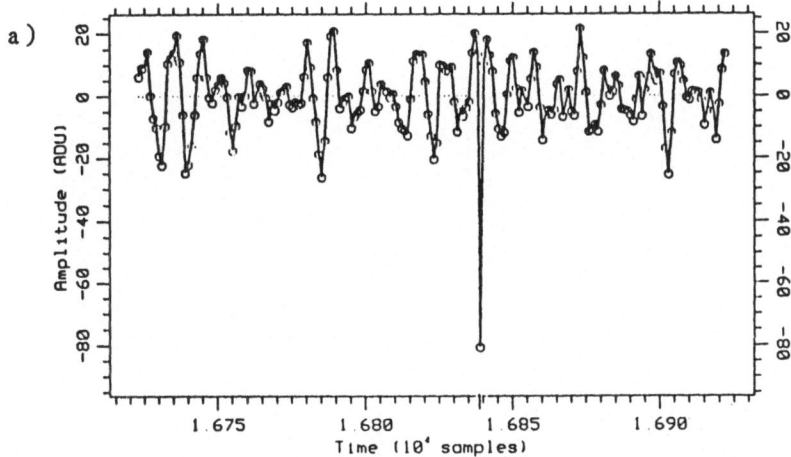

b)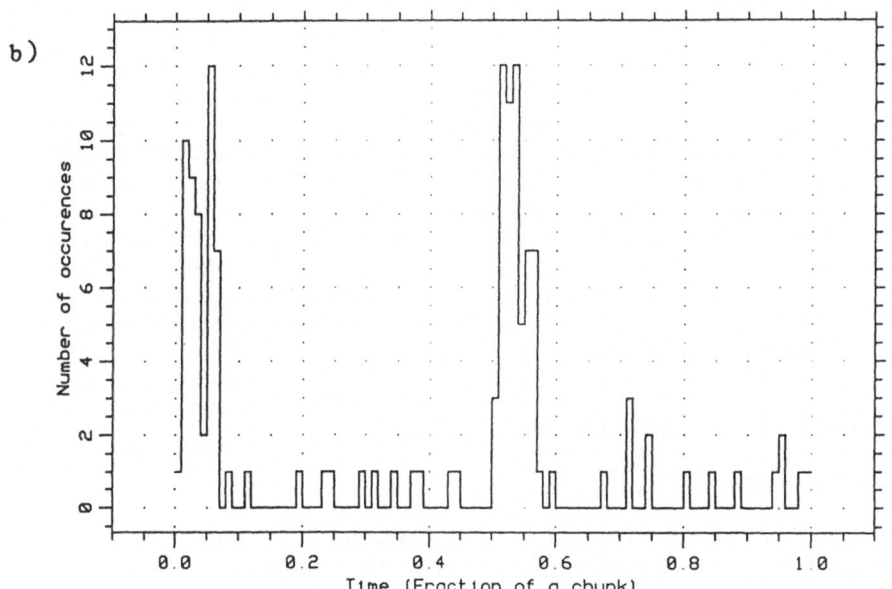

Figure 3. Template-102 events. In a) a typical high-SNR event detected by template 102 is shown. In b) template-102 events have been binned by occurance time within one data taking cycle, a chunk. The two peaks suggest that these events are due to feedthrough originating during write operations to the tape drive which occur twice per chunk.

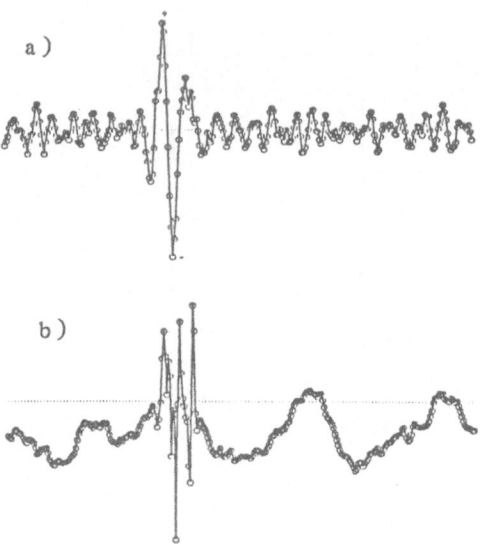

a)

b)

Figure 4. Some transients in the data. Shown in
a') is a pulse that was detected by the 736 template
at an SNR of ≃23.5; this time series is seen after
the digital highpass filter, as it appeared to the
template filters. In b) is shown one of several
glitches which appeared in the output of the interferometer and were excluded from the pulse-height
distribution, in this case shown before digital
highpass filtering.

5. CONCLUSIONS AND FUTURE DIRECTIONS

5.1. Conclusions

The results of the application of matched filters to the MIT
interferometer are encouraging: the instrument operated at a

TABLE 2. Summary of the Results. All non-statistical occurrences in the 1.7 hours of data are
summarized below:

10 Regions excluded from pulse-height binning,
    each is ≃10-20 seconds of data (Sec. 3.2.6)
12 Isolated events ($h_o > 5 \cdot 10^{-14}$)
15 Regions (<1 second) with several events
____

37 Total non-statistical 'events' in 1.7 hours,
    or ≃ 500 per day

constant rms noise level over the course of one night and
the filter output distributions showed agreement with
Gaussian statistics out to an amplitude signal-to-noise
ratio of 5.5. As listed in Table 2 the actual non-statistical event rate (where a generous interpretation of an
event is taken) is of order 500 per day. Finally, the
technique succeeded in detecting, isolating, and identifing
a transient noise source associated with our data collection
hardware, this source would not have been apparent in spectra from the instrument.

5.2. Chirp Sources and Improved Algorithms

The templates described here were chosen to obtain near
optimal detection of many of the predicted burst sources of
gravitational radiation. Templates for a second class of
transient sources, the chirp sources[2], were not explicitly
considered because chirps from, say, a system of two 1.4 $M_\odot$
objects, will consist of only a few cycles above the MIT
antenna corner frequency of 1 kHz and would therefore be
adequately detected by the burst templates.

With interferometric antennas now operating at noise
corner frequencies of 500 Hz or less, and given the exciting
astrophysical information chirp sources can provide[10], much
effort is now directed to detection schemes for these
sources. Unlike the burst souces, chirp sources are expected to emit 100 to 1000 or more cycles of detectable waveform; thus, naive matched templates will be very long, requiring large amounts of processing time or custom digital
filters. With innovative thinking, for example, the short-
time FFT approach[11] or schemes similar to Eq. (3.1), efficient algorithms and hardware will be developed to detect
these sources.

5.3. Diagnostic Philosophy

In the process of searching for real signals, data analysis
techniques can also detect a variety of noise sources. For
example, using a spectrum analyzer we (implicitly) search
for continuous sources and identify some noise sources: shot
noise, seismic noise, mirror resonances, etc.; in this case
the strain noise $h(f)$ is a measure of antenna performance.

With a philosophy similar to that of spectral analysis,
the matched filter analysis described here can be viewed as
a diagnostic for transient noise sources. In this context
very little data-taking hardware is required to acquire
transient statistics (for example, neither accurate absolute
timing nor continous data-taking are needed). Through this
transient analysis one might uncover a variety of potential
transient noise sources; some examples are: scattered light
noise (which may mimic chirp signals), outgassing bursts in

the vacuum, local electrical disturbances, local mechanical disturbances, laser intermittency, and the up-conversion of low frequency motions to high-frequency transients (already seen and important in bar systems[12]). Because bar detector groups have focussed on searches for transient sources there is a chance to learn from and contribute to their research efforts; for example[13] plots of (single-antenna) event arrival-time differences would have identified the template-102 events without requiring the hunch that lead to Fig. 3b.

Finally, it is important to emphasize that by searching for both continous and transient sources at the prototype stage will we ensure that all noise sources are understood in LIGO-scale antennas.

ACKNOWLEDGEMENTS

I'd like to thank Bernard Schutz for his organization of this workshop, the members of the Groupe de Recherches sur les Ondes de Gravitation at Orsay for their pre-workshop hospitality, and Elisabeth Busch at MIT for making travel arrangements. This work was supported in part through NSF grant 85-04836-PHY.

REFERENCES

[1] R.L. Forward, Phys. Rev. D **17**, 379 (1987).
[2] D. Dewey, in *Proceedings of the Fourth Marcel Grossmann Meeting on General Relativity*, edited by R. Ruffini (Elsevier, Amsterdam, 1986),p. 581.
[3] M. Davis, in this volume.
[4] D.O. North, Proc. of the IEEE **51**, 1016 (1963).
[5] C.W. Helstrom, Statistical Theory of Signal Detection (Pergamon Press, Oxford, 1969).
[6] D. Dewey, PhD Thesis, MIT Physics Dept. (1986).
[7] A.R. Whitney, A.E.E. Rogers, H.F. Hinteregger, C.A. Knight, J.I. Levine, S. Lippincott, T.A. Clark, I.I. Shapiro, and D.S. Robertson, Radio Science **11**, 421 (1976).
[8] J. Livas, R. Benford, D. Dewey, A. Jeffries, P. Linsay, P. Saulson, D. Shoemaker, and R. Weiss, in *Proceedings of the Fourth Marcel Grossmann Meeting on General Relativity* (Ref. 2),p. 591.
[9] S.M. Bozic, Digital and Kalman Filtering (Edward Arnold Ltd., London, 1979).
[10] B. Schutz, Nature **323**, 310 (1986); and in this volume.
[11] S. Smith, Phys.Rev.D, 36, 2901 (1987)
[12] P.F. Michelson, J.C. Price, and R.C. Taber, Science **237**, 150 (1987) and references therein.
[13] M.F. Bocks and W. Johnson, this volume

DATA ACQUISITION AND ANALYSIS WITH THE GLASGOW PROTOTYPE DETECTOR

N.L. Mackenzie, H. Ward, J. Hough, G.A. Kerr, J.B. Mangan,
B.J. Meers, G.P. Newton, D.I. Robertson and N.A. Robertson.
Department of Physics and Astronomy
University of Glasgow
Glasgow G12 8QQ

ABSTRACT.   After the discovery of Supernova 1987A,   data was recorded
from the prototype detector at Glasgow, some of which contained
artificially produced signals.   We describe   the   method of data
acquisition and report on the preliminary analysis of the data.

Following the recent Supernova 1987A it was decided to record some
data from the 10 metre prototype detector at Glasgow. It was felt that
this would be a good opportunity to attempt to record data in
coincidence with the detectors at Caltech and MIT. To this end the
three groups recorded around 2.5 hours of data coincidently. During
these runs 5 hours of data were recorded at Glasgow and, allowing for
dropouts, the detector had a duty cycle of around 80%.   So far no
cross-correlation of the data has been performed.   However, the
opportunity has been taken at Glasgow to analyse the characteristics of
the detector noise and, in collaboration with colleagues at University
College Cardiff, to test the effectiveness of matched filters for
detecting coalescing binary 'chirps' in the presence of this noise.
The Glasgow prototype has two Fabry-Perot cavities of 10 metres
formed between mirrors mounted on test masses suspended as pendulums.
The illuminating laser is frequency locked to one cavity, the primary,
and the length of the other cavity, the secondary, is adjusted to
maintain that cavity on resonance. At present, the feedback to the
secondary cavity is obtained by a coil and magnet arrangement which can
apply a force to one of the test masses.   The data recorded at the
secondary error point is the effective gravity wave signal.
The recording system used at Glasgow consists of a mini-computer
connected to two magnetic tape drives (Fig. 1). These can be used
alternately in order to record data continuously. Data is collected
from two 8 bit Analogue to Digital Converters which sample the primary
and secondary error points 6000 times a second. This sampling rate is
derived from a clock which is phase locked to standard frequency
transmissions. In addition a 1 kHz clock signal, derived in the same
manner, is used to provide timing information which is recorded on the
tape with the data.

B. F. Schutz (ed.), Gravitational Wave Data Analysis, 269–273.
© *1989 by Kluwer Academic Publishers.*

270

The servo system used to maintain the resonance of the secondary cavity has a unity gain point of around 1 kHz which means that, although the detector noise is reasonably white, frequencies around 1kHz are amplified and the recorded signal has the characteristic form shown below (Fig. 2). It is therefore necessary to periodically produce calibration peaks at a number of different frequencies if a true measure of strain is to be obtained across the bandwidth of the detector. For the coincidence runs a series of harmonics of equal amplitude (obtained by differentiating a square wave) was applied to the primary cavity for 1 second every 100 seconds. This appears on the secondary error point as a series of harmonics representing constant displacement.

MAGNETIC TAPE UNITS

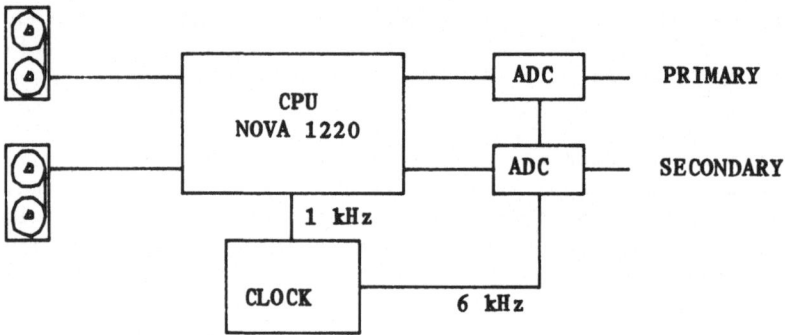

Fig.1. Schematic diagram of the data recording system.

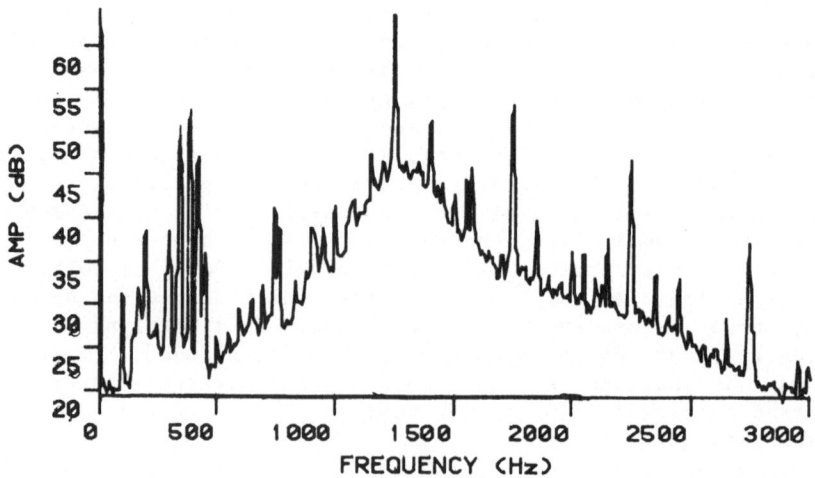

Fig.2. Spectrum of the secondary cavity error point signal.

An additional problem is that of detecting significant pulses in the data. If a pulse contains mainly high frequency information, then it will not be detectable against the relatively larger noise around 1 kHz. In order to overcome this problem the data can be digitally filtered with a filter with a frequency response which is the inverse of that imposed by the servo system. It is known that the noise has a reasonably white character, therefore the filter resonse can be obtained by inverting a smoothed spectrum of the data. A polynomial correction for any variation in sensitivity with frequency can be made using the calibration peaks. The filter can also be used to apply a band-pass to the data and thus exclude resonances around 400Hz due to the wires suspending the test masses. The frequency spectrum of data filtered in this way is shown below (Fig. 3).

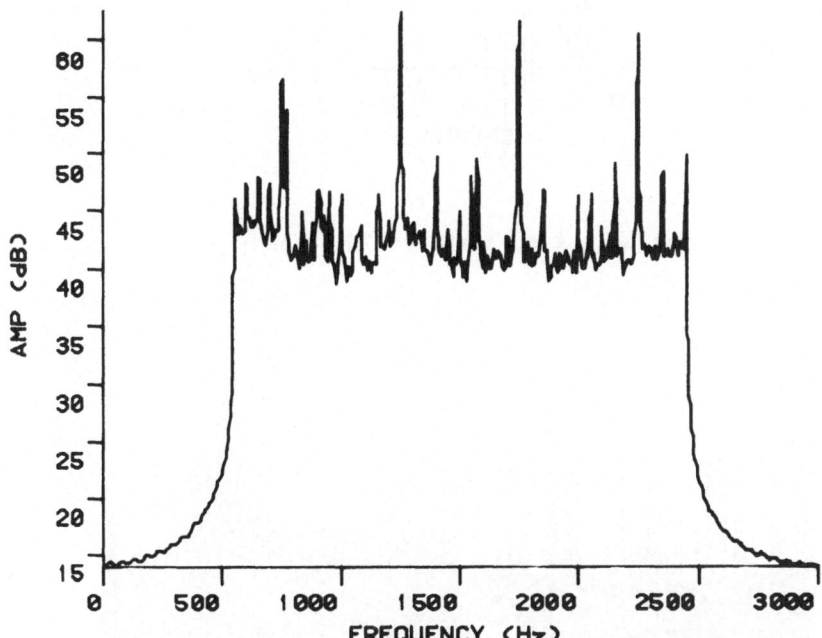

Fig.3. Spectrum of digitally filtered secondary error point.

Another characteristic of the present detector is that variations in the gain of the secondary servo system produce variations in the variance of the recorded data making analysis over long time intervals difficult. However, from the analysis of shorter lengths of data where the gain remains constant it appears that the noise is indeed Gaussian to at least 3 standard deviations. The pulse height spectrum of a period of 75 seconds of data is shown below (Fig. 4). Further analysis of normalised data should allow the limit of 3 sigma to be tightened.

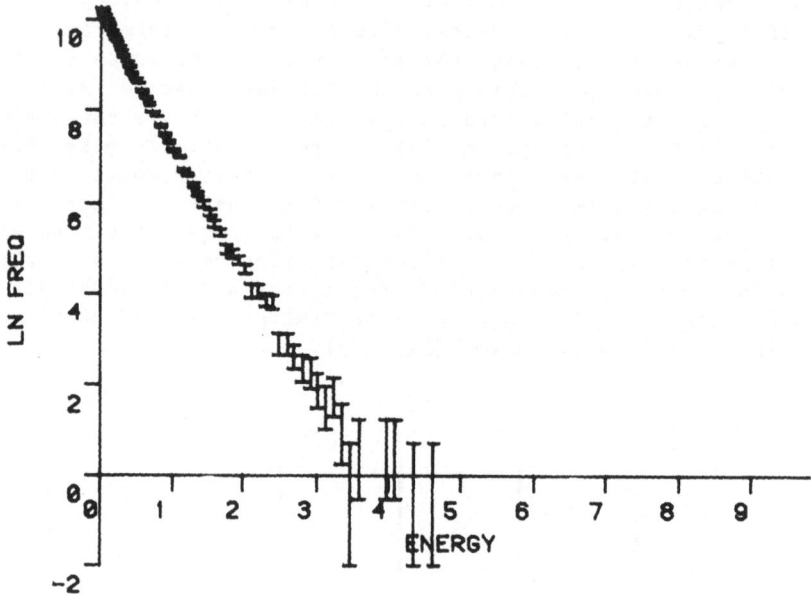

Fig.4. Pulse height spectrum of 75 seconds of data.

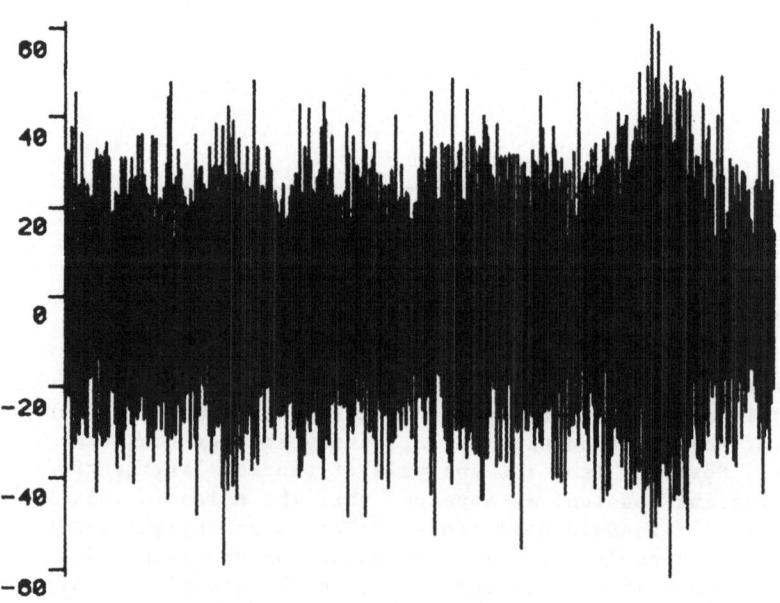

Fig.5. An example of a section of data containing a chirp.

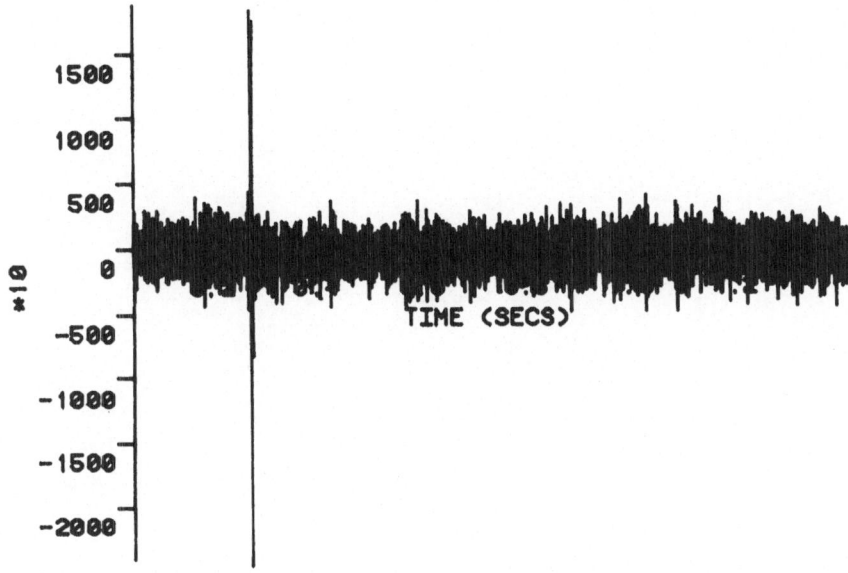

Fig.6. The result of filtering the data of Fig.5. with a suitable
       matched filter.

     Since there has been much interest in the possible detection of
gravitational radiation emitted by coalescing binary stars using
matched filter techniques, it was decided to add such chirps to the
recorded data. The desired waveform was calculated by a computer and
was introduced to the primary cavity via a digital to analogue
converter. In addition the waveform was added directly to the signal
recorded by the primary ADC which will allow timing resolution of
matched filters to be investigated. The chirps were produced at
different levels and examples of the detector output (Fig. 5) and of
the output from a suitable matched filter (Fig. 6) are shown below. A
data tape containing these chirps has been given to University College
Cardiff in order that a blind search may be conducted.

ON THE ANALYSIS OF GRAVITATIONAL WAVE DATA

S. P. Lawrence
E. Nahmad-Achar*
B. F. Schutz

Department of Applied Mathematics & Astronomy
University College, Cardiff
P.O. Box 78, Cardiff, U.K.

ABSTRACT.   An analysis of data was carried out from the interferometric gravitational wave detector prototype at Glasgow, to which artificial signals analogous to those emitted by coalescing binary systems were applied. We report on the ability to detect those signals, and the methods used, as well as on a full account of the detector's output noise analysis.

1.   INTRODUCTION

The possibility of building detectors capable of responding to gravitational waves brings of necessity the development and improvement of techniques to analyse their output. In this paper, we report on the analysis of data from the interferometric gravitational wave detector prototype at Glasgow. The idea was to see if we could detect signals associated with coalescing binaries, that were artificially added to the detector's output by the Glasgow group. Not only do we answer the above affirmatively, but we also give a full account of the output's noise analysis.
  Section 2 below describes the contents of the data sent to us by the Glasgow group.   Section 3 describes the Fourier analysis and compares the results obtained from the Glasgow data to those expected theoretically. The noise analysis is described in Section 4, and we conclude this paper with a few comments.

2.   DESCRIPTION OF THE DATA

The output data from the Glasgow prototype detector consists of two independent streams, each associated with one of the two arms of the

---

*   Present   address: Centro   de   Estudios   Nucleares,   UNAM,   Ciudad Universitaria, Apartado Postal 70-543, 04510 Mexico, D.F., Mexico.

B. F. Schutz (ed.), Gravitational Wave Data Analysis, 275–283.
© *1989 by Kluwer Academic Publishers.*

detector. (For the technical details of how the data is taken, cf. the paper by Norman L. Mackenzie in this volume.) One of the streams, called the *primary data*, is used as housekeeping data, and it contains the evidence of drop-outs (i.e. when the detector is not locked properly; data obtained in these intervals is to be rejected), the time records, and a large-amplitude version of the artificially applied chirps. The other stream, called the *secondary data*, contains the detector's output itself (which is essentially noise in this case), the artificially applied signals corresponding to chirps from coalescing binary systems (see below), and calibration signals which consist of equal-amplitude pulses applied at different frequencies. These calibration "combs" are used to compensate for the fact that the secondary data are related (for instrumental reasons) to the detector's response in a way that is not uniform throughout the frequency spectrum.

The signal in both streams was sampled at 6000 Hz for 25 minutes (to avoid aliasing, the data itself was filtered above 3000 Hz before sampling), and was stored as 1-byte data. When searching for signals in noise one byte data is ideal. However its use does have some restrictions on the analysis of the amplitude statistics of the noise because of saturation or truncation effects.

The artificial signal applied to the detector's output, corresponding to that which one would obtain from a coalescing binary system, is a chirp which increases in amplitude and frequency with time in a well-defined way. If we place the chirp in a standard form in which it has unit amplitude and frequency 100 Hz at $t = 0$, then we have

$$\text{chirp}(t) = \left[1 - 0.34\rho t\right]^{-\frac{1}{4}} \cos\left\{\frac{320\pi\left[1 - (1 - 0.34\rho t)^{\frac{5}{8}}\right]}{0.34\rho} + \phi\right\} \qquad (1)$$

where ρ is the *mass parameter*, which is given by $\rho =$ (reduced mass) (total mass)$^{2/3}$ (in units of solar masses) and characterises the chirp, t is measured in seconds, and ϕ is the phase of the wave when it reaches 100 Hz. If we confine the chirp to the interval (t_0, t_1) during which its frequency rises from f_0 to f_1, then we have

$$t_0 = \frac{\left[1 - (f_0/100)^{-8/3}\right]}{0.34\rho} \quad , \qquad t_1 = \frac{\left[1 - (f_1/100)^{-8/3}\right]}{0.34\rho} \quad ,$$

Figure 1a shows a chirp with $\rho = 10$ in a frequency range 300–2000 Hz (this value for ρ was only chosen to make the graph clearer), while Figure 1b shows the real part of its Fourier transform. Figure 1c shows the power spectrum of a chirp with $\rho = 0.15$; again, this value was chosen for clarity.

The signals actually applied to the detector were chirps in the range 300–2100 Hz. All the applied chirps had the same mass parameter, which was unknown to us, as was their location in the data streams.

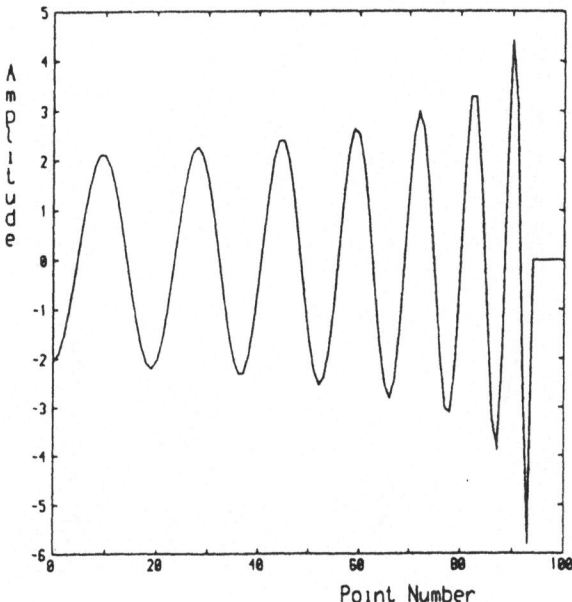

Figure 1a.   Theoretical chirp for ρ = 10.   This large value
        for ρ gives such a short chirp (1/60 s) that the sharp
        increase in frequency towards the end is not noticeable.

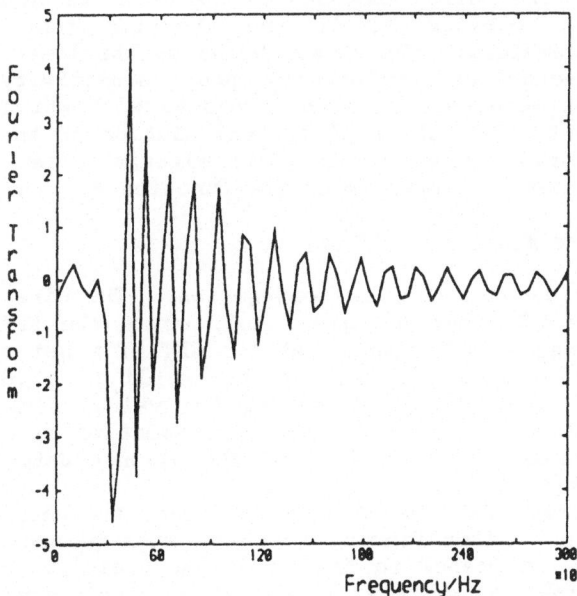

Figure 1b. Real part of the Fourier transform of the chirp in
        Fig.1a.

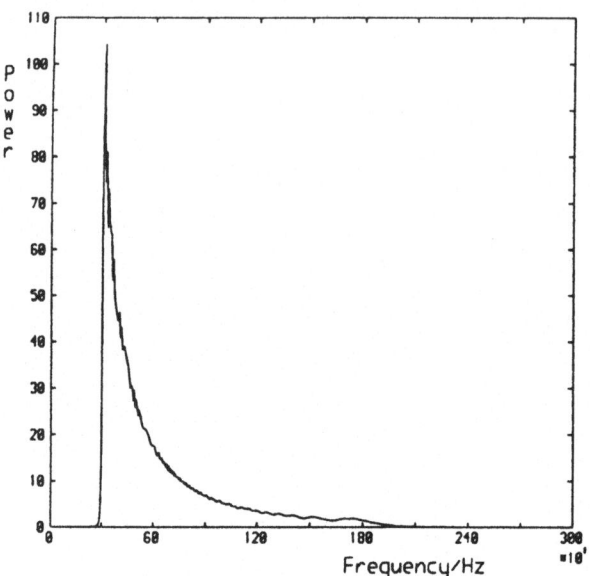

Figure 1c.   Power spectrum of a chirp with $\rho = 0.15$.

3.   FOURIER ANALYSIS

In order to detect the chirps embedded in the data, we built a family of
theoretical chirps obeying Eq.(1) and starting from 300 Hz, with
different mass parameters.   (We chose 300 Hz as the lower limit because
the Glasgow detector has relatively poor sensitivity below this
frequency.)   These were then used as templates with which to correlate
the detector output.   For the explicit calculation of the correlations
we used the fast Fourier transform (FFT) algorithm: we know that, if
F[g] denotes the Fourier transform of the function g, then

$$F[corr(g,h)] = F[g]F^{*}[h] \qquad (2)$$

where an asterisk denotes complex conjugation.   The direct correlation
of N points takes of order N^2 operations, while the FFT of N points
takes of order $Nlog_2N$.   [We found that the FFT of a set of 4096 points
take 0.87 s on a MicroVax II; writing the result to a (direct access)
file takes a further 0.20 s, and reading the values from a file takes
0.01 s, making a total of 1.08 s.   An improvement of speed by a factor
of about 100 is needed if the output of the planned detectors is to be
analysed in real time.]
        Using these methods we were able to detect the chirps embedded in
the secondary data.   After finding that the mass parameter lay in the
range $0 < \rho \lesssim 0.1$, we wished to fix ρ more accurately.   To do this, a
small subset of mass parameters in the above range was chosen.   The
correlations for these mass parameters were calculated and the value of
ρ which yielded the maximum correlation was then used, as the centre of
a smaller interval.   The range was continually narrowed in this way,

until it was felt that no advantage could be gained by having further accuracy; i.e. that any change in the maximum correlation would be masked by fluctuations due to the noise.   In all, ten filters were needed.

The best fit was obtained with $\rho = 0.087 \pm 0.001$.   Discussions with the Glasgow group later revealed that 0.087 was exactly the mass parameter that they had used in constructing their artificial chirps. This initial 'blind' test of the correlation method was therefore successful. (We should note that, because of the very low value of this mass parameter, resulting in a very long chirp, only that portion of the template between 300 and 400 Hz was used for the final correlations. This allowed for more manageable subsets of data.)

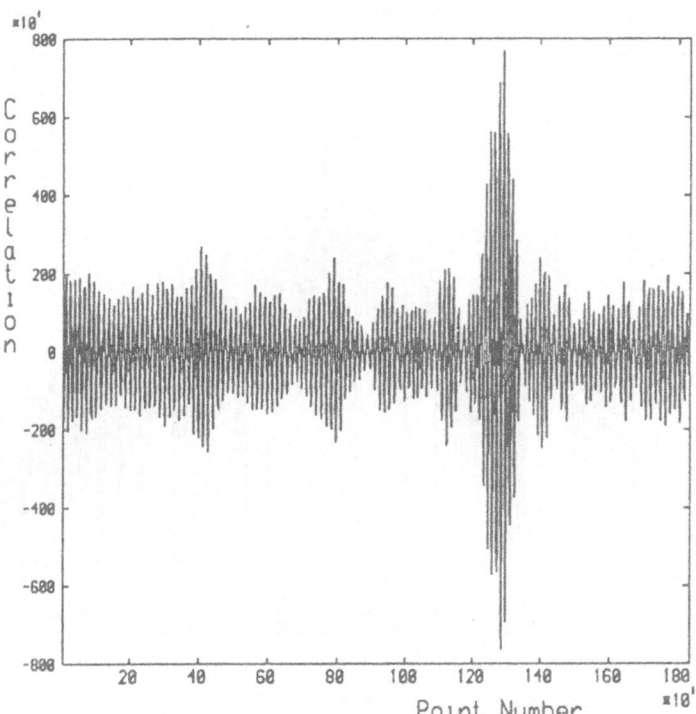

Figure 2.   Correlation of theoretical chirp of $\rho = 0.087$, with
subset of secondary data.

Figure 2 shows the correlation (not normalised) between the theoretical chirp with $\rho = 0.087$ and a section of the secondary data containing an applied chirp.   The plot is made against record point number, and the maximum peak corresponds to the region in the embedded chirp where its frequency is approximately 400 Hz.

Figure 3 shows the log of the power spectrum of the record in which the maximum correlation in Figure 2 occurs. One notes a peak at approximately 380 Hz, and this may be due to the presence of an applied chirp. However, since corresponding peaks occur in other records at similar frequencies (cf. Figure 5), we cannot easily judge whether a chirp is present in Figure 3, or whether such peaks are due to instrumental bias. For this reason, correlations are to be preferred to power spectral methods in detecting chirps in our data.

Finally, one notes that there are equally spaced, narrow peaks in Figure 3, corresponding to interference from the mains supply at 50 Hz and higher harmonics.

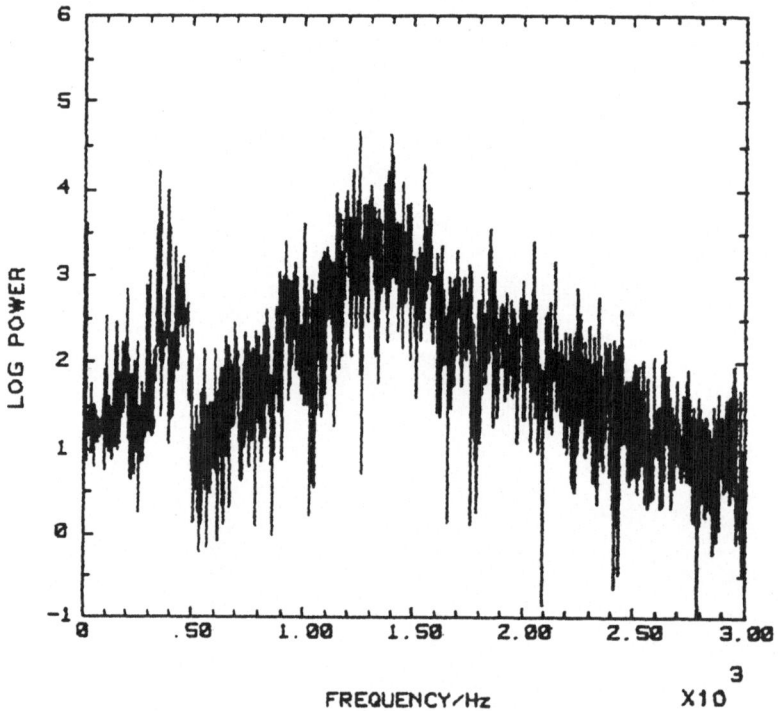

Figure 3.   Log of the power spectrum of a record in which an applied chirp occurs

4.   NOISE ANALYSIS

To examine the performance of the detector, we studied the output data itself, which essentially consisted of noise. Ideally this noise would be Gaussian, with zero mean. However instrumental effects will always cause small zero offsets, and other non-Gaussian noise sources may be present. In fact, taking the whole data set (excluding calibration combs and drop-outs) we obtained

$$\text{mean} = -11.79$$

$$\text{standard deviation} = \sigma = 26.16. \tag{3}$$

This non-zero mean introduces a zero frequency, large amplitude peak in the power spectra of all records of the secondary data, (Figure 3 has been re-scaled so this peak is not present there). This mean has been subtracted from the data before doing the analysis discussed in this section, to account for this problem.

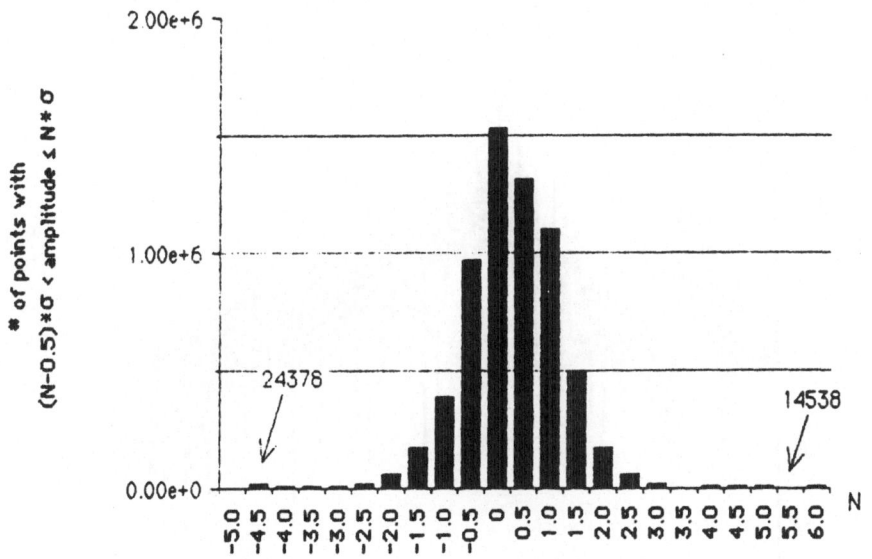

Figure 4. Histogram of time data.

Figure 4 shows a histogram of the centralised (i.e. mean-subtracted) time series. The number of points having amplitude greater than N times σ, but less than (N+0.5) times σ, is plotted against N. Note that, while the central region of the histogram looks roughly Gaussian, two peaks show a relatively large number of points at $-4.5\,\sigma$ and $6\,\sigma$. This is most likely due to the limiting of the dynamic range of the data (cf. Section 2), by which all amplitudes are truncated at the absolute value of 127.

We have also mentioned interference at 50 Hz and higher harmonics from the mains supply. The power spectrum of an arbitrary subset of the data not only shows this interference, but also the non-uniform instrumental response, peaking near 1200 Hz (see Figure 5). This bias can be compensated for by using the calibration signals (combs) applied to the data (cf. Section 2).

However, the data also contains non-white noise and to allow for this one could use a "local estimate" of σ: for every small subset of data (approximately 100 points), calculate its σ value and look for peaks inside the subset, relative to this σ. An analysis along these lines should not only allow a study of the statistics of the power spectrum of the raw data, but would also provide a way to set an upper limit on the detector's sensitivity. Such analysis is currently being carried out.

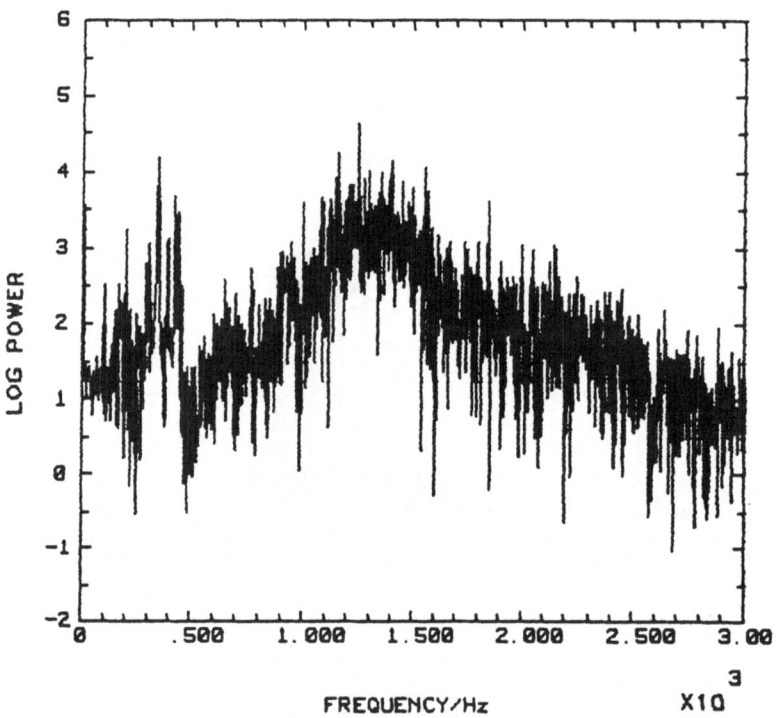

Figure 5.   Log of the power spectrum of arbitrary record of time
series in which a chirp is not present.

5.   COMMENTS

We have shown that, in a blind test of simple filtering techniques, we have been able to detect the artificially applied chirps in the Glasgow data and to determine the mass parameter of the chirp accurately.   The correlation method we used can therefore be expected to be useful in the detection of true gravitational wave signals in future detectors' output. It must be said, however, that a continuously running detector will output vast amounts of data, and since one cannot predict (at least with sufficient accuracy) the kind of event that will take place, all the data has to be correlated with more than one family of filters (for coalescing binaries, for supernovae explosions, for neutron star and black hole collapses, etc.). This asks for both real-time analysis of the output and extremely fast hardware-software combinations that can handle it.

Our experience with this experiment has shown that any attempt to develop an automatic data-analysis system for real-time analysis of long data streams will have to find ways to eliminate or cope with instrumental features in the data. In this case these included interference from the electricity supply, and truncation and centring errors in the 1-byte data, and the non-uniform frequency response of the output data. In the long run, a standard format for data storage would allow the software to run more efficiently (e.g. by choosing the record lengths to be an integer power of 2), and this together with agreement on the information to be stored in the housekeeping data, should facilitate data interchange between the different groups.

ACKNOWLEDGEMENTS

The authors would like to thank the Glasgow group for supplying the data and for valuable discussions. Computations were carried out on a DEC MicroVax II supplied under SERC grant GR/E46158.

GRAVNET, MULTIPLE   ANTENNA COINCIDENCES AND ANTENNA PATTERNS FOR RESONANT
BAR ANTENNAS

David Blair[+] (Parts A and B), Sergio Frasca* (Part B),
Guido Pizzella* (Part B)
+University of Western Australia
*University of Rome "La Sapienza"

ABSTRACT. Reliable dscrimination of individual gravitational wave pulses
against a stochastic background of local events requires a combination of
multiple coincidences between 3 or more antennas, and resolving time less
than 100 ms. GRAVNET has been set up to facilitate coordination of world
wide groups. By changing orientations, four antennas can be optimised
for maximum 2-way coincidence sky coverage, (near 100%) or for greatest
probability of 4-way coincidences for which sky coverage is reduced
(typically to 50%).

1.   INTRODUCTION

This paper is in two parts. Part A discusses the logistics of operating
multiple antennas on a world wide coincidence basis, and considers the
relative advantages of increased bandwidth and increased antenna number
for coincidence operation. To facilitate communication between groups
the GRAVNET network has been set up. The network is an aid to
coordination, optimisation and analysis of the independent antennas.
    It is shown here that coincidences between at least 3 antennas are
essential to reliably distinguish gravitational wave events from
accidental coincidences, even at bandwidths exceeding 30 Hz.
    Part B of this paper gives the results of a collaborative effort
between the UWA and Rome groups to analyse the antenna patterns of the
existing antennas of the four foundation GRAVNET groups Louisiana State
University[2], Stanford University[3], the University of Western
Australia (Perth) [3] and the University of Rome[4]. These results
highlight the fact that with only 4 antennas there is a choice between
orientation to maximise possible 4-way coincidences, or to optimise sky
coverage. Only if more than four antennas are used it is clear that a
high degree of sky coverage can be achieved, combined with the
statistical advantage of multiple coincidences as a means of improving
the discrimination against local noise sources.

B. F. Schutz (ed.), Gravitational Wave Data Analysis, 285–298.

A(1)   GRAVNET

It is well known that Weber used coincidences between a pair of widely
spaced antennas as a means of distinguishing between infrequent
background noise and signals in his antenna.   Gravitational wave signals
should be correlated, whereas locally induced signals should be
uncorrelated.   Unfortunately the rate of background events has been
sufficiently high that so far it has only been possible to consider a
statistical excess of "zero time delay" events as an indication of
excitation by a non local source.   Occasional rare events (e.g. 1 per
month) cannot be distinguished above a substantial accidental coincidence
rate, except at very high thresholds.
      This problem cannot be overcome effectively by simply improving
antenna noise temperatures, especially when the candidate events are
non-Gaussian in origin.   Improvements can be achieved, however, by using
coincidences between more than 2 antennas, and also through improved
antenna bandwidths.   Multiple coincidence analysis requires coordination
of four or more widely spaced antennas, and for this purpose GRAVNET has
been set up.
      We will briefly described GRAVNET and go on to discuss the improved
performance possible with multiple antennas.
      The world gravitational radiaton antenna network, GRAVNET, was
founded in 1986 by the gravitational radiation groups of Louisiana,
Perth, Rome and Stanford.   The purposes of GRAVNET are fourfold:
      (1)   To improve co-operation between groups by providing regular
            communication between the member groups.   Each group has a
            local secretary.   There is a world wide coordinator who
            collates and distributes network information to the groups.
            Meetings are held once per year.
      (2)   To facilitate data exchanges between groups through
            communication of run times and data formats.   In the past it
            has been difficult to synchronise run times.   Easy
            communication can alleviate this problem.
      (3)   To clarify protocols for publication of coincidence data.
      (4)   To maintain a directory and data base on the member groups and
            their antennas.   The directory and data base is maintained by
            the coordinator, periodically updated and distributed to member
            groups.
      The above items are self-explanatory.   Table I gives a page of data
from the data base, listing the antenna parameters and locations of the
four Gravnet groups.   Table II gives a summary of the present data
exchange protocols as they have been agreed by the founder groups.
      Many discussions have been held on the question of biasing
coincidence data through the arbitrary variation of thresholds.   Although
no agreement has been made, many GRAVNET member consider that in the
first instance all thresholds should be set locally and only thresholded
data should be exchanged.

A(2)   BANDWIDTH AND MULTIPLE COINCIDENCES

Candidate events in a single resonant bar antenna occur either due to
environmental perturbation, (seismic events, electromagnetic

TABLE I.   **Antenna Characteristics**

Antenna	Material, mass, frequency	Orientation east-west	Timing	Vetos used & sampling time	Last noise temp. threshold	100 event/day
Rome	Al 5056 2270 kg, 916 Hz	6.1E, 46.22 N	Loran C	Seismic Electromagnetic (offline)	18 mK, 0.3 s	70-100 mK
Stanford	Al 6063, 4660 kg, 840 Hz	122.2 W, 37.4 N north-south	NBS UTC (WWV)	Second harmonic SQUID reset	70 mK, 0.12 s	.89 K
LSU	Al 6061, Al 5056* 2360 kg, 900 Hz	91.1 W, 30.5 N north-south (±5)	WWV (1 ms) SQUID unlock	Amplifier ov'ld Lock-in ov'ld	150 mK, 0.1 s	1.4 K
UWA	Niobium 1500 kg, 710 Hz	115.8 E, 32.0 S north-south (±3) (1 ms)	USN Flying Clock Cosmic ray (offline)	Seismic Electromagnetic		

*next run

Notes (1) the UWA timing can be adapted to give 2µs accuracy

TABLE II.   GRAVNET DATA EXCHANGE PROTOCOLS

**Protocols for Publication of Results from Data Exchange
between groups**

— a "gentleman's agreement"

a)   Publication of results should be by mutual agreement of all parties
b)   Two year rule
    (i)   Publication of results may be vetoed by one party for 2 years
        only, from the date of the data exchange.
    (ii)   If results are published without mutual consent of all
        parties, according to the 2 year rule, there should be
        1. full acknowledgement of the sources of the data
        2. an explanation why it has been published without mutual
        consent

c)   Third parties

    (i)   Data may be analysed by groups or individuals not party to
        the data exchange only by agreement with all groups
        originating the data.
    (ii)   Publication by mutual consent of all parties, or subject to
        the above 2 year rule.

---

interference, intense cosmic ray charged particle showers, or cryostat
noise) or due to rare Gaussian high energy excursions. These may be
idealised as an independent set of background events, occuring at a
constant rate R per unit time. There is evidence that the background
events are not entirely independent but to some extent are clustered.
However as antenna suspension and isolation is improved this will be a
better and better approximation.

The probability of a background event in one antenna during the
antenna resolving time τ (which is generally the optimum sampling time)
is given by

$$P_1 = R\tau \tag{1}$$

Now if there are N independent antennas, the probability of a coincident
excitation of all N antennas in the same sampling time interval is

$$P_N = \tau^N . \prod_{i=1,N} R_i \quad . \tag{2}$$

and if all antennas experience the same background at the rate R,

$$P_N = R^N \tau^N \quad . \tag{3}$$

That is, the probability per unit time, P_N of an accidental coincidence
is

$$P_N = R^N \tau^{N-1} \quad .$$ (4)

It is useful to express this result numerically. We consider a range
of optimum integration times from 1 second to 10^{-2} seconds, and N=1-4.
A realistic value for R is one event per 100 seconds (10^3 events per
day). Table III summarises the resulting probabilities. To detect rare
events we should be looking for $P_N < 10^{-8}$ (1 accidental event per 3
years).

TABLE III. Probability of Accidental Coincidence for R = 10^{-2} s^{-1}

Number of Antennas	τ	1s	0.3s	0.1s	0.01s
1.	P_1	10^{-2}	10^{-2}	10^{-2}	10^{-2}
2.	P_2	10^{-4}	3×10^{-5}	10^{-5}	10^{-6}
3.	P_3	10^{-6}	10^{-7}	10^{-8}	10^{-10}
4.	P_4	10^{-8}	3×10^{-10}	10^{-11}	10^{-14}

It is clear that $P_N < 10^{-8}$ can be achieved by 4 narrow band detectors
(τ = 1 second), or 3 detectors with τ = 0.1 second, but it cannot be
achieved by a pair of detectors even if τ is reduced to 10^{-2} seconds
unless R is reduced to less than 10^{-3} s^{-1}. Table III emphasises that
multiple antenna operation is essential to reduce the background
coincidence rate.

Hamilton has independently reached the same conclusion
(W.O. Hamilton, private communication). Note that if the antennas have
different resolving times, the τ used in equation (3) is generally the
longest resolving time. If the candidate events are non-Gaussian (in
energy distribution) and non-Poisson (in time distribution) it is likely
that accidental coincidences will have a higher probability than
estimated here.

We now go on to consider the present antenna orientations on the
surface of the Earth with regard to possible multiple coincidence
operation.

B(1)   ANTENNA PATTERNS

The antenna patterns for one and two resonant bar antennas have been
analysed previously by Nitti and Frasca[6,7] while Schutz and Tinto
[8,9] have analysed antenna patterns fo pairs of laser interferometers.
Here we consider the case for the four antennas at Perth, CERN,
Louisiana, and Stanford.   In their present orientations all are placed
in a north-south (N-S) direction (within about 1 degree) except for the
antenna at CERN which is oriented east-west (E-W).   Since all antennas
are horizontal, their orientation with respect to the sky is largely
determined by their various locations on the earth, and considered as a
whole this leads to a complex antenna pattern.   Moreover the antenna
pattern is also dependent on the data processing and detection criterion
used, which itself depends on background noise.
      The presence of local sources of background noise lead to the
minimum detection requirement that a two antenna zero time delay
coincidence be observed.   With unknown source direction there can be no
constraint based on the relative signal sizes.   However with four
antennas it is possible to consider much stronger detectionc riteria,
the strongest of which is the presence of four antenna coincident events
with relative amplitudes consistent with aplane gravitational wave of a
particular polarisation originating from some source direction.   (We
note that the present generation of detectors have insufficient time
resolution to be able to use phase information.)
      Here we answer the following questions:
      (1)   What fraction of the sky can be monitored by the existing
            configuration of four antennas, (i.e. what is the fractional
            sky coverage) for different strengths of the detection
            criterion?
      (2)   Would there be any advantage in varying the orientation of any
            of the antennas and what level of sky coverage can be
            achieved?
      The angular dependence of the signal S observed in a single
resonant bar antenna is given by

$$S(\theta,\phi,\varepsilon) = (0.5(1-\varepsilon) + \varepsilon \cos^2 2\phi) \sin^4\theta \qquad (1)$$

where θ is the angle of the incoming plane wave relative to the cylinder
axis of the antenna, and ϕ is the polarisation angle of the wave measured
relative to the plane of the antenna and the source.   The polarisation
fraction ε measures the fraction of linear polarisation of the wave for
e=0 the wave is circularly polarised, whereas for e=1 the wave is 100%
linearly polarised, with polarisation angle ϕ.

The antenna sensitivity can be expressed in geodetic coordinates relative to a source of given hour angle α and declination d, using trigonometric expressions given by Tyson and Douglass[10].

We use two separate criteria for analysing the 4-antenna array. The first consists of a measure of the signal product for all antennas in the array, or for a subset of them. The 4-antenna signal product function, P_4, has the advantage of being analytic at the expense of imprecision in specification of the individual antenna signals:

$$P_4 = \prod S_i, \quad i=1,4 \tag{2}$$

To measure 3 and 2-way coincidences we define the functions P_3 and P_2 as follows:

$$P_3 = P_4/S_{m1} \tag{3}$$

$$P_2 = P_4/S_{m1} \cdot S_{m2}) \tag{4}$$

where S_{m1} is the smallest and S_{m2} is the second smallest of $\{S_i\}$.

Since the antennas considered here are expected to have roughly equal sensitivity, at about the same frequency, 700 – 900 Hz, we make the assumption that all S_i have a maximum value of 1 when the antenna is optimally aligned relative to a source direction. Thus P_N all have a maximum value of 1.0. A value of $P_4 = 1$ implies all antennas optimally aligned, while $P_4 = 0.1$ implies a range of S_i such as (0.1, 1.0, 1.0, 1.0) to (0.6, 0.6, 0.6, 0.6). Since signals can only be expected to appear marginally above the noise (otherwise unambiguous detection would have already occurred) reasonable thresholds for 4-way coincidences vary from 10^{-1} to 10^{-4} corresponding to equal signals in each antenna varying from 0.6 to 0.1.

Since the product functions allow S_i to occupy a range of values (some of which would be buried unmeasurably in the noise), we use a second criterion defined by the locus of points L_N for which 4, 3 or 2 antennas are above a specified threshold.

The loci L_n are useful in determining the fractional sky coverage by 4, 3 or 2 antenna coincidences, and the dependence of sky coverage on threshold. The locus criterion is useful when antenna signals are analysed independently to obtain lists of possible events. However they do not make maximal use of the data available. An optimal filter for multiple antennas would take into account both Gaussian and non-Gaussian noise components, and would be expected to see significantly below the noise thresholds implied by use of the locus criterion. In practice L_n and P_n give very similar results for fractional sky coverage, although the antenna patterns generated differ qualitatively in structure.

The antenna patterns for P_n and L_n can be plotted as a map, and because the 4-antenna array is symmetrical with respect to wave propagation direction, a half-sky map suffices. Figure 1(a) is an example, which shows contours for P_4 in the case of circularly polarised radiation. The map is most usefully expressed in sky coordinates for a given universal time, thus expressing the areas of sky to which the antenna is sensitive. The antenna pattern clealry rotates on the sky with a sidereal day period. The antenna patterns are more complex in the case of linearly polarised radiation, as figure 1(d) illustrates.

The fractional sky coverage C_n can be obtained for P_n as follows:

$$C_{n,P} = \frac{1}{4\pi} \iint_{\alpha\delta} L_n \, ds \quad n = 2,4. \tag{5}$$

Similarly $C_{n,L}$ can be defined for the locus criterion L_n.

Using simple numerical methods we have generated antenna pattern maps and determined fractional sky coverage for the four antennas. The results are discussed below.

B(2)   RESULTS

Figures 1(a)-(e) show a variety of contour maps for both P_4 (for logarithmic contour intervals) and L_2, L_3, L_4 for a fixed threshold. Maps are given both for all antennas in their present (true) positions and for the UWA antenna turned to an E-W direction. By comparing figures 1(a) and 1(b) the difference between P_4 and L_n can be seen, and by comparing figures 1(b) and 1(c) it is clear that sky coverage can be improved if the UWA antenna is placed in a E-W direction.

We have examined the antenna orientations in greater detail, by determining $C_{n,L}$ and $C_{n,P}$ as a function of both threshold and antenna orientation, in the case of circularly polarised radiation. For the minimum detection criterion P_2 or L_2, there is a significant advantage in the antennas being arranged with two oriented E-W and two oriented N-S. The optimum arrangement for 2-way coincidences is for the UWA antenna to be placed E-W insteadof N-S. The optimum is quite flat, however, and sky coverage differs by less than 1% for orientation changes of 10 degrees from the optimum.

Figure 2 shows the dependence of $C_{2,L}$, $C_{3,L}$, and $C_{4,L}$ on threshold for UWA both in the N-S and E-W orientation. The most striking feature is that the UWA antenna oriented E-W allows 100% sky coverage for 2-way coincidences, for thresholds less than 0.3. At this same threshold $C_{2,L}$ is only 69% if the UWA antenna is oriented N-S. The optimisation of C_2 causes negligible degradation of C_4. However C_3 is degraded significantly from 67% to 37% by this optimisation. Thus the optimisation for 2-way coincidences is at the expense of reduced coverage for the more significant 3-way coincidences.

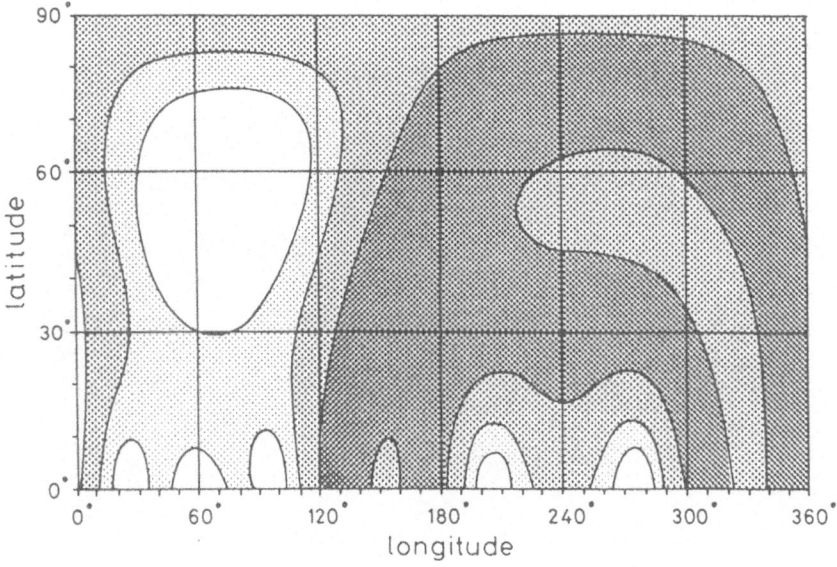

(a)

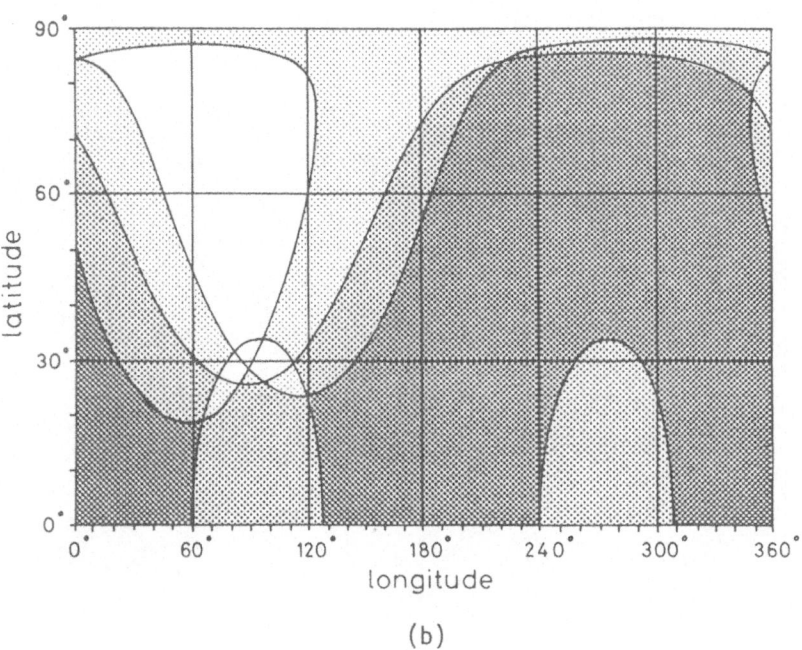

(b)

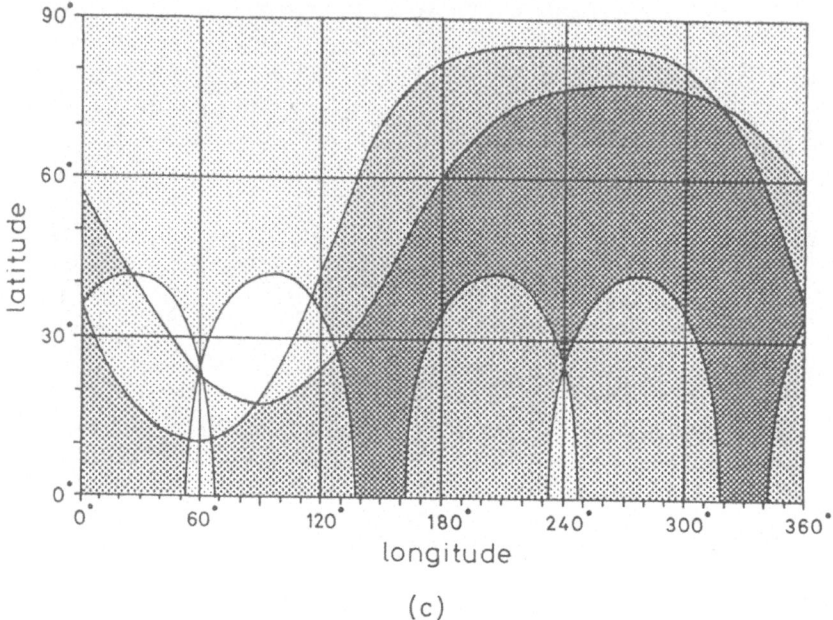

(c)

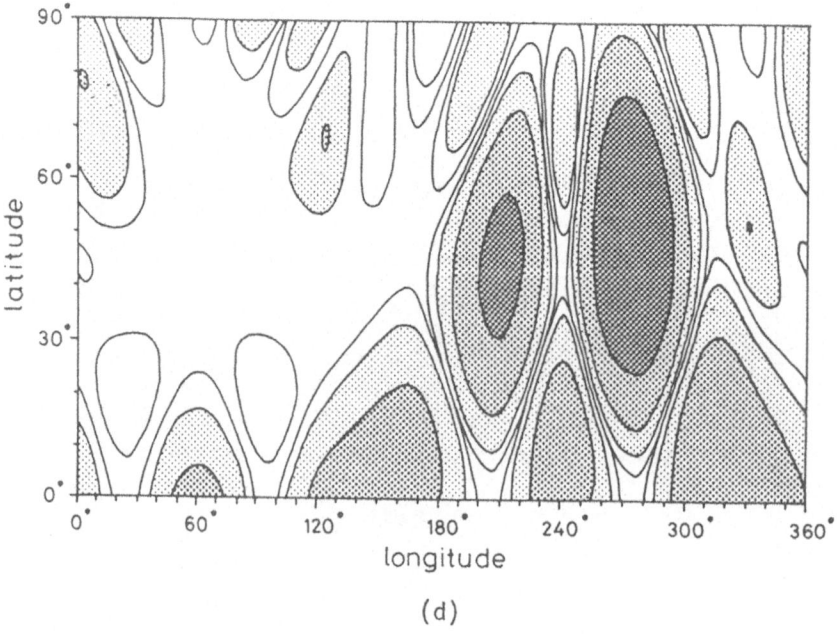

(d)

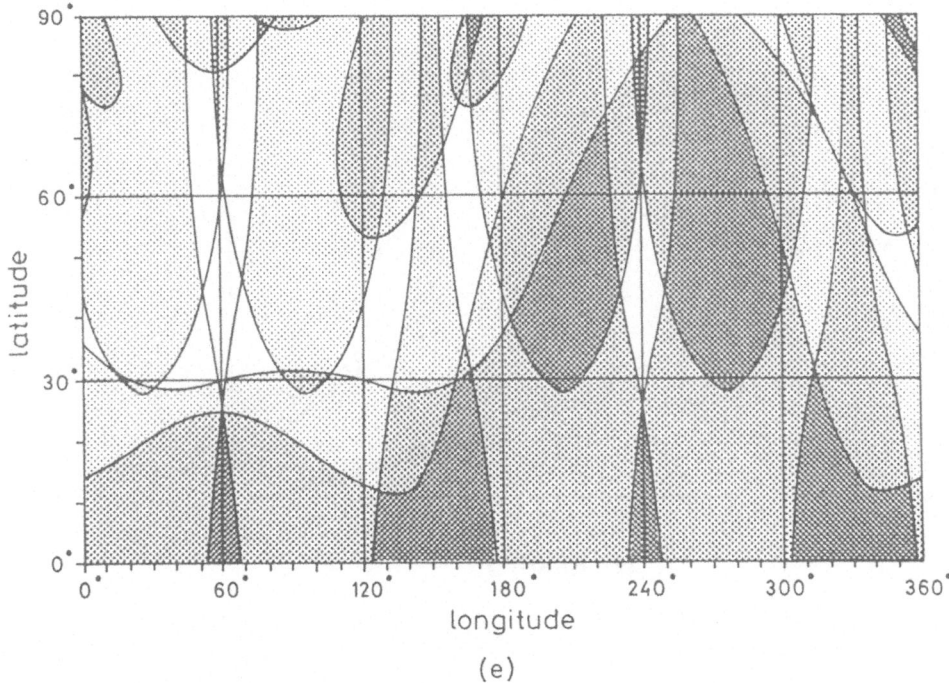

(e)

Figure 1

a) Contour map for product P_4 with circularly polarised radiation. Antenna orientations LSU and Stanford N-S, UWA and CERN E-W. Shading density indicates higher values of P_4. Highest density $P_4 > 10^{-2}$, intermediate density $10^{-3} < P_4 < 10^{-2}$, light density $10^{-4} < P_4 < 10^{-3}$, unshaded $P_4 < 10^{-4}$.

b) Contour map for loci L_2, L_3, L_4, for circularly polarised radiation, and threshold 0.05. Orientations Stanford, LSU and UWA, N-S, CERN E-W. Highest density is the L_4 locus. Intermediate density is the L_3 locus (which includes the L_4 locus) and low density is the locus for two antenna coincidences L_2 (which includes L_3 and L_4).

c) Same as b) except UWA oriented E-W, threshold 0.1.

d) Same as a) but for linearly polarised radiation. ($\varepsilon=1$, $\psi=0$).

e) Same as c) for linearly polarised radiation ($\varepsilon=1$, $\psi=0$), threshold 0.05.

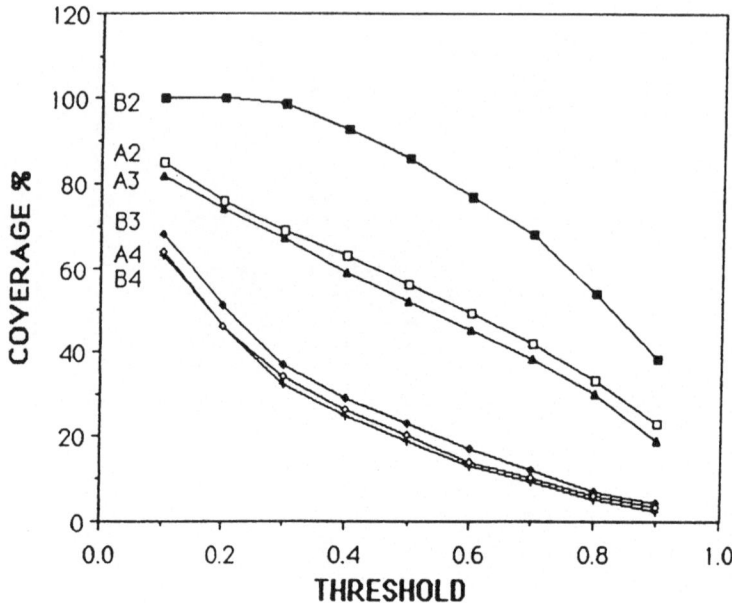

Figure 2

Percent sky coverage versus threshold for present antenna orientations
(all N-S except CERN E-W), compared with two antennas N-S and two (UWA
and CERN) E-W. A2, A3, A4: present orientations B2, B3, B4 UWA and CERN
E-W. (2-way, 3-way and 4-way coincidences respectively.) These curves
are obtained using the locus criterion, for circular polarisation, but
similar results apply to the products P_4, P_3 and P_2.

If for noise reduction it were necessary to use 4-way coincidences, it
would be advantageous to rotate the antenna at CERN to a N-S orientation.
This would increase $C_{4,L}$ from about 30% to about 45% for a threshold of
0.3.

    The strong advantage of using multiple coincidences has been
emphasised in Part A. This must be weighed against the probability of
missing events. Only additional antennas can overcome this dillema.
This is illustrated by a numerical map for 7 antennas, including
Maryland, Guangzhou (China) and Moscow. (See figure 3). The numbers
give the number of antennas above threshold for given source directions.
It is evident that sky coverage is good, and that for most of the sky
multiple coincidences can be expected.

	alfa 0	1	2	3	4	5	6	7	8	9	10	11										
90	4	4	4	4	4	4	4	4	4	4	4	4	4	4	4	4	4	4	4	4	4	4	4
80	4	4	4	4	4	4	4	4	4	4	4	4	4	4	4	4	4	4	4	4	4	4	4
70	4	4	4	4	4	4	4	4	4	4	4	4	4	4	4	4	4	4	4	4	4	4	4
60	5	4	4	4	4	4	4	4	4	4	4	4	4	4	4	4	4	4	4	4	4	4	4
50	7	7	5	4	4	4	4	4	4	4	4	4	4	4	4	4	4	4	4	4	4	4	4
40	3	3	3	3	0	4	4	4	4	4	4	4	4	4	4	4	4	4	4	4	4	4	0
30	3	3	3	3	3	3	0	4	4	4	4	4	4	4	4	4	4	4	4	4	0	3	3
20	3	3	3	3	3	3	3	6	4	4	4	4	4	4	4	4	4	4	4	6	3	3	3
10	3	3	3	3	3	3	3	3	7	7	6	6	6	6	6	6	6	7	7	3	3	3	3
0	3	3	3	3	3	3	3	3	7	7	7	7	7	7	7	7	7	7	3	3	3	3	
-10	3	3	3	3	3	3	3	3	7	7	7	7	7	7	7	7	7	7	3	3	3	3	
-20	3	3	3	3	3	3	3	7	7	7	7	7	7	7	7	7	7	7	3	3	3		
-30	3	3	3	3	3	3	3	7	7	7	7	7	7	7	7	7	7	7	3	3	3		
-40	3	3	3	3	3	7	7	7	7	7	7	7	7	7	7	7	7	7	7	7	7	7	3
-50	5	7	7	7	7	7	7	7	7	7	7	7	7	7	7	7	7	7	7	7	7	7	7
-60	4	4	5	6	7	7	7	7	7	7	7	7	7	7	7	7	7	7	7	7	7	7	7
-70	4	4	4	4	5	5	5	6	7	7	7	7	7	7	7	7	7	7	7	6	5	5	5
-80	4	4	4	4	4	4	4	4	4	4	4	4	4	4	4	4	4	4	4	4	4	4	4
-90	4	4	4	4	4	4	4	4	4	4	4	4	4	4	4	4	4	4	4	4	4	4	4

Figure 3

Antenna coincidence map for a threshold of 0.3, with 3 N-S antennas at
Stanford, LSU and Guanzhou, and 4 E-W antennas at Perth, CERN, Moscow and
Maryland.

CONCLUSION

The need for multiple coincidences as a means of positively identifying
gravitational waves has been emphasised. Two way coincidences cannot
positively discriminate against accidental coincidence as for accessible
resolving times and typical background event rates. The present array of
resonant bar antennas is optimised neither for all sky coverage, nor for
multiple coincidences. Sky coverage for unknown and random source
directions can be increased substantially by reorienting one antenna in
an E-W direction, and for thresholds less than 0.3 this gives practically
100% sky coverage for 2-way coincidences. However additional antennas
will be required to allow all sky multiple coincidence detection.

ACKNOWLEDGEMENTS

We thank the members of GRAVNET for contributing data on their antennas.
DGB thanks members of the Rome gravitational radiation group and the
Institute of Astronomy, Cambridge for hospitality during the preparation
of part of this manuscript, and Bernard Schutz, Massimo Tinto and Bill
Hamilton for useful discussions.

REFERENCES

(1)   W.O. Hamilton et al. Cryogenics **22**, 107, (1982).
(2)   P.F. Michelson and R.C. Taber Phys. Rev. D **29**, 2149 (1984).
(3)   P.J. Veitch et al. Proc 17th Int Conf on Low Temp Phys LT-17.
      ed. U. Eckern et al. Pt 2 941.   North Holland (1984).
(4)   E. Amaldi et al Il Nuovo Cimento 7C 338 (1984).
(5)   J. Weber in Proc. Sir Arthur Eddington Centenary Symp. Vol. 3 ed.
      J. Weber and T.M. Karade World Scientific, Singapore (1986).
(6)   G. Nitti Il Nuovo Cimento 3C, 420 (1980).
(7)   S. Frasca Il Nuovo Cimento 3C, 237 (1980).
(8)   B.F. Schutz and M. Tinto Mon. Not. R. Astr. Soc. **224,** 131 (1987).
(9)   M.Tinto Preprint Univ. College, Cardiff (1987).
(10) J.A. Tyson and D.H. Douglass Phys. Rev. Let. 28, 991 (1972).

COINCIDENCE PROBABILITIES FOR NETWORKS OF LASER INTERFEROMETRIC DETECTORS OBSERVING COALESCING COMPACT BINARIES

Massimo Tinto
Department of Applied Mathematics & Astronomy,
University College, Cardiff CF1 1XL, Wales, UK.

ABSTRACT.   The threshold averaged coincidence probability and the coincidence probability as a function of the thresholds of two or more laser interferometers, are applied to the case of waves from coalescing compact binaries.   These are thought to be the most likely sources of gravitational waves to be detected by broad-band detectors.

We obtain the various coincidence probabilities as functions of the distance to the source relative to the maximum distance a network will be able to look at.   After deducing the probability distribution for binaries located inside the observable volume, we calculate the detection efficiency, defined as the averaged value of the coincidence probability over the volume.   By assuming the figure of three events per year for neutron-star binaries out to 100 Mpc, we calculate the event rate that a network of interferometers will be able to register over a given observation time.

We find that the currently proposed four detectors in California, Maine, Scotland and Germany, working with light recycling will be able to observe in coincidence 2000 events per year out to 2.1 Gpc.

1.   INTRODUCTION

Independent detections of any given gravitational wave event by different detectors is vital for its reliable identification as a gravitational wave.   This is because the events may not be much larger than the noise level.   More detectors on the same site provide one way of increasing signal to noise, but by spacing them across the globe one automatically gains much more information about the wave and its source.

With a network of three interferometers, for instance, one has in principle optimal directional information for good signal to noise ratios (SNRs).   If the events are not corroborated by an electromagnetic detection (optical, X-ray or radio telescopes) this could be crucial.   However, even if there is an optical activity, an all-sky survey is nearly impossible optically.   With three detectors, from the two independent time delays and the three amplitude observations, one has in principle enough information to narrow down the position of the source in the sky, the two independent polarization amplitudes of the wave, and the phase lag between them (Schutz, 1986) to two possible solutions.

With four detectors it would be possible to resolve the ambiguity

B. F. Schutz (ed.), Gravitational Wave Data Analysis, 299–313.
© 1989 by Kluwer Academic Publishers.

of the 3-detector solution and to test Einstein's predictions regarding gravitational wave polarization: four detections over determine the solution for a transversely polarized quadrupole wave, so any inconsistency among the data would be evidence for other polarization states.

Among all the theoretically predicted gravitational wave forms, those from coalescing compact binaries (containing neutron stars or black holes) have a unique signature that enables them to be extracted from wide-band data by digital filtering techniques (Thorne, 1987; Schutz, 1988; Dewey, 1986). This signature is their accelerating sweep upwards in frequency as the binary orbit decays. Coalescing binaries have the advantage over other sources in signal to noise ratio by a factor depending on the square-root of the ratio between the corresponding number of cycles in the wave trains (Thorne, 1987). This is in fact the enhancement in effective signal that the experimenters will achieve by optimal signal processing in their search for these frequency-sweeping bursts. The key to being able to take advantage of this is the confidence with which we can predict the wave form. Because the binary system spends more time in the low-frequency part of the sweep than in the high frequency part or in the final coalescence, and because laser interferometers should eventually have less noise amplitude at low frequency (~100 Hz) than at high (> 100 Hz) (Hough *et al.*, 1986), it will be easier for the detectors to see the Newtonian regime of the sweep than the post-Newtonian one.

In the Newtonian regime, if we orient the wave's polarization axes along the axes of the projection of the orbital plane on the sky, then the wave's amplitudes assume the following form: (units c = G = 1)

$$h_+ = 2(1 + \cos^2 i)(\mu/r)(\pi M f)^{2/3} \cos(2\pi f t)$$

$$h_\times = \pm 4\cos i \ (\mu/r)(\pi M f)^{2/3} \sin(2\pi f t)$$

$$(1.1)$$

where i is the angle of the inclination of the orbit to the line of sight of the detector; M and μ are the total and reduced masses, r the absolute distance to the binary and f the frequency of the waves. We observe that the orbit can be assumed to be circular because of the radiation reaction effect on the original eccentricity (Thorne, 1987; Krolak, 1988; Schutz, 1988).

From a study of the wave form (1.1), by using a network of broad-band detectors located at different places on the Earth, we can in principle deduce the following information: (i) the direction to the source; (ii) the inclination of the orbit to the line of sight; (iii) the direction the stars move in their orbit; (iv) the combination $\mu M^{2/3}$ of the reduced and total masses; (v) the distance r to the source.

Besides these important points, Schutz (1986) has shown that, by observing bursts from coalescing compact binaries, four detectors of sufficiently large SNRs would allow us to obtain a significantly better value of the Hubble constant than now we have.

Although it is possible to predict the shape of the signal with great confidence, there is considerable uncertainty about the rate of coalescence events. Clark, van den Heuvel and Sutantyo (1979) have estimated, from neutron star observations in our own galaxy, that to see three coalescences of neutron-star binaries per year one should look out to a distance of ~ 100 Mpc. Recent improvements in the statistics of neutron-star binaries (Schutz, 1988a) reinforces this conclusion but, at the same time, shows that this number might be drastically changed by plausible astrophysical scenarios. In what follows we shall adopt the figure of three per year out to 100 Mpc, bearing in mind its uncertainty.

At the present there are four groups around the world operating prototype Laser Interferometric Gravitational Wave Observatories (LIGOs), at the California Institute of Technology, the Massachusetts Institute of Technology, Glasgow University and the Max Planck Institute for Quantum Optics at Garching, near Munich. These groups, with two others (at Orsay, near Paris, and at Pisa) are proposing to build detectors of large size (at least 1 km arm-lengths). If these proposals are funded soon, then by the end of the century they could be operating with full design sensitivity.

Once a detector is built, it will be difficult to move it or even to change its orientation. These detectors have quadrupolar antenna patterns, so their location and orientation on Earth affect their sensitivity to gravitational wave bursts and especially the likelihood of coincidences with two or more detectors. Since, as we said earlier, the events may be rare and because directional antennae miss valuable data, it comes natural to ask the following questions:

1)   How should we orient two or more detectors, sited in different places on the Earth, in order to optimize the likelihood of coincidences?

2)   How much further away will a network of interferometers be able to look than single detectors?

3)   What is the fraction of the number of events, out to this distance, that they will be able to register in coincidence?

In this paper we answer these questions. However, for a more detailed analysis we refer the reader to Schutz & Tinto (1987) and Tinto (1987a,b,c,d). Here we briefly summarize our method.

In Section 2 we write the analytical expression of the *antenna pattern* ⋂ for waves from coalescing compact binaries. This function depends on the source angle $(\theta;\phi)$, the polarization angle of the wave ψ, the inclination i of the angular momentum of the binary to the line of sight of the detector and the angles $(\alpha;\ \beta;\ \gamma)$ representing the orientation, latitude and longitude of the detector (see Fig. 1). After observing that the response of a laser interferometric detector depends not only on its relative geometrical disposition with respect to the source of the wave but also on the level of noise and its statistical distribution, we turn to coincidence experiments. We are able to prove a simple relationship between the geometrical factors affecting the response of the detectors and their thresholds relative to the maximum amplitude of the wave.

Let C be the coincidence probability for two detectors observing the same event. Clearly C depends upon the thresholds of the two instruments. We have been able to prove that the averaged value of C over all possible thresholds is equal to the mean value (over random arrival time and orientation of the wave's polarization ellipse) of the squared-modulus product of the antenna patterns of the two instruments. By treating this as a function of the detector orientations, for several pairs of likely locations, we can obtain the values optimizing coincidence rates.

In Section 3, after observing that the threshold averaged coincidence probability does not give a fair indication of the typical coincidence probability, we deduce a general algorithm to calculate the coincidence probability C, for an arbitrary number of detectors, as a function of the distance to a binary system relative to the maximum distance a network will be able to look at. This is determined in fact by the largest of the thresholds that the detectors will assume.

Since binary systems are homogeneously and isotropically distributed in the Universe, we can calculate the detection efficiency, defined as the averaged value of the coincidence probability over the volume. By assuming the figure of three events per year for neutron-star binaries out to 100 Mpc and defining the binary masses and the wave frequency, we calculate the event rate that a network of interferometers will be able to register over a given observation time.

In Section 4 we discuss the results and conclude that the proposed four detectors working with light recycling will be able to observe in coincidence 2000 events per year.

2. ANTENNA PATTERNS AND MEAN COINCIDENCE PROBABILITIES

The response of an interferometric detector of gravitational waves consists basically of the change $\delta\ell/\ell_0$ in the relative length of the two arms. Its analytical expression has been deduced by Schutz & Tinto (1987) for plane waves in the long-wave length approximation (reduced wavelength $\lambda/2\pi \gg$ arm length ℓ_0), and it may be written as follows:

$$\frac{\delta\ell}{\ell_0} = \left[E(\alpha;\beta:\gamma - \Phi;\theta;\psi)h_+ + E(\alpha;\beta;\gamma - \Phi;\theta;\psi + 45^{\circ})h_{\times}e^{j\delta} \right] \tag{2.1}$$

where j is the imaginary unit.

The function E and its properties have been discussed by Schutz & Tinto (1987) and we refer the reader to them entirely. Here we only point out that:

(i) the parameters α, β and γ are respectively the angle between the direction of the detector bisector and the local meridian, the latitude and longitude of the detector on the Earth (see Fig. 1).

(ii) $(\theta; \Phi)$ and ψ are associated, respectively, to the direction of the incoming wave and the inclination of the axes of the polarization ellipse.

(iii) h_+ and $h_{\times}$ are the amplitudes of the two independent polarization states (referred to the orientation angle ψ) and ξ is the phase lag between them.

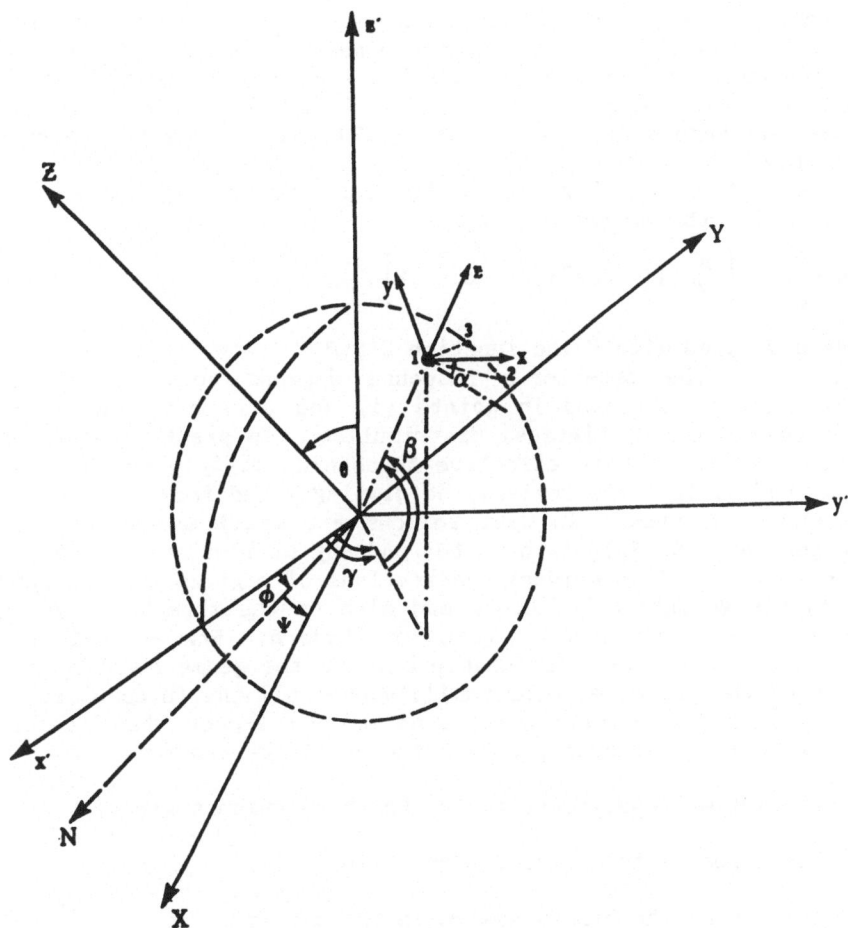

Figure 1. The relations among the detector's axes (x,y,z), the Earth's axes (x',y',z') and the wave's axes (X,Y,Z). Here β and γ are the detector's latitude and longitude respectively, and α its orientation, defined as the angle between the x-axis and the east-west direction.

In the case of a chirp-burst the wave's two amplitudes h_+, $h_\times$ and their relative phase ξ assume the following form (Thorne, 1987):

$$\xi = \pm \pi/2; \quad h_+ = \frac{1}{2}(1 + \cos^2 i)h; \quad h_\times = \cos(i)h \qquad (2.2)$$

where i is the inclination of the angular momentum of the source to the line of sight of the detector and h is equal to

$$h = \frac{4G^{5/3}}{c^4} \left[\frac{\mu}{r} \right] \left[\pi M f \right]^{2/3} = 2.7 \times 10^{-23} \left[\frac{\mu_{\odot}}{r_{100}} \right] \left[M_{\odot} f_{100} \right]^{2/3} \tag{2.3}$$

where $M_{\odot}$ amd $\mu_{\odot}$ are the total and reduced masses in units of solar masses, r_{100} the absolute distance to the binary in units of 100 Mpc and f_{100} the frequency of the wave in 100 Hz. We notice that h is the maximum possible amplitude one can receive from a binary system; in a general case the angular factors in Eq. (2.2) will reduce the observed amplitude below h.

By substituting Eq.(2.2) into Eq.(2.1) we easily deduce the following form for the *antenna pattern* η:

$$\eta = \frac{\delta\ell}{\ell_0 h} = \left[\frac{E}{2} (1 + \cos^2 i) \pm j\cos i\ \overline{E} \right] \tag{2.4}$$

where we have denoted with $\overline{E}$ the function $E(\alpha; \beta; \gamma - \phi; \psi + 45^{\circ})$.

The size of the detector's response depends not only on the geometrical factors mentioned in points (i) and (ii) but also on the level of noise and its statistical distribution. In practice a variety of noise sources will set an effective threshold $(\delta\ell/\ell_0)_*$ on the level of response $(\delta\ell/\ell_0)$ that can reliably be distinguished from noise during a given observation time. In what follows are shall assume that any $(\delta\ell/\ell_0)$ larger than $(\delta\ell/\ell_0)_*$ is detected and any smaller one is lost.

The probability of coincident observations of a given event depends therefore on the geometrical factors and also on the thresholds of the detectors relative to the maximum wave amplitude h. Now we shall show that there exists a simple relationship between the geometrical factors and the mean of the coincidence probability over all the thresholds.

Over a large number of observations we can expect the following variables, affectig the response of a detector, to be random:

(i)   the arrival-time (equivalent to the Earth rotation angle ϕ);

(ii) the orientation of the polarization ellipse (ψ);

(iii) the location of the binary system in the sky (θ);

(iv) the inclination of the orbit to the line of sight (i).

Let us assume h to be given and let $P(\phi, \psi, \theta, i)$ be the probability distribution for (ϕ, ψ, θ, i). Let us denote with X_* the threshold of the squared-modulus amplitude, the minimum detectable value of $|\eta|^2$. The single-antenna detection probability for a given h is:

$$S(X_*) = \int P(\phi, \psi, \theta, i)\ d\phi\ d\psi\ d\theta\ di \tag{2.5}$$

$$X > X_*$$

where the integral is over the region of the (ϕ, ψ, θ, i) space in which

X exceeds X_*. The coincidence probability is therefore given by the following expression:

$$C(X_{1*}; X_{2*}) = \int P(\phi,\psi.\theta,i)d\phi \; d\psi \; d\theta \; di \qquad (2.6)$$

$$X_1 > X_{1*}$$

$$X_2 > X_{2*}$$

where the indices 1, 2 refer to different detectors. Now let us consider the mean of the product of the antenna power functions:

$$\langle X_1 X_2 \rangle = \int_0^{2\pi} \int_0^{2\pi} \int_0^{\pi} \int_0^{\pi} X_1 X_2 \; P \; d\phi \; d\psi \; d\theta \; di \qquad (2.7)$$

Changing variables of integration to (X_1, X_2, θ, i) we may write Eq.(2.7) as follows:

$$\langle X_1 X_2 \rangle = \int_0^1 \int_0^1 X_1 X_2 \; P|J| \; dX_1 dX_2 \qquad (2.8)$$

where $J = \partial(\phi,\psi)/\partial(X_1,X_2)$ is the Jacobian of the coordinates transformation, the limits of integration are the extreme values that X can assume and the symbol " $\quad$ " means "integrated over θ and i".

Integrating Eq.(2.8) by parts on X_1 we get:

$$\langle X_1 X_2 \rangle = \int_0^1 X_2 \left\{ -\left[X_1 \int_{X_1}^1 P|J|d\bar{X}_1 \right]_{X_1=0}^{X_1=1} + \right.$$

$$\left. + \int_0^1 \left[\int_{X_1}^1 P|J| \; d\bar{X}_1 \right] dX_1 \right\} dX_2$$

The integrated term vanishes at both limits. After a similar integration on X_2 and a change of notation $X \to X_*$, $\bar{X} \to X$ for both the variables we obtain:

$$\langle X_1 X_2 \rangle = \int_0^1 \int_0^1 \left[\int_{X_{1*}}^1 \int_{X_{2*}}^1 P|J| \; dX_1 \; X_2 \right] dX_{1*} \; dX_{2*} \qquad (2.9)$$

Comparing this equation with Eq.(2.6) we see that this is equal to:

$$\langle X_1 X_2 \rangle = \int_0^1 \int_0^1 C(X_{1*}; X_{2*}) dX_{1*} \; dX_{2*} \qquad (2.10)$$

Thus, the *mean overlap* of the antenna power patterns $\langle X_1 X_2 \rangle$ is the same as the *average coincidence probability*. If we assume the locations of the detectors to be in California, Maine, Scotland and Bavaria, we can plot $\langle X_1 X_2 \rangle$ as a function of the orientations α_1, α_2 and deduce therefore the values optimizing coincidence rates, (see for details Schutz &

Tinto, 1987; Tinto, 1987a).

Finally we observe that the mean coincidence probability and its geometrical implications can easily be extended to resonant bar-detectors.

3. REAL COINCIDENCE PROBABILITIES AND DETECTION EFFICIENCY

The mean coincidence probability $\langle X_1 X_2 \rangle$ tells us how we should orient two detectors in order to maximise the chance of detecting the same event, but however does not contain quantitatively exact information of the probability itself, because it is only a mean value. In order to answer this question we should calculate the integrals written in Eq.(2.6).

The coincidence probability as a function of the thresholds of a network of n detectors can be written as follows:

$$C(X_{1*}^{\frac{1}{2}};\ X_{2*}^{\frac{1}{2}};\ldots;\ X_{n*}^{\frac{1}{2}}) = \int_{\substack{X_1^{\frac{1}{2}} > X_{1*}^{\frac{1}{2}} \\ X_2^{\frac{1}{2}} > X_{2*}^{\frac{1}{2}} \\ \cdot \\ \cdot \\ \cdot \\ X_n^{\frac{1}{2}} > X_{2*}^{\frac{1}{2}}}} \frac{1}{16\pi^2}\ \sin i\ \sin\theta\ d\phi\ d\psi\ d\theta\ di \qquad (3.1)$$

where we have explicitly assumed ϕ, ψ to be uniformly distributed in the interval $(0,\ 2\pi)$ and θ, i to be uniformly distributed over the sphere. The single-antenna detection probability as well as the double, triple and quadruple coincidence rates, in terms of the detectors' thresholds relative to the maximum wave amplitude, have been calculated by Tinto (1987c) using a Monte Carlo method and assuming the detectors' orientations to be equal to the values optimizing mean coincidence rates. There we observed that probably, for geographical reasons, not all detectors will assume the configurations suggested by the threshold averaged coincidence probability calculations. In order to quantify by how much the coincidence probabilities would be reduced by a different choice of the orientations, we calculated coincidence rates in terms of the detectors' orientations for characteristic thresholds. We obtained that the orientations optimizing mean coincidence probabilities are also optimum for real coincidence rates, which appear to be not significantly affected by the angles we choose (see for details Tinto, 1987c,d).

This point should be remembered when a decision for a suitable orientation of the detectors in Europe will be taken. Since they will lie essentially on the same tangent plane to the Earth, by orienting them with 45⁰ of difference we would be able to get useful information about the degree of elliptical polarization of the wave by looking at the time delays in the various baselines without sacrificing many events.

In what follows we shall calculate the single-antenna detection probabilities and quadruple coincidence rates in the hypothesis that the detectors will be working with light recycling and will have their proposed arm-lengths (4 km for the detectors in North America, an arm-length equivalent to about 2.5 km for the triangular configuration of the detector in South Germany and a 1 km arm-length for the British antenna). This study was carried out before details were available of the planned detector in Italy and hence could not incorporate them. As far as the results for double and triple coincidences are concerned, we refer the reader to Tinto (1987d).

For this configuration the values of the thresholds in such a network will be in the following proportions:

$$T_{AM} = 0.5 \ T_B; \quad T_{AM} = 0.8 \ T_G; \quad T_G = 0.6 \ T_B \tag{3.2}$$

where T_{AM}, T_G and T_B are the thresholds in the detectors in America, Germany and Britain respectively.

According to the assumption made in Section 2, we deduce that a gravitational event will be regarded as a coincidence only if its maximum wave amplitude h will be greater or equal to the largest threshold in the network (which we shall refer to as T). By normalizing h to T and remembering that for binaries their analytical expressions are (units c = G = 1):

$$h = 4 \ \left[\ \frac{\mu}{r} \ \right] \left[\pi M f \right]^{2/3} \ ; \quad T = \frac{r}{R} \ h \tag{3.3}$$

we can calculate real coincidence probabilities as function of the ratio between the distance to the source r and the maximum distance R a network will be able to look in coincidence. As expected each coincidence probability C(r/R) is equal to one when r is equal to zero and goes to zero when r goes to R (see Figs. 2, 3).

In order to calculate the fraction of the total number of events occurring within a certain distance and that the antennae will be able to register, we have to take into account the distribution of the sources within this region. Since binary systems are homogeneously and isotropically distributed over the sky, the probability density of having a source at a distance r is simply equal to:

$$f \ \left[\ \frac{r}{R} \ \right] = \frac{3r^2}{R^3} \tag{3.4}$$

where we have normalized to the largest distance R. However, when we average the coincidence rates over smaller regions of the Universe, we should substitute the corresponding distance we want to consider for R.

If we define Y(r/R) to be the product of the coincidence probability C with the probability density f, we find that this function has a maximum. In other words there exists a value of the distance,

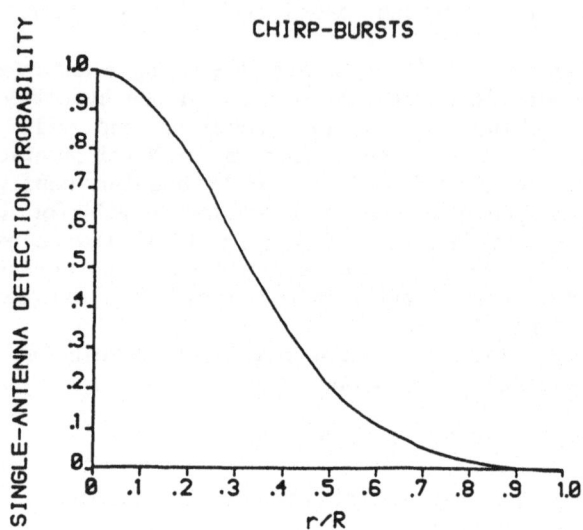

Figure 2. Single antenna detection probability plotted against the ratio between the distance to the source r and maximum distance R the antennae will be able to look at (see text for full discussion).

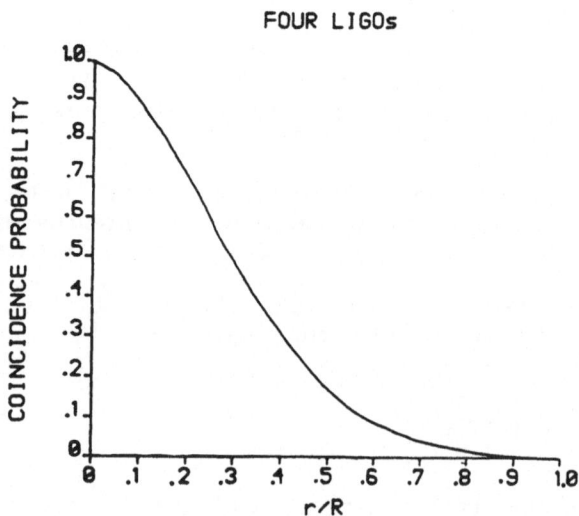

Figure 3. As Fig. 2, but for a network of four LIGOs. R here is given by the largest threshold in the network (see text).

309

distributed in the interval (r, r+Δr) with probability density f(r/R), which maximizes the coincidence probability. From now on we shall refer to it as r´ (see Figs. 4, 5).

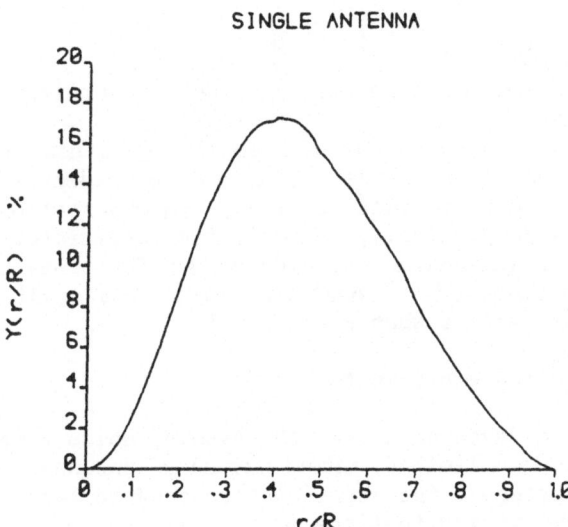

Figure 4. Plot of the function Y(r/R), introduced in the text, for a single antenna.

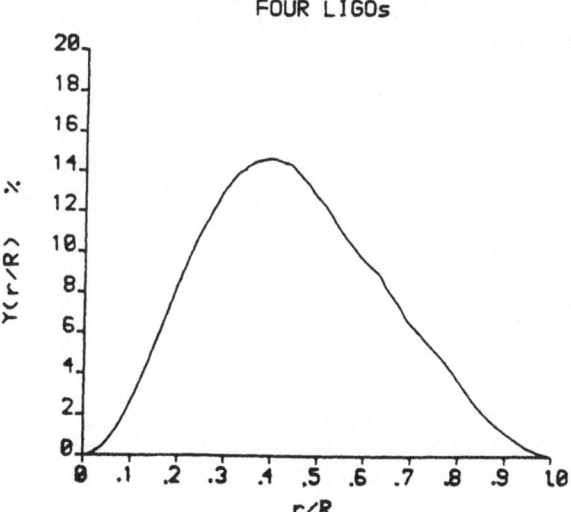

Figure 5. As for Fig. 4, but for four LIGOs.

In order to evaluate the fraction of the number of events occurring per year, within a volume of radius r, that will be observable by a network of LIGOs, we can calculate the detection efficiency, defined as the average of the coincidence probability over that volume. Its expression is the following:

$$\langle C \rangle = \left[\frac{R}{r} \right]^3 \int_0^{r/R} C\left[\frac{X}{R} \right] \, 3 \left[\frac{X}{R} \right]^2 d\left[\frac{X}{R} \right] \tag{3.5}$$

where the factor $(R/r)^3$ takes into account the proper normalization for the probability density $f(r/R)$.

Multiplying $\langle C \rangle$ by the number of events occurring within the space region of radius r, we may deduce the number of coincidences that a certain network will be able to register over a given observation time.

Figures 6 and 7 show respectively, the detection efficiency for the single antenna and for a quadruple coincidence as functions of r/R. These are decreasing functions of r, equal to 1 when r is equal to zero and going to their minimum values when r goes to R.

3.1 Single—Antenna detection experiments

In the assumption of white—Gaussian noise, the type of noise expected at the output of a shot noise limited interferometric antenna, one can calculate the level of threshold the experimenters need to take in order to ensure that the number of fluctuations above threshold due to noise will be acceptably small. Schutz (1988b) finds that, with a sample time of 10^{-3} s, there will be a random noise fluctuation of amplitude greater than 7.4 times the rms noise level σ only once a year. In what follows we shall assume this value as an appropriate threshold.

For an antenna with 1 km arm—length, working with light—recycling and optimally filtered for signals from two 1.4 solar masses neutron—star binaries, at 100 Hz, the expected attainable rms noise level is equal to: (Hough et al. 1986):

$$\sigma_B = 7.8 \times 10^{-25}$$

Taking one noise event per year to be an acceptable threshold, we may use Eq.(3.2) to deduce the threshold levels for the planned antennae:

$$T_B = 57.7 \times 10^{-25}$$

$$T_G = 34.6 \times 10^{-25} \tag{3.6}$$

$$T_{AM} = 28.8 \times 10^{-25}$$

By substituting the values of the masses and of the wave's frequency into Eq.(3.3) we obtain the following relation between the threshold T and the maximum distance the antennae will be able to look at:

$$T = 3.8 \times 10^{-23} \left[\frac{100\text{Mpc}}{R} \right] \tag{3.7}$$

From Eq.(3.6) we find the following maximum distances:

$$R_B = 0.7 \text{ Gpc}$$

$$R_G = 1.1 \text{ Gpc} \qquad (3.8)$$

$$R_{AM} = 1.3 \text{ Gpc}.$$

We have calculated numerically the integrals in Eq.(3.5) for the detection efficiency and the results are plotted in Figs. 6 and 7. For a single antenna and at $r = R$ we find a detection efficiency $\langle C \rangle$ of 8.6%. This allows us to evaluate the number of events per year, N, that each antenna will be able to register if we assume the rate of three events per year, for systems out to 100 Mpc, suggested by Clark *et al.* (1979). The values we get are as follows:

$$N_B \simeq 90 \text{ yr}^{-1}$$

$$N_G \simeq 345 \text{ yr}^{-1} \qquad (3.9)$$

$$N_{AM} \simeq 570 \text{ yr}^{-1}$$

Even if the estimate of the rate of formation of binaries of this type is too high by a factor 100 because of small number statistics or other uncertainties, from Eq.(3.9) we see that the antennae would have a reasonable event rate.

In Fig. 4, where we plot the function $Y(r/R)$ for a single antenna, we deduce a value of r' equal to 0.41 times the maximum distances as given in Eq. (3.8). At this r', from Fig. 6 we obtain a detection efficiency $\langle C \rangle$ equal to 55%. With these data we deduce the following distances and event rates for the various antennae:

$$r_B' = 0.3 \text{ Gpc} \qquad n_B \simeq 45 \text{ yr}^{-1}$$

$$r_G' = 0.4 \text{ Gpc} \qquad n_G \simeq 105 \text{ yr}^{-1} \qquad (3.10)$$

$$r_{AM}' = 0.5 \text{ Gpc} \qquad n_{AM} \simeq 205 \text{ yr}^{-1}.$$

3.2   Quadruple Coincidences

For a quadruple configuration a threshold of 2.3σ in Scotland will be such as to reduce the number of coincident noise events to one per year (Schutz, 1988b). This gives the following theshold T:

$$T = T_B = 18.17 \times 10^{-25} \qquad (3.11)$$

which takes us out to:

$$R = 2.1 \text{ Gpc} \qquad (3.12)$$

and with a $\langle C \rangle$ of 7.2% we get:

312

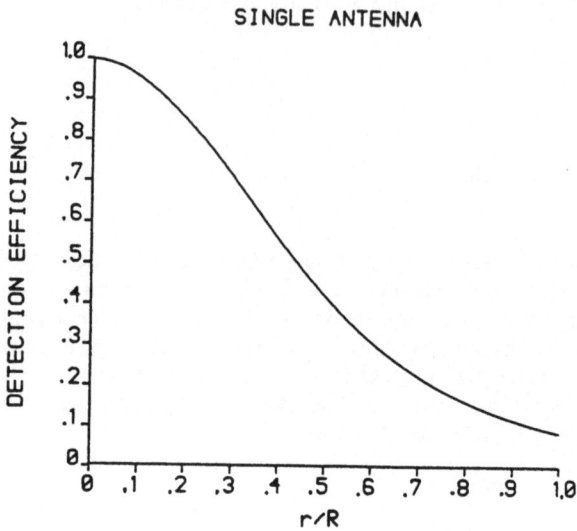

Figure 6.   Detection efficiency for a single antenna as a function of
r/R (see Eq.(3.5) for its definition).

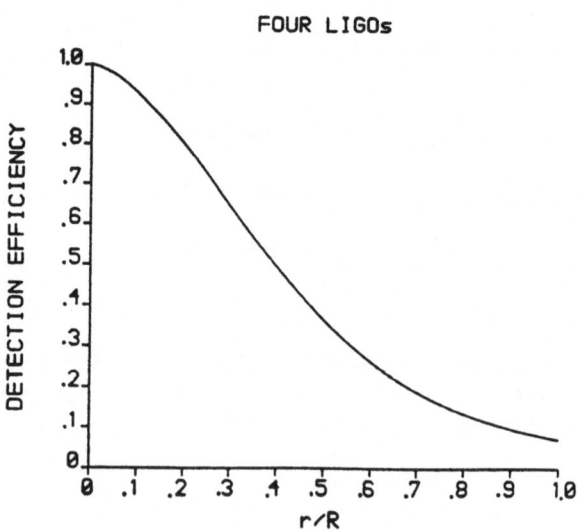

Figure 7.   As for Fig. 6, but for four LIGOs.

$$N \simeq 2000 \ yr^{-1} \tag{3.13}$$

With r´ equal to 0.40 R and a corresponding detection efficient of 50% we obtain the following results:

$$r´ = 0.8 \ Gpc; \quad n \simeq 770 \ yr^{-1} \tag{3.14}$$

4. CONCLUSIONS

A coincident experiment with two or more detectors enhances the probability of detection of gravitational bursts because it singles them out from uncorrelated noise pulses.

The principal results presented in this review paper are the threshold averaged coincidence probability (Eq.(2.10)), the calculation of the coincidence probabilities and of the detection efficiencies. We found that the network of four planned detectors in USA and Europe, working at their optimum sensitivities, would be able to register in coincidence 2000 events per year when looking for signals from two 1.4 solar masses neutron-star binaries and at 100 Hz.

This would allow us a number of astrophysical observations of extraordinary relevance, otherwise impossible, and a test of Einstein's theory of Relativity.

REFERENCES

Clarke, J.P.A., van den Heuvel, E.P.J. & Sutantyo, W., 1979.
    Astron.Astrophys., 72, 120
Dewey, D., 1986. in *Proceedings of the Fourth Marcel Grossmann Meeting
    on General Relativity*, ed. R. Ruffini, Elsevier
Hough, Meers, B.J., Newton, G.P., Robertson, N.A., Ward, H., Schutz,
    B.F., Drever, R.W.P., Tolcher, R. & Corbett, I.F., (1986) *A
    British Long Baseline Gravitational Wave Observatory*, Rutherford
    Appleton Laboratory Report GWD/RAL 86-001, Chilton, Oxon. U.K.
Krolak, A. 1988. 'Post-Newtonian Coalescing Binaries', this volume.
Schutz, B.F., 1986. *Nature*, 323, 310
Schutz, B.F., 1988. in *Proceedings of the XIV Yamada Conference on
    Gravitational Collapse and Relativity*, ed. Sato, H., Nakamura,
    T., World Scientific, Singapore
Schutz, B.F., 1988a. 'Sources of gravitational radiation', this volume
Schutz, B.F., 1988b. 'Data analysis requirements of networks of
    detectors', this volume
Schutz, B.F. & Tinto, M. 1987. *Mon.Not.R.astr.Soc.*, 224, 131
Thorne, K.S., 1987. in *300 Years of Gravitation*, eds. Hawking, S.W. &
    Israel, W. Cambridge University Press
Tinto, M., 1987a. *Mon.Not.R.astr.Soc.*, 226, 829
Tinto, M., 1987b. in *Proceedings of the 7th Italian Conference on
    General Relativity and Gravitational Physics*, eds. Bruzzo, U.,
    Cianci, R. & Massa, E. World Scientific, Singapore
Tinto, M. 1987c. *Mon.Not.R.astr.Soc.*, to appear
Tinto, M., 1987d. Ph.D. thesis, University of Wales, unpublished

DATA ANALYSIS REQUIREMENTS OF NETWORKS OF DETECTORS

Bernard F. Schutz
Department of Applied Mathematics and Astronomy
University College Cardiff
P.O. Box 78
Cardiff, Wales, UK

ABSTRACT. It is generally accepted that gravitational wave detectors must work together for successful gravitational wave observations. The most elementary reason is to gain confidence: rare and unmodelled sources of noise in single detectors can be eliminated by demanding that a gravitational wave event must be seen in two or more detectors on different sites. But there are other aspects of joint observation that I discuss here. A complete solution for the gravitational wave requires observations by 4 laser interferometers or 5 bar detectors. There are at least three likely data analysis modes: threshold, summation, and correlation. The summation and correlation modes require the exchange of raw data, the volume of which makes stringent demands of any data storage and transmission methods. The threshold mode will almost certainly be the main mode of operation, and I discuss the question of how to set a threshold for observations against a background of gaussian noise, when one allows for the fact that there are time-delay windows within which coincidences will be accepted, and also that the data will be run through a large number of digital filters. Finally I raise a number of issues that detector groups need to address in planning for laser-interferometric detectors, such as establishing standards for data exchange and storage media, for software, and for the means by which data will be analyzed.

1. BASIC PROBLEMS OF DATA ANALYSIS FOR NETWORKS

Observations by networks of detectors will be necessary to extract the full astrophysical information from any detected gravitational wave. The following simple counting argument shows that a minimum desirable size of network is four if the instruments are broadband and five if they are narrow band. Each detector basically gives only one number, the amplitude of its response at any time. Broadband detectors have in addition sufficient time resolution to determine the delay-time between events in two detectors (typically tens of milliseconds for detectors distributed around the globe). A network of n broadband detectors therefore can produce 2n-1 data (n amplitudes and n-1 independent time delays) for the wave, while a network of n narrowband detectors can produce n data. Now, a wave is characterized by five numbers: two angles giving its direction, two amplitudes for its independent polarizations, and one angle giving the orientation of its polarization ellipse on the plane of the sky. In

B. F. Schutz (ed.), Gravitational Wave Data Analysis, 315–326.
© *1989 by Kluwer Academic Publishers.*

principle, then, we need five narrowband detectors or three broadband detectors for a complete solution. In fact, however, the solution for three broadband detectors is ambiguous (two-valued), so four are necessary.

The management of a network of narrowband detectors is probably not a technically difficult problem. GRAVNET (Blair, et al, 1988) has already begun to establish data-exchange agreements and protocols. The volume of data produced will not be particularly large: at a data rate of, say, 10 bytes per second one year's data from a detector would occupy a single 300 Mbyte optical disk. Analysis -- at least for bursts -- will not require sophisticated filtering algorithms and can be accomplished on ordinary computers.

Broadband detectors present a completely different scale of problem. If each detector samples at 10 kHz, uses 2-byte data words, and takes four channels of data (one 'real' output and three housekeeping channels), then it will produce 160 kbytes/sec, and it will fill up that 300 Mbyte optical disc in half an hour. To look for coalescing binaries, the output data needs to be filtered through perhaps 100 filters with different values of the mass parameter; special computing equipment may be needed to keep up with the data rates. Moreover, in order to look for rare or unexpected events it may not be enough simply to analyze each data stream for events and then to exchange lists of them, looking for coincidences: it will be important to cross-correlate the full raw data among all pairs of detectors. It is clear that the problems of data storage and exchange need considerable planning, and one of the reasons for arranging this Workshop was to initiate that process.

Because the broadband detectors present such formidable problems, I will concentrate on them in this paper, although much of what I say will be applicable to narrowband systems as well. In the next section I will identify and discuss three possible modes of data analysis in networks: threshold mode, summation mode, and correlation mode. In the third section I will calculate the thresholds that detectors must set in order to ensure that no more than, say, one spurious noise-generated detection occurs per year. These thresholds depend on the number and relative sensitivities of the detectors in the network, and on such complicating factors as how much of a time delay must be allowed for between detectors, and how many filters will be applied to the data. In turn, they determine how far a network can see and what the rate of events it detects will be. In the final section I will make a few remarks on the data-analysis questions that the current planning exercise (Corbett 1988) might try to answer.

2.   METHODS OF COORDINATED DATA ANLAYSIS

In this section I will examine three modes in which data from networks may be analysed for burst events. Each seems to have certain advantages and disadvantages. I will not deal with the thornier problem of finding continuous wave signals in long stretches of data, which is discussed by Livas (1988).

2.1. Threshold mode

In this mode, each detector's output is examined (perhaps after filtering) to determine when it crosses a predetermined threshold. Lists of these 'events' are exchanged among the groups and mutually consistent time-delays are sought in these lists. This is the way that bar detectors currently operate, and it will certainly be the first way that laser interferometers will look for coincidences too.

This has several clear advantages:

(i) It is *fast*. Lists of events for reasonably high thresholds will not be very long, and can be exchanged by electronic mail or telephone line, completely automatically. It allows for quick ($<$ 1 day) recognition of significant events, and consequently quick notification of other astronomers who may wish to look at the position of suspected events.

(ii) It is *easy*. The necessary filtering can be done on-line, so that events can be picked up in each detector almost instantly. All the groups can afford to do the coincidence analysis with exchanged lists of events: it requires no great computing or programming overheads.

(iii) It is *versatile*. It is suitable for diverse antennas: bars can operate in coincidence with laser interferometers in this mode with no difficulty.

Against this are some disadvantages of the method:

(i) It can miss relatively significant events which are unanticipated, that is for which no filter has been constructed to pull them out of noise.

(ii) For proper operation, the filters and threshold tests used at different sites must be consistent, and preferably identical. It requires agreements on standardization around the network.

These disadvantages are relatively minor, so this method is likely to become the most important data analysis mode, at least at first. The thresholds must be set by taking into account the number of 'accidental' coincidences that one would expect (false-alarm rate). I shall consider this question in some detail in §3 below, because it determines how far away events can be detected and hence the expected number of events that we can anticipate in our network.

2.2. Summation mode

Given two identical (or at least comparable) antennas, one takes the raw data and simply adds them together before applying any filters or threshold tests. One advantage is that it clearly improves signal-to-noise (S/N) ratios. For example, if two laser interferometers are limited by shot noise, then this is like adding the two beams, doubling the light, and reducing the noise by $\sqrt{2}$. Presumably this gives a potentially better S/N than the threshold mode.

It has at least two disadvantages:

(i) The same event arrives in different detectors at different times, so simply adding data taken at the same instant will not pick it up. One has to allow for these time-delays, as well as for possible differences in waveforms due to the fact that different detectors are sensitive to different polarization states of the waves because of their different orientations.

(ii) To operate this mode one has to exchange or at least pool the raw data: there are large volumes involved and higher costs in manpower and computing resources dedicated to analysis.

(iii) To operate with different kinds of detectors, say bars and laser interferometers, the data would have to be filtered before addition in order to equalize bandwidths and sampling rates.

This method has not received much attention from the gravitational wave community, and it deserves further study. Despite its disadvantages, the improvement it gives in S/N may make it useful in special circumstances.

2.3. Correlation mode

This method is likely to become a very important adjunct to the threshold mode of analysis. In this, one forms the correlation of two raw data streams,

$$c_{12}(\tau) = \int y_1(t) \, y_2(t+\tau) \, dt. \qquad (2.1)$$

It is useful to think of this as using detector 2 as a *filter* on the output of detector 1. Unlike the ideal filters one would use in the threshold mode, this filter is noisy, and so this introduces extra noise into the system. In compensation, one doesn't have to know ahead of time what to expect: *every* wave event is in the output of detector 2. Thus, its advantages are:

(i) It is *robust*. It can find events that had not been anticipated.

(ii) It is *unbiased*. It does not filter only for our preconceptions, but gives us anything that has occurred.

Its disadvantages are also clear:

(i) Because the filter is noisy, the S/N is worse than in the threshold mode, roughly by $1/\sqrt{2}$ for each pair correlated. Moreover, since the amplitude of the event in the 'filter' is unknown, the amplitude of the response cannot be determined if there are only two detectors. (But it can be if there are three or more.)

(ii) The time delay between different detectors is no problem here, but the possibility that events will not look the same in the two detectors due to their different sensitivities to the two polarizations of the wave is a serious difficulty.

(iii) As in the summation mode, raw data must be exchanged or pooled, with consequent overheads and delays; also the method is hard to apply between detectors of different types.

This is likely to be a network's 'discovery mode': it is the only way to find unanticipated events of moderate to small S/N, provided they have enough structure to be enhanced by filtering. If we theorists are not clever enough, it is even possible that the *first* gravitational waves will be seen this way! More work on this is necessary to understand the expected noise levels and the statistics of detection, especially when there are more than two detectors in the network. (See, *e.g.*, Armstrong 1977.)

3.   THRESHOLDS IN NETWORKS

In this section I will concentrate on an important aspect of the threshold mode, namely deciding what thresholds produce an acceptable false-alarm rate in a given network for a given type of observation. This is important even in the present stage of planning for the detectors, because the threshold determines how far away detectors will be able to see. Random coincidences among four detectors are obviously less likely than among two or three, so that every time another detector is added to a network, the threshold goes down and the network can see farther and therefore gather a larger number of events.

3.1. Gaussian Noise

Let us assume that in the output, x, of our detector, the only noise source is Gaussian with zero mean and standard deviation σ, so that its probability distribution function is

$$p(\sigma;x) = \frac{1}{(2\pi)^{1/2}\,\sigma}\, e^{-x^2/2\sigma^2} \quad . \tag{3.1}$$

The probability that x will exceed some threshold X in either the positive or the negative direction is given by

$$p(\sigma;|x|>X) = 2 \int_X^\infty p(\sigma;x)\,dx$$

$$\simeq \left(\frac{2}{\pi}\right)^{1/2} \frac{\sigma}{X}\, e^{-X^2/2\sigma^2}\, \left[1 - \left(\frac{\sigma}{X}\right)^2 + \ldots\right]. \tag{3.2}$$

Equation (3.2) is an asymptotic approximation for large X; it is good to 10% for X > 2.5 σ, and its first term is good to 10% for X > 3.2 σ.

3.2. Simple false alarms

Suppose we define an 'event' in any detector as a time when the response x exceeds the threshold X in either the positive or negative direction. The single-detector probability of a spurious, noise-generated event is therefore $p(\sigma;|x|>X)$. Our aim is to calculate the threshold X required to ensure that the number of spurious gravitational wave 'events' is acceptably small. Clearly, if we determine that an acceptable probability for spurious events ('false alarms') is f, then we must solve the transcendental equation

$$p(\sigma;|x|>X) = f$$

for X. For a detector that samples at 1 kHz, we might want to choose f so that the expected one-detector false alarm rate is once per year, i.e. f = 3 x 10^{-11}. The solution is X = 6.6 σ.
     If two detectors are operating together, then the simplest coincidence experiment is to look for both detectors to be above threshold at the same time. If the detectors have independent noise (which we shall always

assume) and are identical (we shall drop this assumption later), then the false alarm probability is $[p(\sigma;|x|>X)]^2$. Similarly, for n detectors the n-way false alarm probability is $[p(\sigma;|x|>X)]^n$, and the appropriate threshold is given approximately by the solution to

$$(\frac{2}{\pi})^{\frac{1}{2}} \frac{\sigma}{X} e^{-X^2/2\sigma^2} = f^{1/n}. \qquad (3.3)$$

Since the left-hand side is dominated by the exponential term, a rough approximation is that the solution should behave like $X \propto n^{-1/2}$. Thus, if the single-detector threshold is 6.6, then we expect that two detectors can operate at $X \simeq 4.7$, three at $X \simeq 3.8$, four at $X \simeq 3.3$, and five at $X \simeq 3$. The actual solutions to Eq.(3.3) are, for identical detectors with a false alarm probability of 3×10^{-11},

number of detectors n	1	2	3	4	5
threshold X/σ	6.6	4.5	3.6	3.0	2.6

The importance of this decrease of the threshold is that the volume of space that becomes accessible to the network increases dramatically, as $1/X^3$. Roughly, this goes as $n^{3/2}$, so that five detectors can survey a volume of space 11 times as large as one detector can. In fact, the solutions given in the table above give an even more rapid increase in the volume than this: if V_n is the volume accessible to n detectors, then we have

number of detectors n	1	2	3	4	5
V_n / V_1	1.	3.1	6.3	11.	15.

Provided that the first detector can see at least to the Virgo cluster (about 15 Mpc), then the rough homogeneity of the distribution of galaxies further away guarantees that the true gravitational wave event rate will increase by the same factors.

While these numbers are instructive, they leave out some important features of real networks:

(i) real detectors will not all be identical;

(ii) the necessity of allowing for an unknown travel-time delay between correlated events in separated detectors will increase the false alarm rate at a given threshold; and

(iii) the need to filter the data through a hundred or more statistically independent filters to look for coalescences and other events can also increase the false alarm rate.

We shall now include each of these effects in turn into our threshold calculations.

3.3. Non-identical detectors

What matters for the false-alarm probability is the signal-to-noise (S/N) ratio x/σ in each detector. Suppose we number our detectors (1, ..., n) and normalize the sensitivity of detectors to detector number 1. Suppose that, for reasons of different intrinsic noise or different size, a given gravitational wave will produce a different S/N in detector j than in detector 1. Let r_j be the ratio of S/N's:

$$r_j = \frac{(S/N)_j}{(S/N)_1}.$$

Then if X is the threshold in detector 1, one will have to set a threshold $r_j X$ in detector j if one wants to detect the same gravitational wave. [This ignores the fact that different detectors are oriented differently and would not respond to a given gravitational wave identically even if they had the same sensitivity. For waves arriving from random directions, this effect presumably averages out, but to take it into account fully would require a Monte-Carlo calculation of the type perfomed by Tinto (1988).] The equation governing the false alarm rate [replacing Eq. (3.3)] is then

$$(\frac{2}{\pi})^{n/2} \; (\frac{\sigma}{X})^n \; e^{-SX^2/2\sigma^2} \; \prod_{j=1}^{n} \{ \; \frac{1}{r_j} \; [1 - (\frac{\sigma}{r_j X})^2] \} = f, \qquad (3.4)$$

where S is defined by

$$S = \sum_{j=1}^{n} r_j^2.$$

We will not recalculate the thresholds for assumed networks of non-identical detectors until after we have included the effects of time delays and filtering.

3.4. Time-delay windows

Since we must allow for the light-travel time between detectors (some tens of milliseconds in experiments with millisecond or better time resolution), noise-generated events that occur within a certain window of time in separated detectors will contribute to the false alarm rate.

For only *two* detectors operating a coincidence experiment, suppose the time-delay window is W sampling times long, *i.e.* given an event in one detector, any event that occurs within $\pm W/2$ sampling times in the second detector will be accepted. Then since W will be very much smaller than the observation time (which will be days or even years), the false alarm probability just increases by a factor of W to $W[p(\sigma; |x|>X)]^2$. One sets this equal to f and solves for X as above.

If there are three or more detectors the situation is more complex. It is helpful to think in terms of an (n-1)-dimensional 'time-delay' lattice T_{n-1} for n detectors. If we take detector number 1 as the reference, then any event is located in T_{n-1} by the delays to the other detectors (t_2-t_1, t_3-t_1, ...). The space is a lattice because of the finite sampling time: the time delays are integer multiples of this time. In this space there is a region around the origin within which real events must lie, and the 'volume' R_{n-1} (number of lattice cells) of this region is the n-detector analogue of W. We must multiply the left-hand-side of Eq. (3.4) by R_{n-1} and then solve for X.

Unfortunately, this volume is not straightforward to calculate. For *three* detectors it depends on the relative positions of the detectors. If W is the largest window between any pair, then we overestimate R_2 as W^2. This errs in the conservative direction, producing slightly larger thresholds than optimum. For *four* detectors, further complications set in because the inverse problem is overdetermined. Therefore, any noise-generated event would have to lie in a region of T_3 which was consistent with the amplitudes of the events in the four detectors. Given the amplitudes, we require in principle only one time delay to determine the solution, and we can reject noise-generated events whose second and third time delays do not fit the solution. In practice, there will inevitably be some uncertainty in the measurements of the event amplitudes which will allow some room in T_3 for the noise-generated events. If again we take the largest window to be W, then we shall crudely approximate the error-width of R_3 in the second and third dimensions as εW, where $0 < \varepsilon < 1$. Then we have $R_3 = \varepsilon^2 W^3$. Finally, for *five* detectors the various amplitudes completely determine the solution, so the only room in T_4 is error-generated: $R_4 = \varepsilon^4 W^4$.

The conclusions of this section are summarized in the following array:

number of detectors n	2	3	4	5
window volume R_{n-1}	W	W^2	$\varepsilon^2 W^3$	$\varepsilon^4 W^4$

where W is the maximum window in units of the sampling time and ε is defined above.

3.5. The effect of filtering

Filtering the raw data numerically has two effects on the setting of thresholds. On the one hand, it is clear that if two filters are statistically independent (no correlation in their outputs when applied to white noise), then they offer twice as much opportunity for false alarms as one. On the other hand, filters reduce the effective noise bandwidth, which has much the same effect as increasing the sampling time and therefore decreasing the false-alarm probability. These two effects tend to compensate each other. In my study of the coalescing binary filters (Schutz 1987), I found that a typical filter had zero correlation with itself when shifted by about 2 msec, so this might indicate an effective sampling time of 2 msec for this problem. The same calculation showed that a typical filter had very small correlation with another whose mass parameter differed by perhaps 2%, which suggests that something like 200 or so filters will be needed to span a reasonable range of mass parameters. But these conclusions are preliminary and require further study. For the present, we will simply take the number N_F of filters and multiply it by R_{n-1} on the left-hand-side of Eq. (3.4). We will take the sampling rate effect into account in setting the false-alarm probability f.

3.6. Realistic thresholds for networks

We are now in a position to redo the calculations of §3.2, taking into account windows and filters in networks of non-identical detectors. In this section I will assume that the sampling time is 1 msec and an acceptable false-alarm rate is once per year, so that f is the same as before, 3×10^{-11}. I will then take W to be 50 (msec), appropriate to a baseline of 7500 km, and I will assume 200 filters. For ε I will take 0.1, but this may be too small, particularly for low thresholds, where the signal-to-noise ratio is small and the errors in the inverse problem are large. The following numbers are therefore illustrative of what the importance of these effects may be: they are not definitive.

First let us assume a network of *identical detectors*, so that we can compare with §3.2. For the thresholds we find

number of detectors n	1	2	3	4	5
threshold X/σ	7.4	5.4	4.6	3.9	3.4

and for the ratios by which the observable volume increases

number of detectors n	1	2	3	4	5
V_n / V_1	1.	2.5	4.1	6.9	11.

We see that windows and filters can reduce the volume of space that a network of four or five detectors can see by about 30%.

Next we turn to networks of *non-identical detectors*. First consider a network consisting of the four detectors that had been proposed as of the end of 1986, which is the network on which Tinto (1988) has based his calculations of detector efficiency. If we assume that the laser, isolation, and mirror technologies will be comparable in each detector, then their relative sensitivities in recycling mode will scale as the square root of their effective arm lengths. Taking the US 4-km detectors as the standard, then the 3-km German detector with an included angle of 60° has relative sensitivity r = 0.81, and the 1-km Glasgow proposal has r = 0.5. The threshold in the US detectors works out to be 4.7σ, while the German detector would operate at 3.8σ and Glasgow at 2.3σ. Tinto (1988) has used these figures to show that such a network could observe 50% of all coalescing binary events out to 1.7 Gpc, with an estimated event rate of almost two per hour!

We shall also consider other possible networks. The recent Pisa-Orsay collaborative proposal has opened the possiblity of a five-detector network, perhaps with larger European detectors. Let us consider two 4-km US detectors observing with either two or three 3-km European detectors (r = 0.87). When operating with two European detectors, the US threshold is 4.1σ and the European threshold is 3.6σ. This network would be able to see a 50% larger volume of space than the one considered in the previous paragraph, with a corresponding increase in the event rate. Even better, of course, would be the network with five detectors. In it, the US detectors would be set at 3.6σ and the European ones at 3.2σ; this addition of a fifth European detector increases the accessible volume by a further 50%.

There is also now a possiblity that a detector will be built by a group in Tokyo. While I have not included this in my calculations, it would be straightforward to do so. Such a detector has the great advantage that its baseline from the others is very large, permitting good directional sensitivity. This large baseline increases W as well, of course, but this will not prevent the sensitivity of a network from improving when such a detector is added.

3.7. Large-statistics surveys: the false-alarm rate as a fraction of the event rate

In what I have described so far, I have kept the false-alarm rate to a fixed number of events per year (one, if the sampling time is 1 msec). But if the event rate for coalescing binaries turns out to be as large as Tinto (1988) estimates (see the discussion in Schutz 1988), then much of the data will be used to make good-statistics surveys of such things as the homogeneity of the universe and the mass function of neutron stars. In such surveys it would be more appropriate to choose the threshold to guarantee that the false-alarm rate is a specified fraction (say 1%) of the true event rate. This would lower the threshold and allow more of the Universe to be surveyed.

The true event rate will be proportional to the accessible volume, which in turn is proportional to X^{-3}. We want f to be proportional to this. It follows that we need to replace the right-hand-side of Eq. (3.4) by α/X^3, where α is a constant that depends on the true event rate and the fraction of this that can be allowed to be false alarms. If for example we take Tinto's figure of 2000 true events for $X = 4.7\sigma$, and if we accept a 1% false-alarm rate for data sampled at 1 msec intervals, then it is straightfoward to show that

$$\alpha = 0.01 \times 2000 \times 4.7^3 / 3 \times 10^{10} = 6.9 \times 10^{-8}.$$

The four-detector network as proposed by the end of 1986 would then be able to operate with a US threshold of 4.4σ; the German threshold would be 3.5σ and Glasgow 2.1σ. This raises its event rate by perhaps 20% over the rate it would have if it allowed only one false alarm per year. Again, this increase is only illustrative, especially since it depends on the assumed coalescence rate, which is very uncertain.

4. CONCLUSIONS

In this paper I have discussed a few topics that are germane to networks of gravitational wave detectors: the number of detectors needed to reconstruct the wave, the amount of data they will produce, three different modes of data analysis, and the thresholds that are necessary to keep the false-alarm rate at a reasonable level in a network.

I would like to conclude by making some observations about the sort of planning that needs to be done for data analysis in networks of laser interferometers before they become operational. If the data are to be analyzed thoroughly, each detector will have to be equipped with a system to filter the stream of data as it comes out, at a rate that can keep up with the data. Since the most demanding filter for bursts is likely to be for coalescing binaries, for which a typical filter would be about two seconds in duration and sampled at 1 kHz, and since one would want to be able to apply perhaps 200 filters to each sample of data, the on-line analysis system will need to be able to perform a 2048-point Fourier transform in 10 msec maximum. This does not seem to be a difficult goal: it is already achievable with special-purpose hardware (array processors, small arrays of digital signal processing chips, or small arrays of transputers), and in five years it may be attainable in inexpensive general-purpose computers. Another problem that will have to be solved, but which does not seem intractable, is to arrange for the data-analysis computers to exchange lists of events automatically, then to process the lists for coincidences (allowing for time delays), and finally to solve the inverse problem and produce lists of gravitational waves with their positions, polarizations, and amplitudes.

A more difficult job may be storing the data. If networks operate in the correlation or summation modes of analysis, then they will have to store and exchange raw data. Even if they operate only in the threshold mode, there is a strong argument for archiving the raw data so that it can be searched later if other observations in astronomy make it seem likely that a gravitational wave event may have occurred at a certain time. At the present time there are optical disc and videotape storage systems available that not only have large capacity but are also relatively easy to store and to transport. Unfortunately there are no international standards for either of these media, and if this situation persists then the network will presumably have to settle on one standard for everyone.

The analysis of data in the threshold mode need not be very demanding, but the joint analysis of raw data, looking for correlations among different antennas, presents problems mainly of getting the data to a single site where the calculation can be performed. The actual calculations are not significantly different from the filtering that will be done on-line, but groups will have to decide whether they want universal data exchange (each group in an n-detector network making n-1 copies of its data and shipping it off to each other group), or alternatively a small number of data-pooling centres where archiving, correlation, and distribution of data will be organized. It may be that in five years there will be relatively inexpensive high-bandwidth fibre-optic or satellite data transmission services that will make the distribution of data easy. But 'universal' data analysis also makes manpower and storage demands, and groups will have to decide whether they wish to meet these or to displace them to data-pooling centres.

In any case, wherever the analysis is done, the network will need to fix certain standards for data formats and for interfaces with data-analysis software. The software has to be designed, at least to the stage of being able to handle the initial data rates and to be able to effect data exchanges. In order to make software transportable and useful everywhere, guidelines need to be agreed as to language, special extensions, I/O formats, and so on. And, very importantly, they will have to agree protocols on the use of data: vetoes on the publication of one's own data, access of third parties to the data, and so on. (GRAVNET already has a set of agreed protocols.) None of these difficulties is unique to gravitational wave research, but they will have to be addressed before the networks can be fully operational.   The working party set up by this meeting (Corbett 1988) will begin this process for networks involving laser interferometers.

REFERENCES

Armstrong, J.W. 1977, *Astron. Astrophys.*, **61**, 313.
Blair, D. 1988, this volume.
Corbett, I 1988, this volume.
Livas, J. 1988, this volume.
Schutz, B.F. 1987, in *Gravitational Collapse and Relativity*, eds. H. Sato & T. Nakamura (World Scientific, Singapore).
Schutz, B.F. 1988, this volume.
Tinto, M. 1988, this volume.

ROUND-TABLE ON DATA EXCHANGE

I F Corbett
Rutherford Appleton Laboratory
Chilton
Didcot
United Kingdom

This was a discussion session on data exchange protocols, hardware
and software standards and problems of coordination and
collaboration. It concentrated on the initiatives that should be
taken in the short term in order to ensure that the community was
well prepared for effective collaboration once the proposed new
generation of detectors were operational.

Prior to the workshop, a questionnaire had been circulated to all
active experimental groups represented. From the replies it is
possible to summarise the present status and future plans of the
community as a whole.

The present data acquisition systems are fairly simple, as befits
prototype instruments, and all groups have plans for modest
improvements, for the most part concentrating on improved data
storage. Even at the present, prototype, stage it is clear that the
interferometric detectors have state-of-the-art storage requirements.

On data sharing needs and protocols, there was gratifying unanimity,
with all groups declaring their willingness to make raw data,
corrected only for instrumental artefacts, available essentially
immediately. There was a preference for collaborative data analysis,
and in the event of separate analysis the group supplying the data
would wish to retain a veto on publication for up to 2 years or so.
To a very large extent these protocols are the same as those of
'GRAVNET', already described by D G Blair (University of Western
Australia) at an earlier session.

Turning to the future, the following possibilities for
standardisation between detector systems were noted:

B. F. Schutz (ed.), Gravitational Wave Data Analysis, 327–329.
© *1989 by Kluwer Academic Publishers.*

COMPONENT	STANDARDISATION ELEMENT
The detector itself	Almost nothing
A to D conversion	Sample Frequency Number of Bits Time Standard
Raw Data Store	Hardware medium Formats
"First Pass" Processing	Algorithms Selection Criteria
Selected Data Store	Hardware Medium Formats
Full Analysis	Algorithms Computer Operating System Input/Output Mediums Storage Mediums

The discussion ranged widely over all these topics. Some of the more salient points made are summarised below.

Several contributors stressed the impossibility of data being analysed by people without an intimate knowledge of the detector producing it. It was agreed that a large part, certainly initially, of the software would be 'detector dependent' and would be contributed by the detector group. The 'detector independent' software could be written and developed elsewhere, for example by signal processing specialists, and used on the data from all groups provided the software interfaces and hardware media were compatible. This needs careful planning.

It would be desirable to have relative calibration between detectors good to (say) 20%. Currently this is probably very hard to achieve, and longer term drifts in relative calibration are not well studied. It will be very important for groups to work together on calibration data, and on 'housekeeping' and monitoring data. This should be recorded to enable subsequent analysis to take into account how the detector was actually performing, and to allow periods of doubtful data to be either devalued or discarded.

In this context, powerful on-line monitoring and analysis facilities were regarded as important, to enable the experimenters a) to develop their equipment, b) to understand what was going on during a data taking session, and c) to give early-warning of potential problems eg. slow drifts driving feed-back loops out of range.

It was generally felt that good data should be stored indefinitely: astronomy is an observational science and data from many centuries ago is still valid and useful.

In the discussion on the likely evolution of computing power, memory size and mass storage systems, it was made clear by the representatives of Sun, Masscomp and Honeywell present that they anticipated a doubling at roughly 12-18 month intervals of what was available in all these sectors. Thus in 5 years time 20 Mips, 32 Mbyte and more than 5 Gbyte storage should be readily available at reasonable prices. One problem, already raising its head, is the absence of a true industry standard for mass storage which can absorb new technologies as they come on stream. With groups currently thinking of investing in new mass storage devices, this now needs careful consideration.

In the future, very high speed data links (approx 20 Mbaud) will come down in price, so real-time data transfer to computing centres with very powerful processors could be a reality. This would be ext-remely significant once gravitational radiation had been detected and observation in conjunction with optical or radio telescopes was established. Such links would also allow users remote from such a powerful centre to perform sophisticated interactive analysis. It is quite probable that some forms of model dependent analysis will require CRAY type computing power, which can only be provided by a large centre.

The need for professional software and signal analysis specialists, who would be important and integral members of the experimental teams, was stressed by many people.

Finally, in recognition of the importance of the analysis chain and the need to start effective collaboration now, it was agreed that Prof B F Schutz, University College, Cardiff, would seek the cooper-ation of all active groups in setting up a representative panel to assist the community in coordinating its data handling and data analysis efforts. Each group would be invited to nominate a contact person who would, preferably, be someone actively involved in the evolution of their data acquisition and analysis system. This panel would aim to start work soon, and would hope to present some recommendations for discussion at, for example, the 5th Marcel Grossmann meeting to be held in Perth, Australia, in August 1988.